Blood Cells and Vessel Walls: functional interactions

The Ciba Foundation for the promotion of international cooperation in medical and chemical research is a scientific and educational charity established by CIBA Limited—now CIBA-GEIGY Limited—of Basle. The Foundation operates independently in London under English trust law.

Ciba Foundation Symposia are published in collaboration with Excerpta Medica in Amsterdam.

Excerpta Medica, P.O. Box 211, Amsterdam

J.L. Gowans, CBE, FRCP, FRS
(Photograph reproduced by permission of the Medical Research Council)

Symposium on

Blood Cells and Vessel Walls: functional interactions

Ciba Foundation Symposium 71 (new series)

In honour of Dr J L Gowans FRS

1980

Excerpta Medica

Amsterdam · Oxford · New York

ISBN Excerpta Medica 90 219 4077 9
ISBN Elsevier/North-Holland 0 444 90112 4

Published in January 1980 by Excerpta Medica, P.O. Box 211, Amsterdam
and Elsevier/North-Holland, Inc., 52 Vanderbilt Avenue, New York, N.Y. 10017.

Suggested series entry for library catalogues: Ciba Foundation Symposia.
Suggested publisher's entry for library catalogues: Excerpta Medica.

Ciba Foundation Symposium 71 (new series)
369 pages, 56 figures, 23 tables

Library of Congress Cataloging in Publication Data

Main entry under title:

Blood cells and vessel walls.
    (Ciba Foundation symposium; 71 (new ser.))
    Includes bibliographical references and indexes.
    1. Blood cells—Congresses. 2. Endothelium—Congresses. 3. Blood-vessels—Congresses.
4. Gowans, J.L. I. Gowans, J.L. II. Series: Ciba Foundation. Symposium; new ser., 71.
QP94.B55   591.1'1   79–26528
ISBN 0-444-90112-4 (Elsevier/North Holland)

Printed in The Netherlands by Casparie, Amsterdam

Contents

J.L. GOWANS Chairman's introduction

LEON WEISS The haemopoietic microenvironment of bone marrow: an ultra-structural study of the interactions of blood cells, stroma and blood vessels 3
Discussion 13

WILLEM VAN EWIJK Immunoelectron-microscopic characterization of lymphoid microenvironments in the lymph node and thymus 21
Discussion 33

MAYA SIMIONESCU Structural and functional differentiation of microvascular endothelium 39
Discussion 51

G.V.R. BORN Haemodynamic and biochemical interactions in intravascular platelet aggregation 61
Discussion 73

J.R. VANE and S. MONCADA Prostacyclin 79
Discussion 90

General Discussion I Endothelial products 99
Ia antigens 100
Permeability of small vessels to cells 104

J.L. GOWANS and H.W. STEER The function and pathways of lymphocyte recirculation 113
Discussion 122

WENDY TREVELLA and BEDE MORRIS Reassortment of cell populations within the lymphoid apparatus of the sheep 127
Discussion 140

R.N.P. CAHILL, I.HERON, D.C. POSKITT and Z. TRNKA Lymphocyte recirculation in the sheep fetus 145
Discussion 157

IAN McCONNELL, JOHN HOPKINS and PETER LACHMANN   Lymphocyte traffic through lymph nodes during cell shutdown   167
Discussion   190

J.G. HALL Effect of skin painting with oxazolone on the local extravasation of mononuclear cells in sheep   197
Discussion   205

P. ANDREWS, W.L. FORD and R.W. STODDART   Metabolic studies of high-walled endothelium of postcapillary venules in rat lymph nodes   211
Discussion   227

General Discussion II   231

JUDITH J WOODRUFF and BARRY J. KUTTNER   Adherence of lymphocytes to the high endothelium of lymph nodes *in vitro*   243
Discussion   257

EUGENE C. BUTCHER and IRVING L. WEISSMAN   Cellular, genetic and evolutionary aspects of lymphocyte interactions with high-endothelial venules   265
Discussion   281

J.H. HUMPHREY Macrophages and the differential migration of lymphocytes   287
Discussion   296

SALLY H. ZIGMOND Polymorphonuclear leucocyte chemotaxis: detection of the gradient and development of cell polarity   299
Discussion   307

PETER D. RICHARDSON   Dynamic theory of leucocyte adhesion   313
Discussion   329

Final General Discussion   'Receptors' and adhesion   327

IRVING L. WEISSMAN Summing up   343

LEON WEISS Epilogue   349

Index to contributors   351

Subject index   353

Participants

Symposium on Blood Cells and Vessel Walls: functional interactions, held at the Ciba Foundation, London, 6–8 March 1979

J.L. GOWANS *(Chairman)* Medical Research Council, 20 Park Crescent, London WlN 4AL, UK

P.ANDREWS Department of Pathology, University of Manchester, Stopford Building, Oxford Road, Manchester M13 9PT, UK

G.V.R. BORN Department of Pharmacology, University of London, King's College, Strand, London WC2R 2LS, UK

E.C. BUTCHER Department of Pathology, Stanford University Medical Center, Stanford, California 94305, USA

R.N.P. CAHILL Basel Institute for Immunology, 487 Grenzacherstrasse, CH-4005 Basel 5, Switzerland

A.J.S. DAVIES Division of Biology, Chester Beatty Research Institute, Institute of Cancer Research, Royal Cancer Hospital, Fulham Road, London SW3 6JB, UK

W.L. FORD Department of Pathology, University of Manchester, Stopford Building, Oxford Road, Manchester M13 9PT, UK

R.J. GRYGLEWSKI Department of Pharmacology, Copernicus Academy of Medicine, Grzegorzecka 16, Cracow, Poland

J.G. HALL Chester Beatty Research Institute, Institute of Cancer Research, Royal Cancer Hospital, Clifton Avenue, Belmont, Sutton, Surrey SM2 5PX, UK

J.C. HOWARD Department of Immunology, Agricultural Research Council Institute of Animal Physiology, Babraham, Cambridge CB2 4AT, UK

J.H. HUMPHREY   Department of Immunology, Royal Postgraduate Medical School, Ducane Road, London W12 0HS, UK

I. McCONNELL   MRC Group on Mechanisms in Tumour Immunity, c/o Laboratory of Molecular Biology, The Medical School, Hills Road, Cambridge CB2 2QH

A.J. MARCUS   Hematology Section, 13 West, New York VA Hospital, 408 First Avenue, New York, NY 10010, USA

B. MORRIS   Department of Immunology, The John Curtin School of Medical Research, The Australian National University, P.O. Box 334, Canberra City, ACT 2601, Australia

J.J.T. OWEN   Department of Anatomy, Medical School, The University of Birmingham, Vincent Drive, Birmingham B15 2TJ, UK

P.D. RICHARDSON   Division of Engineering, Brown University, Providence, Rhode Island 02912, USA

M. SIMIONESCU   The Institute of Cellular Biology and Pathology, 8 Hasdeu Street, Bucharest 70646, Romania

J.R. VANE   The Wellcome Research Laboratories, Langley Court, South Eden Park Road, Beckenham, Kent BR3 3BS, UK

W. VAN EWIJK   Department of Cell Biology and Genetics, Erasmus University of Rotterdam, P.O. Box 1738, Rotterdam, The Netherlands

L. WEISS   Department of Animal Biology, University of Pennsylvania, School of Veterinary Medicine, 3800 Spruce Street, Philadelphia, Pennsylvania 19104, USA

I.L. WEISSMAN   Department of Pathology, Stanford University Medical Center, Stanford, California 94305, USA

A.F. WILLIAMS   MCR Cellular Immunology Unit, Sir William Dunn School of Pathology, University of Oxford, South Parks Road, Oxford OX1 3RE, UK

J.J. WOODRUFF   Department of Microbiology and Immunology, Downstate Medical Center, State University of New York, 450 Clarkson Avenue, Brooklyn, New York 11203, USA

M.B.H. YOUDIM   Department of Pharmacology, Israel Institute of Technology, Faculty of Medicine, 12 Haalyah Street, Bat-Galim, P.O. Box 9649, Haifa, Israel

S.H. ZIGMOND   Department of Biology, Faculty of Arts & Sciences, Joseph Leidy Laboratory of Biology G7, University of Pennsylvania, Philadelphia, Pennsylvania 19104, USA

Editors: RUTH PORTER *(Organizer),* MAEVE O'CONNOR and JULIE WHELAN

Chairman's introduction

J.L. GOWANS

Medical Research Council, 20 Park Crescent, London W1N 4AL

The theme of this meeting is the interaction of blood cells with vessel walls. The major, but not the only, component of the vessel wall with which we will be concerned is the endothelium, which we now know is not simply a bland lining but a layer of cells with a complex ultrastructure and special metabolic activities, including the secretion of pharmacologically active substances, notably the prostacyclins. The blood cells under consideration will be mainly platelets and leucocytes. Thus, the study of platelet–vessel wall interactions will include an examination of the phenomena of haemostasis and platelet aggregation and deaggregation.

Leucocyte–vessel wall interactions lead to the adhesion of leucocytes to endothelium and their subsequent migration across the walls of fine blood vessels. The most familiar example is the emigration of polymorphonuclear leucocytes during acute inflammation and the hope here is that the role of chemotactic factors *in vivo* will be illuminated.

Studies on lymphocyte migration have generated a mass of descriptive data with rather little understanding, as yet, of basic mechanisms or function. A list of the preoccupations of those in this field would include the following:

(a) Lymphocytes normally recirculate in large numbers through the lymph nodes by way of the vascular endothelium of what used to be called 'postcapillary venules' but for which the term 'high-endothelial venules' (HEV) is now preferred. Other leucocytes do not migrate across these vessels unless the nodes are inflamed. What are the components of the interacting cell membranes which lead specifically to the adhesion of lymphocytes to this particular endothelium and why is adhesion followed by migration across the vessel wall?

(b) Lymphocytes circulate from blood to lymph by way of tissues other than lymph nodes, e.g. normally through the lamina propria of the small intestine;

© *Excerpta Medica 1980*
Blood cells and vessel walls: functional interactions
(Ciba Foundation symposium 71) p 1-2

abnormally through granulomata. Is this explained by the presence of tissue-specific or regionally modified endothelia which possess the properties, if not the appearance, of the HEV of lymph nodes?

(c) Are there subpopulations of lymphocytes within the total recirculating pool which recirculate preferentially through certain regional lymphoid beds or certain tissues?

(d) What is the functional significance of lymphocyte recirculation? There has been little speculation on this point, apart from the suggestion that it increases the efficiency of regional immune responses by allowing antigen-induced selection of precursors from a pool of cells larger than that accommodated by the regional nodes alone. This argument turns on the extent to which the frequency of specific B and T lymphocytes limits the size of immune responses *in vivo*. The discovery of suppressor T cells raises the further possibility that recirculation may provide a mechanism for the regulation of immune responses *in vivo*. But immunological theories cannot ignore the fact that lymphocytes recirculate from blood to lymph in the fetus before conventional immune responses are called upon to operate. What is the role of this intrinsic recirculation?

This partial list of topics illustrates the wide range of problems in pathology, physiology and cell biology which are brought together when we begin to consider the common focus of our different interests—the properties of vascular endothelium. We are particularly grateful to Dr R. Porter and Professor W.L. Ford for initiating this ambitious meeting and, as ever, to the Ciba Foundation for its excellent hospitality.

The haemopoietic microenvironment of bone marrow: an ultrastructural study of the interactions of blood cells, stroma and blood vessels

LEON WEISS

University of Pennsylvania Department of Animal Biology, School of Veterinary Medicine, 3800 Spruce Street, Philadelphia, PA 19104

Abstract   This paper draws on studies of normal, depressed and enhanced haemopoiesis in mutant mice (Wv/Wv, Sl/Sl^d, ob/ob, ha/ha, sph/sph, MRL lpr/lpr, BXSB and NZB × BXSB) and in bone marrow experimentally modified by bleeding, hypertransfusion, the administration of phenylhydrazine, busulphan, vinblastine, methylcellulose, X-irradiation or saponin, and in protozoal and helminthic infections. Mice, rats, dogs and chicks have been studied.

Differentiation, maturation and proliferation of haemopoietic cells in bone marrow occur in cords between vascular sinuses. On maturation, blood cells cross the blood vessel wall and reach the circulation. Several cell types are associated with the differentiation and maturation of blood cells and their transmural passage. Macrophages are associated with haemopoiesis in each of the blood cell types. They phagocytize defective cells or discarded cell parts and secrete factors controlling haemopoiesis. Lymphocytes may stimulate or suppress haemopoiesis. Reticular cells are the principal cell type of the stroma of the haemopoietic cords. They appear to have roles in the induction of differentiation, proliferation, cell sorting and the delivery of blood cells to the blood. Reticular cells are a component of the wall of vascular sinuses. They form an adventitial layer and cover, in normally haemopoietic marrow in mice, approximately 65% of the adventitial surface of the vascular sinuses. They also branch into the haemopoietic cords, forming a meshwork which holds haemopoietic cells. Reticular cells or related branched cells appear to be fibroblastic, synthesizing reticular fibres. They may accumulate fat, markedly so in hypoplastic marrow. In regenerating marrow putative stem cells are associated with reticular cells at the outset of haemopoiesis. In intense haemopoiesis dense branched cells appear, displaying large numbers of lysosomes and broad continuities of nuclear cisternae and endoplasmic reticulum. They are probably a variant of reticular cells.

The reticular cell cover of vascular sinuses decreases with increasing haemopoietic cell-crossing. Cells cross the endothelium near cell junctions. The endothelium normally remains without apertures except during cell-crossing. With large-scale crossing, however, endothelium may be disrupted.

3

© *Excerpta Medica 1980*

Blood cells and vessel walls: functional interactions
(Ciba Foundation symposium 71) p 3-19

The demonstration of the recirculation of small lymphocytes by Dr Gowans and his colleagues (Gowans & Knight 1964, Marchesi & Gowans 1964) brought to a close the classic descriptive period in the study of haemopoietic tissue and initiated the modern period in which it has become evident that the structure of haemopoietic tissue is the morphological expression of the interactions of migratory cells, stroma and vasculature. For Dr Gowans not merely showed that lymphocytes recirculated; he also demonstrated their histological pathways through complex lymphatic tissues and, in the process, revealed patterns and tempi of cellular migrations and of cellular interactions upon which the functions of the haemopoietic system depend and through which regulation is effected.

In seeking to honour Dr Gowans and his continuing contributions to knowledge of the haemopoietic system, I should like to present work from my laboratory on the bone marrow. The bone marrow may well be regarded as the central haemopoietic tissue. It possesses an unrivalled capacity to sequester stem cells, it initiates the differentiation of each of the blood cell types, and it is uncommonly rich in interactions between blood cells and blood vessels. In an attempt to convey the kind of histological information provided by Dr Gowans and his colleagues in their work on the recirculation of small lymphocytes I shall present not only observations of the direct interplay of blood cells and blood vessels, but also information on the organization of bone marrow and on the cell types associated with the differentiation, maturation and proliferation of blood cells. I believe I shall be able to make clear, at this symposium dealing with functional interactions of blood cells and blood vessel walls, that the full sequence of stem cell sequestration, differentiation and delivery, and not merely the obvious event in which a vessel is crossed by a blood cell, represents the interaction of blood cells and blood vessels.

Haemopoiesis in mammalian marrow occurs in branching cords separated by anastomosing venous sinuses. The idea of a haemopoietic inductive microenvironment was developed by Trentin (Curry & Trentin 1967, Trentin 1970) largely on the basis of the cellular make-up of 5- to 6-day-old stem-cell-derived haemopoietic colonies in spleen and bone marrow of irradiated mice. Bone marrow normally displays all the various haemopoietic microenvironments because it normally initiates the differentiation of stem cells into each of the blood cell types. Bone marrow-derived fibroblasts harvested from tissue culture and grafted beneath the renal capsule have been shown by Friedenstein and his colleagues (1974) to possess haemopoietic inductive capacity. There have been demonstrations in a number of different systems of the inductive capacities of fibroblasts, stemming from Grobstein's work (1954) on the induction of differentiation of mouse submandibular gland.

Thus, odontoblasts are required for the differentiation of ameloblasts (Croissant et al 1975), fibroblasts for the formation of seminal vesicle, and collagen for the differentiation of myoblasts. In several aplastic marrows displaying incipient haemopoiesis—as in the recovery phase of myelofibrosis induced by saponin (Hoshi & Weiss 1978) in the vascularized fetal marrow, and in regenerating autografts of bone marrow—an association of reticular cells, the distinctive fibroblastic cells in haemopoietic tissue, with putative stem cells immediately precedes the appearance of haemopoiesis.

The haemopoietic microenvironment may well be fashioned partly of reticular cells or related branched stromal cells, as discussed below. It is probable that the first step in the differentiation of a haemopoietic stem cell which commits it to one blood cell type may be induced by local influences in the haemopoietic microenvironment. But as soon as this commitment is made, powerful systemic influences force further differentiation and proliferation on a large scale. In illustration of this hypothesis, a haemopoietic stem cell conditioned by an erythropoietic inductive microenvironment, say in the central red pulp of mouse spleen, differentiates into an erythropoietin-responsive cell (ERC). Erythropoietin reaches ERCs and induces their further differentiation and proliferation into burst-forming units (BFU_es) and colony-forming units of the erythroid type (CFU_es) and then on into the long-recognized basophilic and polychromatophilic erythroblasts. A population of working proliferating and differentiating cells is thereby created, permitting stem cells to rest in G_0.

In our work on bone marrow, my colleagues and I are exploiting mutant and experimental mouse models. We have studied stroma-deficient (Sl/Sl^d) and stem cell-deficient (Wv/Wv) aplastic anaemias (Bernstein et al 1968), spectrin-deficient (ha/ha and sph/sph) haemolytic anaemias, lympho-proliferative autoimmune syndromes (MRL lpr/lpr, BXSB and NZB × BXSB) and obese mice (ob/ob). We have perturbed the bone marrow by bleeding and by hypertransfusion, and by the administration of vinblastine, of busulphan and, in rabbits, of saponin. We have seen the initiation of haemopoiesis in irradiated marrow regenerating after endocloning, in fetal human marrow, in embryonic chick femurs, and in autografts of bone in rats. We have induced haemolytic anaemia with methylcellulose, by phenylhydrazine, and in experimental rodent malaria *(Plasmodium berghei)*. We have studied the eosinophilopoiesis in ascaris and trichinella infections. All in all, we have been provided with virtually every type of haemopoiesis for our studies. While most of our work is in inbred mice, it also includes rats, dogs, cats, birds, rabbits and invertebrates. Out of this material I should like to discuss the cellular concomitants of haemopoiesis and the interaction of blood cells and blood vessels.

Macrophages are consistently associated with haemopoiesis. A well-studied association characterized by Bessis (1958) is the erythroblastic islet which consists of erythroblasts, often in concentric tiers in which the more mature cells are peripheral, surrounding a macrophage whose velamentous processes extend among and surround the erythroblasts. The degree of organization of the islet varies, depending on the level of haemopoiesis and the species. Among the functions of the erythroblast-associated macrophage are phagocytosis of defective erythroblasts (ineffective erythropoiesis) and of nuclear poles during normal—erythroblast to erythrocyte—maturation. But this kind of coupling of macrophage and haemopoietic cell is not restricted to erythropoiesis. Metcalf (1966) described a periodic acid–Schiff-positive (PAS) cell in the lymphoproliferative zones in thymic cortex. In fact this is a macrophage surrounded by dividing lymphocytes—a 'lymphoblastic islet' where the macrophage has phagocytized lymphocytes (the PAS-positive inclusions) which may be imperfect or, perhaps, among the forbidden clones. In eosinophilopoiesis macrophages are surrounded by late-stage eosinophils which they entwine in cell processes. The association of macrophages and leucocytes may be diffuse or organized into tight little islands. I believe such association is a regular feature of haemopoiesis, regardless of blood cell type, and accords to macrophages a monitoring role which may be simply the elimination of defective cells or discarded cell parts or, more complexly, the establishment of identity by destruction of immunologically competent non-self. Macrophages may have a stimulatory role early in haemopoiesis by elaborating colony-stimulating factors (CSF) such as occur in granulocytic tissue culture colonies. Macrophages, moreover, lie on the outside surface of vascular sinuses of bone marrow, as they do in vessels of other tissues, and exercise surveillance over the traffic of material to and from the vessels. It is not unusual for such perivascular macrophages in the marrow to extend tongue-like processes through the vessel wall into the lumen, and engage in phagocytosis there.

Lymphocytes are a cellular concomitant of haemopoiesis. It will prove to be the case, I believe, that the production of each blood cell type is regulated through the interplay of enhancer and suppressor lymphocytes. The best-studied sequence is that of plasma cell differentiation where, except in the case of T cell-independent antigens, the maturation of the plasma cell precursor (B lymphocyte) through intermediate forms (transitional cells) requires the help of a T lymphocyte subtype (T_H cell). The differentiation of the B lymphocyte may be suppressed by another T lymphocyte subtype (T_S cell). Beeson, Basten and the Oxford group (Basten et al 1970, Beeson & Bass 1977) and Speirs and his colleagues (1973) in Brooklyn showed that the eosinophil response to antigen is dependent upon thymic lymphocytes. The eosinophil response to encysted larvae is deficient in thymus-deficient mice. The production of basophils probably depends on thymic lymphocytes. In cultures of thymus, Ginsburg & Sachs (1963) found massive mast cell formation after about two weeks incubation. Ishizaka et al (1976) showed that these incipient mast cells possess IgE receptors in their plasma membrane at about the time their granules appear and that a thymic lymphocyte is associated with the differentiation. In erythropoiesis differentiation may be depressed by a thymic-derived lymphocyte, with resultant aplastic anaemia (the Diamond-Blackfan syndrome) (Hoffman et al 1976, 1977). The same mechanism may be the basis of the aplastic anaemia in the Wv/Wv mutant mouse. In my laboratory D. Brookoff has observed that lymphocytes are regularly in association with erythroblasts during erythropoiesis and N. Sakai has seen lymphocytes in close physical associaton with immature eosinophils and with megakaryocyte precursors in experimental ascaris infections. The lymphocytes we see in haemopoiesis may be contiguous to haemopoietic cells or separated from them by several cells. Their influence evidently does not depend on specialized intercellular junctions or bridges but rather, perhaps, on diffusible locally acting microhumoral substances of the lymphokine sort.

Reticular cells—characteristic, dendritic stromal cells—are regularly associated with haemopoiesis. Their role may be to support the haemopoietic microenvironments inducing stem-cell differentiation and it therefore may parallel that of other fibroblasts in other systems. Reticular cells form a meshwork supporting haemopoietic cells and the vasculature. They envelop or lie upon argyrophilic reticular fibres and, because they appear to synthesize them, constitute a type of fibroblast. Reticular cells vary in apparent fibroblastic activity. Those in the red pulp of spleen are associated with heavy well-developed reticular fibres while those in haemopoietic compartments of bone marrow lie upon attenuated fibres. There may in fact be scarce evidence of reticular fibres in marrow, although a thick ruthenium red-positive layer on the surface of reticular cells is characteristic. In saponin-treated rabbits, on the other hand, there occurs a pronounced myelofibrosis and reticular cells can be associated with slender reticular fibres and with heavy bands of collagen. In the period of haematology in which Aschoff, Maximow, Sabin, Downey, Dantschakoff and their colleagues were active, reticular cells were accorded stem cell functions and were regarded as phagocytic (the fixed macrophages). In fact they are neither stem cells nor phagocytes. When the nature of the reticuloendothelial system is definitively understood, I believe that reticular cells will be recognized as possessing a central role not, as first supposed, as macrophages or other such effector cells, but as regulator cells controlling the differentiation, proliferation, compartmentalization and sorting of the blood cells.

Reticular cells may be part of the vasculature. In the marrow, they may lie on the outside surface of blood vessels and constitute an adventitial layer. In normal red marrow of mice these adventitial reticular cells cover about 60% of the adventitial surface of vascular sinuses. They branch away from the sinus into the perivascular haemopoietic space. Here their processes, as broad but thin curving sheets, envelop haemopoietic cells or, put another way, haemopoietic cells lie in the interstices of the meshwork formed by these processes. It is the association of such reticular cells and putative stem cells that presages haemopoiesis. In intense haemopoiesis, as in the erythropoiesis of the spectrin-deficient mouse or in the eosinophilopoiesis of the ascaris-infected mouse or rat, there appear dark, branched multinucleate cells with dilated nuclear cisternae which extend into rough endoplasmic reticulum. They lie among haemopoietic cells, often among the more immature of them (in the case of the eosinophil), covering them with broad cytoplasmic processes. They form a reticulum and may be a variant of the normal, less active reticular cells, or, perhaps, of macrophage-derived multinucleate cells.

Adventitial reticular cells display additional conformations in the vascular wall. They interact with maturing haemopoietic cells that approach the adventitial surface of the vessel en route to the circulation. The reticular cells move away from the wall, baring the basement membrane which is readily solvated and then the basal endothelial surface. In active transmural cell transport, the adventitial cover may be reduced to less than 20%. Microfilaments and microtubules are abundant in adventitial reticular cells and, in places, may criss-cross heavily to form felted plaques. These cells may develop junctional complexes of the adherent type with endothelial cells and with other reticular cells.

Adventitial cells may become quite voluminous and lucent, perhaps due to hydration, and if this change is large-scale the marrow becomes white and gelatinous. The haemopoietic space is reduced in such marrows because the expanded adventitial cells lie in the space. Adventitial cells may become fatty and even murine bone marrow, which is virtually all haemopoietic, contains a few fat cells. In the *ob/ob* mouse, a mutant which is quite obese, weighing 85 g instead of the normal 25 at eight months of age, there is a slight to moderate increase in fat in the marrow as observed microscopically but, as D. Brookoff has shown in my laboratory, a 100-fold increase in fat as shown by chemical extraction and spectrophotometry. Fat is regularly present in haemopoietic marrow in rats and human beings. When the fatty transformation of adventitial cells is large scale, the marrow is yellow and fatty. While it is true that fat may expand to occupy space in the marrow when haemopoiesis is reduced—an adaptation of a tissue that cannot shrink because of the rigidity of its bony capsule—marrow fat should not be regarded as merely bulk. Sahebekhtiari & Tavassoli (1978) showed that fat from haemopoietic marrow is unsaturated while that from characteristically yellow marrow is saturated. Marrow fat is capable of aromatizing oestrogens to androgens, as is being demonstrated by Dr Rose Frisch and her colleagues. Dexter et al (1977) have shown that adipocytes are associated with granulocyte formation in cultured marrow. I believe it probable that adventitial reticular cells are the source of the adipocytes seen in culture and that these reticular cells, on becoming fatty, can induce granulocytopoiesis in intact marrow. In fact certain experiments in my laboratory—by D. Brookoff in the hypertransfused mouse and by P. McManus in the busulphan-treated mouse—suggest that this may occur. If this is so, perhaps those cells which have considerable increases in fat, as shown chemically, but not enough to be seen microscopically, may be sufficiently modified to possess granulocyte-inductive capacities. Reticular cells may also become the fibroblastic cells of myelofibrotic marrows, as after treatment of rabbits with saponin. As fatty marrow has been incorrectly assumed to be inert in haemopoiesis, fibrous marrow has been assumed to be an end-stage in the shut-down of haemopoiesis. In fact, however, haemopoiesis often flourishes in such marrow, as revealed in recovering saponin-treated animals. In the light of the evidence that fibroblasts have inductive roles in many systems, fibrous marrow may represent not a burnt-out marrow but one primed for the induction of haemopoiesis. It may be aplastic only because of failure in stem cells or dysfunction in helper or suppressor lymphocytes or in other factors essential to haemopoiesis.

While the adventitial layer of the vascular sinuses of marrow faces out towards the perivascular haemopoietic tissue and regulates its differentiation and access to the endothelium, the endothelium faces the circulation and constitutes the major element in the barrier between blood and haemopoiesis. It is a slender vesiculated active barrier which may be quite flat and thin, or thrown up into irregular luminal projections. Colloidal materials cross this barrier readily. Campbell (1972) has shown that apertures in the wall occur only during blood cell passage, and that blood cells cross the wall near endothelial cell junctions.

Vascular sinuses have distinctive functions other than cell passage. R.Lambertsen, in my laboratory, has found that within 24 hours of lethal irradiation in the mouse, the haemopoietic compartment is drastically reduced in volume and the venous vasculature takes up this sudden slack, becoming so widely dilated as to occupy almost the whole of the marrow cavity. Paradoxically, this congested bright red marrow is an aplastic marrow. I believe that this substantive increase in venous vascular surface is made possible by both the presence of adventitial cells and the nature of interendothelial junctions. In vascular sinuses it is not unusual, especially where transmural passage is active, to find intervals in the venous wall where adventitial cells appear on the lumen as endothelium. I believe, moreover, that the interendothelial junctions are sliding ones. Tavassoli's freeze-etch and lanthanum studies (Tavassoli & Shaklai 1979) have indicated that these junctions are not tight and that they permit colloidal material to penetrate the endothelium. Junctions, moreover, are not at right-angles to the surface but are oblique and quite long. Therefore, sudden dilatation consequent to the melting away of haemopoiesis can be achieved by the endothelial cells sliding apart on one another at their extended oblique junctions, so reducing endothelial overlap and presenting more endothelial surface to the lumen. At the same time, where adventitial cells appear on the lumen, the luminal surface is greatly augmented, as an adventitial cell can slide more of itself into the endothelium. The wall now has less adventitial cell cover, shorter interendothelial cell junctions, thinner endothelium and an enormously increased surface area.

The events in the delivery of blood cells from marrow to blood culminate in blood cells crossing the wall of venous sinuses and entering the lumen. As blood cells approach the wall the adventitial cells may move aside, lift away from the wall, or become pierced by the blood cells. A common arrangement is that adventitial cells separate from the endothelium and the blood-bound blood cells lie between them and the endothelium. There is thereby created a pocket environment, preparatory to transendothelial passage, in which adventitial and endothelial cells, representing fixed elements of stroma and vasculature, enclose the migratory blood cells and, not infrequently, a macrophage with them. The vascular basement membrane disappears, probably being depolymerized. The attack on the basal surface of the endothelium may take different forms. In some instances, especially in large-scale granulocyte transfer, as is evident after endotoxin has been given, lucent areas in overlying endothelium suggest an intracytoplasmic lytic process (the plasma membrane remains intact). Such lucent swelling may spread through a whole endothelial cell, or a series of cells, stopping sharply at interendothelial junctions. In more typical cases, however, the blood cell presses upon the endothelium, whose basal and luminal plasma membranes meet and break down. An aperture is formed and the blood cell squeezes through it. The process may be not unlike a type of endocytosis. (Emperipolesis is quite evident in the microenvironment of the endothelium, notably in megakaryocytes.) But in the endocytosis involving a blood cell and the venous endothelium, the thinness of the endothelium relative to the bulk of the blood cell prevents the blood cell from being encompassed by the endothelium as pinocytotic vesicles are. Instead the endocytotic vesicle, if the endothelial aperture may be referred to in that way, is open on both the basal and luminal surfaces and the relatively massive blood cell, in its mid-passage, sticks out on both sides. When the migrant completes its passage, the endothelium is healed. Microfilaments, according to Campbell (1972), may be arranged radially about the aperture, lying in the plane of the endothelial surfaces. Within a leucocyte in passage, microfilaments may be grouped in a corset-like peripheral ring, pinched in the endothelial aperture.

Whenever cellular passage becomes quite brisk, as in the two to three-day period in an ascaris-induced eosinophilia during which the number of blood eosinophils mounts from several hundred per cubic millimetre to near 1000, the vessel wall is profoundly changed. The endothelium may be quite rarefied and swollen, and many blebs, folds, other protrusions and invaginations may occur, particularly at the luminal surface. The concentration and size of coated and non-coated vesicles may be increased. The endothelium may be disrupted, moreover. Large gaps occur in the wall even in the absence of cells in passage. The changes may be so pronounced that some sinuses are difficult to recognize. Sinuses at the periphery of the marrow are affected more markedly than those more central. Even at the periphery, however, the changes are patchy, with many sinuses unaffected. It appears that the changes in the vascular sinuses are greatest in those places where blood cell delivery is greatest.

In work in progress in my laboratory, I have found that the marrow in the late stages of *Plasmodium berghei* murine malaria is intensely erythropoietic and active in the delivery of red cells to the circulation. The erythropoiesis in these stages appears commensurate with the massive haemolytic anaemia (haematocrit $\cong$ 8) that is due to parasite-induced damage to erythrocytes, opsonins and hypersplenism. By Day 21 of the infection, in the terminal stages of the disease, reticulocytes and polychromatic erythroblasts (but no orthochromatic erythrocytes) are in the circulation and, as *P. berghei* has a propensity for immature erythroid cells, virtually all circulating red cells are parasitized.

But in the early phases of the disease the level of erythropoiesis appears out of proportion to the anaemia. This stimulates two speculations. The first is that the action of *P. berghei* on immature erythroid cells (including those in the proliferating compartment) may be mitogenic—perhaps acting through erythropoietin receptors. The second relates to the possibility that a selective association between circulating red cells and the endothelium of the vascular sinuses in the marrow may induce endothelial breakdown and the release of red cells from the haemopoietic compartment to the lumen of the sinus. I have occasionally observed in red cell delivery from the marrow in other anaemias, without giving it special note, that circulating red cells on the luminal surface may touch and indent the endothelium of vascular sinuses at a place where red cells ready for release in the haemopoietic compartment would indent the basal surface. In our malaria-infected mice, circulating cells must frequently adhere to the endothelium, as can be inferred from the presence of intercellular ground substance between red cell and endothelium, at the sites associated with delivery of cells to the circulation. At such sites, moreover, the endothelium often appears attenuated. We must consider whether this is peculiar to malaria (endothelial adhesion of parasitized erythrocytes is characteristic of certain malarias) or whether it is a clue to a normal feedback mechanism that regulates erythrocyte release.

ACKNOWLEDGEMENT

This work was supported by US Public Health Service Grant AM - 19920 - 03.

References

Basten A, Markley MB, Boyer H, Beeson PB 1970 Mechanism of eosinophilia. I. Factors affecting the eosinophil response of rats to Trichinella spiralis. J Exp Med 131:1271-1287

Bernstein SE, Russell ES, Keighley G 1968 Two hereditary mouse anemias (Sl/Sl^d and W/W^v) deficient in response to erythropoietin. Ann NY Acad Sci 149:475-485

Beeson PB, Bass DA 1977 The eosinophil. Saunders, Philadelphia (Major problems in internal medicine vol 14)

Bessis M 1958 L'îlot érythroblastique, unité fonctionelle de la moelle osseuse. Rev Hématol 13:8-11

Campbell FR 1972 Ultrastructural studies of the transmural migration of blood cells in the bone marrow of rats, mice and guinea pigs. Am J Anat 135: 521-535

Croissant R, Guenther H, Slavkin HC 1975 How are embryonic preameloblasts instructed by odontoblasts to synthesize enamel? In: Slavkin HC, Greulich RC (eds) Extracellular matrix influences on gene expressions. Academic Press, New York, p 515-522

Curry JL, Trentin JJ 1967 Hemopoietic spleen colony studies, I: Growth and differentiation. Dev Biol 15:395-413

Dexter TM, Moore MAS, Sheridan APC 1977 Maintenance of hemopoietic stem cells and production of differentiated progeny in allogeneic and semiallogeneic bone marrow chimeras in vitro. J Exp Med 145:1612-1616

Friedenstein AJ, Chailakhyan RK, Latsinik NV, Pansyuk AF, Keiliss-Borok IV 1974 Stromal cells responsible for transferring the microenvironment of the hematopoietic tissues. Transplantation 17:331-340

Ginsburg H, Sachs L 1963 Formation of pure suspensions of mast cells in tissue culture by differentiation of lymphoid cells from the mouse thymus. J Natl Cancer Inst 31:1-21

Gowans JL, Knight EJ 1964 The route of recirculation of lymphocytes in the rat. Proc R Soc Lond B Biol Sci 159:257-282

Grobstein C 1954 Tissue interaction in the morphogenesis of mouse embryonic rudiments in vitro. In: Aspects of synthesis and order in growth. Princeton University Press, Princeton, NJ p 233-256

Hoffman R, Zanjani ED, Zalusky JVR, Lutton JD, Wasserman LR 1976 Diamond-Blackfan syndrome: lymphocyte-mediated suppression of erythropoiesis. Science (Wash DC) 193:899-900

Hoffman R, Zanjani ED, Lutton JD, Zalusky R, Wasserman LR 1977 Suppression of erythroid-colony formation by lymphocytes from patients with aplastic anemia. New Engl J Med 296: 10-13

Hoshi H, Weiss L 1978 Rabbit bone marrow after administration of saponin: an electron microscopic study. Lab Invest 38:67-80

Ishizaka T, Okudaira H, Mauser LE, Ishizaka K 1976 Development of rat mast cells in vitro, I: Differentiation of mast cells from thymus cells J Immunol 116:747-754

Marchesi VT, Gowans JL 1964 The migration of lymphocytes through the endothelium of venules in lymph nodes: an electron microscopic study. Proc R Soc Lond B Biol Sci 159:283-290

Metcalf D 1966 The thymus: its role in immune responses, leukaemia development and carcinogenesis. Springer-Verlag, Berlin (Recent Results Cancer Res vol 5)

Sahebekhtiari HA, Tavassoli M 1978 Studies on bone marrow histogenesis: morphometric and autoradiographic studies of regenerating marrow stroma in extramedullary autoimplants and after evacuation of marrow cavity. Cell Tissue Res 192:437-450

Speirs RS, Gallagher MT, Rauchwerger J et al 1973 Lymphoid cell dependence of eosinophil response to antigen. Exp Hematol 1:150-158

Tavassoli M, Shaklai M 1979 Absence of tight junctions in endothelium of marrow sinuses: possible significance for marrow cell egress. Br J Haematol 41:303-307.

Trentin JJ 1970 Influence of hematopoietic organ stroma (hematopoietic inductive microenvironments) on stem cell differentiation. In: Gordon AS (ed) Regulation of hematopoiesis. Appleton-Century-Crofts, New York, p 161-186

Discussion

Gowans: You have suggested that particular cells in the marrow provide a microenvironment which is essential for the differentiation of blood cells. After extramedullary haemopoiesis has regressed in, for example, the spleen of a young animal or an embryo, does some element which you think is a necessary component of the microenvironment disappear concomitantly with the disappearance of the haemopoietic cells?

Weiss: The activity of these dendritic cells seems to be correlated with the level of haemopoiesis in the bone marrow. During intense haemopoiesis large numbers of dark branched reticular cells are present. When haemopoiesis slows down, these cells are no longer present. They may be present in another form but their density, their activity, their lysosomes, the connections between the perinuclear space and the rest of the cytoplasm, are not evident. The numbers of lymphocytes and of macrophages also seem to decrease as haemopoiesis falls off. Erythropoiesis in the mouse spleen, heightened by haemolysis, is under study in my laboratory but we do not yet have data on the cellular associations in splenic haemopoiesis.

Davies: Do you regard the adventitial cells as fixed or can they have a haematogenous origin?

Weiss: The question of fixed versus free cells needs to be discussed prudently. The literature of haematology is littered with inappropriate answers. Many cells that we tend to regard as fixed are not fixed at all. Thus, macrophages move about more than we anticipated; endothelial cells may come off the wall and flow in the circulation; and megakaryocytes regularly circulate. Although the adventitial cells can move away from a wall and into the haemopoietic space, I have not seen them in any other location.

Davies: In nude mice, or mice which you injected first with an anti-nude antiserum, are there any major defects in haemopoiesis apart from those involving lymphocytes?

Weiss: The bone marrow of nude mice is not too different from normal bone marrow except that the dark cells associated with the intense eosinophil production after trichinella infection are absent. Although the nude mouse has a normal base number of eosinophils this does not rise in eosinophilia in response to helminthic infection. That may be the case for haemopoiesis in general in the nude mouse. It may turn out that the capacity of the nude mouse to respond to any type of haemopoiesis under stress—as heightened erythropoiesis after bleeding or haemolysis—is impaired.

Davies: In some earlier experiments (Walls et al 1971) the resting level of eosinophils was the same in normal and T-cell-deprived mice. Although on stimulation with *Trichinella* only the normal mice became eosinophilic, I am still not totally convinced that the physical presence of lymphocytes adjacent to eosinophiloblasts in the marrow is a *sine qua non* for eosinophil production. But you may well be right. Your presentation was very seductive in this respect.

Weiss: Lymphocytes are very difficult cells to characterize morphologically. We are not supposed to call the stem cell a lymphocyte but I think in the electron microscope it is morphologically a lymphocyte. When we consider lymphocytes as seen in light or electron micrographs without special markers, we are talking about T cells, B cells, stem cells and so on.

There are, moreover, the stages of development of erythroblasts, including the stem cell, the burst-forming unit and the colony-forming unit—and I suspect that they all look like lymphocytes. They are all, of course, quite different cells functionally.

Davies: The question of whether lymphocytes should be regarded as haemopoietic stem cells is, as you know, an old one. One thing that seems strongly against it is that if one tries to reconstitute an animal after irradiation with lymphocytes alone nothing happens. That suggests that it is either a lousy experiment or that lymphocytes cannot easily be regarded as haemopoietic stem cells. Of course the lymphocytes in these circumstances are derived from lymphoid organs. Lymphocytes derived specifically from bone marrow may be a different kind of cell.

Weiss: We mustn't fall into a historical and semantic trap. Dr Gowans pointed out to Professor Yoffey in 1961 that an irradiated mouse cannot be reconstituted with lymphocytes from lymphoid organs. We know now that an irradiated mouse can be reconstituted with a suspension of bone marrow cells, so stem cells do lie in bone marrow. These stem cells look like the candidate stem cell in Dicke's picture (Dicke et al 1973, Fig 4 p 60) and if one looks at it dispassionately that cell looks somewhat like a lymphocyte. It is in that very restricted morphological sense that I am using the term lymphocyte. It is simply to recognize that cells which are of different cell type—different in function, in cell surface receptors, etc—can look alike by conventional light and electron microscopy. I don't want to summon up old battles which I think are over.

Williams: It is absolutely clear that virtually none of what you call lymphocytes in the marrow share markers with those in the periphery. They are completely different sets of cells. I think most of them are likely to be immature B cells (Hunt et al 1977).

Weiss: Yes. We have reached that stage of haematology where instead of having some remarkable pigment in the cell that marks a red cell, or instead of having granules in a cell that mark a granulocyte, or instead of having a deformed nucleus, we have to go to the realm of the cell surface markers to delineate different cell types that otherwise look alike. We have not yet, to my knowledge, characterized the cell surface markers on a stem cell that are distinctive to the stem cell.

Williams: The stem cell in the rat clearly has Thy-l antigen on it. Most of the lymphocytes in the marrow can be distinguished from those in the periphery, both B and T cells, by at least four marker antigens at the cell surface.

Weiss: I understand. We must remember that stem cells are not a common cell in the bone marrow. Dicke et al (1973) increased the number of stem cells by a factor of 40 to 60, by density gradient centrifugation and by lightly irradiating the donor. Even in the bone marrow where they occur in greatest number they occur in a very low concentration and are often not in the cell cycle. There are certain circumstances, such as in patients recovering from aplastic anaemias and in the establishment of bone marrow in the fetus, where we think we see putative stem cells. So far the only way we have been able to quantitate stem cells is by spleen colony assay. It now appears that we have a means by tissue culture as well. I hope we shall soon be able to recognize a stem cell by a distinctive cell surface marker or set of markers not shared by other haemopoietic cells. Without those markers, I suspect that stem cells, T cells, B cells, cells in burst-forming units and cells in colony-forming units all look monotonously alike.

Weissman: Dicke et al (1973) enriched the cells 40−60-fold but they had about 100-fold or 1000-fold more to go. Thus, one doesn't know whether any particular cell they happen to show you is in fact a haemopoietic stem cell.

You said that lymphocytes are associated with haemopoietic clusters. If you propose that T cells are involved in regulation—an interesting possibility—do these putative T cells express any of the surface markers characteristic of peripheral T cells in an appropriate animal (not the rat), e.g. Thy-l or Lyt markers in the mouse?

We regularly see from 0.5% to as much as 4% of small round cells in mouse marrow with distinctive T cell markers. Could those be the cells you are seeing associated with haemopoietic clusters or are these T cells arranged otherwise?

Weiss: We are planning to look at that.

Weissman: You said that you think that the dendritic cells or adventitial cells are fibroblastic and that they can give rise to the fat-filled cells which you can see later. Have you any markers or any intermediate stages to back that up?

Weiss: Yes. After the irradiation of bone marrow or the creation of aplastic anaemia these adventitial cells begin to accumulate droplets of fat, and after a time they become quite fatty. And in the obese mouse (*ob/ob*), which at eight months of age weighs about 85−90 g instead of 25 g, the bone marrow morphologically contains just a bit more fat, though chemically it contains about 100-fold more fat. That fat is in cells in an adventitial position in the wall of venous sinuses.

With regard to the possible fibroblastic role of these branched stromal cells, normally the marrow is not very fibroblastic. There is not much collagen about, although there is basement membrane which is made of collagen. These cells have on their cell surface a very heavy ruthenium red-staining coat, which is a type of matrix. In some circumstances, however, very dense fibroblastic cultures can be grown from bone marrow. In an aplastic marrow induced by saponin, moreover, the marrow can become myelofibrotic. I believe, though my evidence is only tentative, that these branched cells become fibroblastic in such circumstances. We should be prepared to learn, moreover, that the dendritic cells of the bone marrow may be separable into subgroups, just as lymphocytes have been shown to be of different subtypes.

Marcus: You mentioned that the presence of fat might act as a stimulus. Could it therefore be speculated that in aplastic anaemia one has an accumulation of fat that is acting as a stimulus, but there are no cells present to respond to this stimulus?

Weiss: Yes, in the saponin-treated and the irradiated animals one can get a marrow that is fatty and fibrous. When those marrows begin to recover, as a certain percentage of them do, one finds an interesting association of branched stromal cells, many of which contain fat, with putative stem cells. A fatty and fibrous marrow may not necessarily be a burnt-out marrow. It may be prepared to become haemopoietic but be lacking stem cells or other essential elements. When one induces an aplastic marrow by hypertransfusion, moreover, within days the red cells disappear, fat appears and soon afterwards granulocytopoiesis occurs. Further, there are cultures of bone marrow in which granulocytic colonies can be seen on the aprons of fat cells.

Marcus: In a study that we did in collaboration with Dr Dorothea Zucker-Franklin, cultured monocytes were studied by both electron microscopy and biochemical analysis. Such monocytes accumulated large quantities of fat in about 5–7 days, and Dr Zucker-Franklin was able to identify them as still retaining the properties of monocytes (Zucker-Franklin et al 1978). A lipid profile of these cells was made, and it did indeed change. Dr Zucker-Franklin believes that this might represent a situation analogous to aplastic anaemia. Is it possible that the dendritic cells you showed are comparable to this cultured monocyte system?

Weiss: I don't think so, although the cultured monocyte is one step away from the macrophage, which can be dendritic as well. The composition of the fat varies interestingly in the marrow. Tavassoli (1976) showed that in marrow which is established as fatty marrow, as in limb bones, the fat is saturated. But where fat occurs in haemopoietic marrow it is unsaturated, as shown by the performic acid-Schiff histochemical method.

Marcus: With regard to your comments on megakaryocytes, I have had the opportunity to observe a patient with profound chronic thrombocytopenia. This is an otherwise healthy antique dealer who has survived for more than 20 years with platelet counts varying from zero to 15 000/mm^3. When 425 ml of whole blood from this patient was processed in a manner that theoretically should yield platelet-rich plasma, and when this plasma was centrifuged as if it contained platelets, a thin film was obtained. When this film was examined by electron microscopy by Dr Dorothea Zucker-Franklin, it seemed to consist of segments of megakaryocytes—which one ordinarily does not see in the circulation. This is of course only one subject; but the process we may be observing could be an abortive attempt on the part of the bone marrow to deliver platelets into the circulation. It is also of interest that the patient has been essentially asymptomatic, with an occasional ecchymotic lesion after trauma. Apparently this patient's vascular integrity has in some way compensated; alternatively, the unknown substance that platelets theoretically possess and which maintains vascular integrity is somehow being produced and utilized in this patient (this is a highly speculative explanation).

Weiss: There is a paradox in this. Some years ago Kitchens and I (1975) showed that defects in the endothelium of blood vessels occur in experimental thrombocytopenia in rabbits, and that normal endothelium is restored with restoration of platelets. These results suggest that normal circulating platelets are responsible for endothelial maintenance. Yet megakaryocytes located on the adventitial surface of vascular sinuses in marrow and producing platelets lie on defects in the endothelium.

Marcus: I think it is controversial as to whether platelets physically maintain vascular integrity, or whether they supply some chemical agent to the vascular wall. Such a situation may be analogous to what happens with serotonin, wherein the latter is synthesized in the gut and is picked up and transported by the platelets.

Weiss: I didn't speak to that. I am not sure how platelets work. Prednisone, for example, can substitute for platelets. In an experimental platelet-depleted rabbit the endothelium loses redundancy and becomes defective. But if prednisone is administered during the depletion of platelets the endothelium remains normal despite the absence of platelets.

Owen: Have you a hypothesis to explain the migration of the extravascular elements towards the endothelium as they mature? Is a chemotactic influence involved? Is there some change in the amoeboid capabilities of the cells as they mature?

Weiss: I think it varies. For example, it is known that as granulocytes mature their cytoplasm becomes capable of the sol-gel change associated with locomotion. They show abundant microtubules and microfilaments in the metamyelocyte stage and they are motile. In the promyelocyte and myelocyte stages they just sit still. I find that the earlier stages, promyelocytes and myelocytes, tend to occur away from the vascular wall. My impression is that as the granulocytes mature to the metamyelocyte stage they move over to the vascular wall and then go through. How red cells manage to reach and cross the vascular wall and thereby enter the circulation, I don't know. There are two intriguing possibilities. One is that the macrophages that are always closely associated with erythropoiesis help. Secondly, in heightened erythropoiesis one may observe circulating reticulocytes already in the lumen of vascular sinuses adherent to the luminal surface of endothelium. At sites nearby, apertures develop in the endothelium and reticulocytes enter the circulation. Thus erythrocytes circulating in the vascular sinuses of the marrow may, by interaction with the endothelium, control the delivery of perivascular reticulocytes into the lumen of those vascular sinuses.

Humphrey: If you have done serial studies of lethally irradiated animals, can you say which cells disappear first and at what rate? How far do the cells that remain depend on continuing haemopoiesis in that marrow? I am thinking about what happens when bone marrow is transplanted into a recipient after lethal irradiation.

Weiss: In the lethal irradiation that R.Lambertsen has studied in my laboratory, haemopoiesis in the marrow melts away in a few days. The cells that persist are associated with the vasculature—endothelial cells and the adventitial cells. There are in addition always moderate numbers of macrophages. The adventitial cells seem to undergo fat transformation. Most of the haemopoietic elements go. The megakaryocytes tend to be relatively resistant, as are macrophages. Numbers of stem cells survive the irradiation since moderate numbers of haemopoietic colonies develop in lethally irradiated marrow—obviously in adequate numbers to allow survival—even without endocloning or exocloning.

References

Dicke KA, van Noord MJ, Maat B, Schaefer UW, van Bekkum DW 1973 Attempts at morphological identification of the haemopoietic stem cell in primates and rodents. In: Haemopoietic stem cells. Excerpta Medica, Amsterdam (Ciba Found Symp 13) p 53-62

Gowans JL 1961 In: Biological activity of the leucocyte. Churchill, London (Ciba Found Study Grp 10) p 58-59

Hunt SV, Mason DW, Williams AF 1977 In rat bone marrow Thy-1 antigen is present on cells with membrane immunoglobulin and on precursors of peripheral B lymphocytes. Eur J Immunol 7:817-823

Kitchens C, Weiss L 1975 The dependence of endothelium on blood platelets as demonstrated in experimental thrombocytopenia. Blood 46:567-578

Tavassoli M 1976 Marrow adipose cells: histochemical identification of labile and stable components. Arch Pathol Lab Med 100:16-23

Walls RS, Basten A, Leuchars E, Davies AJS 1971 Contrasting mechanisms for eosinophilic and neutrophilic leucocytoses. Br Med J 3:157-159

Zucker-Franklin D, Grusky G, Marcus A 1978 Transformation of monocytes into 'fat' cells. Lab Invest 38:620-628

Immunoelectron-microscopic characterization of lymphoid microenvironments in the lymph node and thymus

WILLEM VAN EWIJK

Department of Cell Biology & Genetics, Erasmus University, P.O. Box 1738, Rotterdam, The Netherlands, and Department of Pathology, Stanford University, Stanford CA 94305, USA

Abstract   The cell surface morphology of B and T lymphocytes *in vivo* has been investigated by scanning electron microscopy (SEM) in mice experimentally enriched for B cells, and mice experimentally enriched for T cells.   The general findings were: (1) it is *not* possible with SEM to distinguish between T and B cells *in vivo* using morphological criteria; (2) both T and B cells display two types of surface morphologies: lymphocytes homing in their respective microenvironments have a smooth cell surface, whereas cells in recirculation compartments exhibit numerous microvilli; (3) cells attached to the walls of high-endothelial venules (HEV) in lymph nodes show microvilli and apparently use these microvilli to support the cell during the initial attachment phase; (4) cells in the process of passing the endothelial lining of HEV undergo morphological changes from villous to smooth; (5) migrating cells cross the endothelial lining through gaps between the endothelial cells.   No evidence was found for emperipolesis. *In vitro* studies of cell suspensions of thymocytes, dexamethasone-resistant thymocytes and lymph node cells from mice enriched for B cells also showed that lymphocytes have a heterogeneous cell surface morphology. Many cells showed long microvilli oriented towards the substratum, indicating that these organelles have a supporting function. For a more specific characterization of cell types at the ultrastructural level immunospecific methods are required.   We developed an immunoperoxidase method for the detection of cell surface antigens on lymphoid and non-lymphoid cells *in vivo*. With this method we studied the distribution of B lymphocytes in the lymph node and spleen and the distribution of major histocompatibility antigens (MHC) on cells in the mouse thymus.

In peripheral lymphoid organs, T and B lymphocytes each localize in specific microenvironments.   Gutman & Weissman (1972) and Weissman (1976), using immunofluorescence techniques on frozen tissue sections, have shown that B cells localize in follicles in the outer cortex, whereas T cells are distributed throughout the paracortex of the lymph node and the periarteriolar lymphoid sheath in the spleen.   Histologically these specific lymphoid microenviron-

© *Excerpta Medica 1980*
Blood cells and vessel walls: functional interactions
(Ciba Foundation symposium 71) p 21-37

ments are characterized by specific non-lymphoid cells. T cell microenvironments contain interdigitating cells, whereas B cell microenvironments contain dendritic cells (Veerman & van Ewijk 1975). The intimate cell surface contact between lymphoid cells and interdigitating or dendritic cells, as observed with the electron microscope, suggests that these reticular type cells are involved in the homing of lymphoid subpopulations (van Ewijk et al 1974). So far we have not been able to distinguish the two major lymphoid subpopulations at the ultrastructural level. A major problem here is that T cell microenvironments are not exclusively occupied by T cells. Nieuwenhuis & Ford (1976) have shown that migrating B cells may traverse T cell areas on their way to follicles. Also, in immunofluorescence studies using monoclonal rat anti-Thy-1 hybridoma antibodies, we have shown that some T cells occur in follicles (W. van Ewijk, J. Ledbetter & I.L. Weissman, unpublished observation). Another problem in electron microscopic investigation of lymphoid subpopulations is that the ultrastructure of both T and B lymphocytes changes dramatically after antigenic stimulation (van Ewijk et al 1977).

In 1973, Polliack and coworkers and Lin et al (1973b) presented evidence that scanning electron microscopy (SEM) might be a suitable tool for the identification of lymphoid subpopulations. In SEM studies of lymphoid cells isolated from peripheral blood, these workers demonstrated a correlation between the cell surface structure and the presence of class-specific markers on lymphoid cells. T cells generally showed a smooth cell surface, while B cells showed many microvilli. These early observations prompted us to investigate the surface morphology of lymphoid cells *in vivo,* in order to analyse the ultrastructure of cell–cell interactions in lymphoid microenvironments. Furthermore, SEM might provide insight into the nature of cell–cell interactions and migration pathways in high-endothelial venules (HEV) in the lymph node.

To this purpose we analysed the topography of cells of the lymphoid system *in situ* in mice enriched for B lymphocytes, or enriched for T lymphocytes. In studies of the cell surface of B cells we examined the lymphoid organs of two types of mice: (a) (DBA/2 × C57BL/Rij) F1 mice, deprived of T cells by adult thymectomy followed by lethal irradiation and reconstitution with anti-Thy-1-treated bone marrow (T × BM mice), and (b) B10.LP/JPh nude mice. The topography of T lymphocytes was studied in the thymus of 6-week-old (DBA/2 × C57BL/Rij) F1 mice and 6-week-old AKR mice. The peripheral lymphoid organs of (DBA/2 × C57BL/Rij) F1 mice, lethally irradiated and reconstituted with syngeneic thymocytes (T mice) or cortisone-resistant thymocytes (CRT mice), were also examined. Immunofluorescence control studies on cell suspensions of the mesenteric node of T × BM mice 4 weeks after reconstitution showed 82% immunoglobulin-positive lymphocytes and 10% T cells, as determined with heterologous antisera. In nude mice these values were 90% and 5% respectively. The mesenteric node of T mice was totally devoid of immunoglobulin-positive lymphocytes; 98% of the lymphoid cells were positive for the anti-T-cell serum (van Ewijk et al 1975).

All mice were killed by total body perfusion with Dulbecco's phosphate buffer (D-PBS), followed by 1.5% glutaraldehyde in D-PBS. This method has the advantage that the lymphoid tissues are fixed very rapidly: glutaraldehyde was introduced 30 s after perfusion began. Secondly, perfusion clears the blood cells from the lumen of the vessels, which allows the vascular wall to be inspected by scanning electron microscopy (SEM). Details of this fixation method have been published (van Ewijk et al 1974). Next, the lymphoid organs were excised and thin slices of the fixed tissue were dehydrated and critical-point-dried. Gold-sputtered specimens were examined in a Cambridge MK II A scanning electron microscope.

For SEM studies of lymphoid cells *in vitro* we seeded suspensions of lymphoid cells from the mesenteric node of T × BM mice and nude mice, and cell suspensions of the thymus of normal (DBA/2 × C57BL/Rij) F1 mice and cortisone-treated mice, into a plastic film dish. The preparation of cells in plastic film dishes for SEM studies has been described in detail elsewhere (van Ewijk & Mulder 1976). We allowed the lymphoid cells to attach to the plastic (Teflon) film for 30 min. Subsequently, the cells were fixed and dehydrated *in situ* under carefully controlled conditions, using a semiautomatic fixa-tion−dehydration device (van Ewijk & Hösli 1975). Small pieces of Teflon film were taken from the dish, critical-point-dried and sputter-coated with gold. This preparation method guarantees a minimal loss (<5%) of the lymphoid cells during the preparation for SEM.

RESULTS AND DISCUSSION

Topography of lymphoid cells in recirculation pathways

In 1964 Gowans & Knight demonstrated autoradiographically that intravenously injected radiolabelled lymphoid cells enter the lymph node parenchyma through the vascular wall of HEV. These venules are distributed throughout the paracortex of the lymph node. On entering the parenchyma, T and B cells segregate and migrate towards their respective micro-environments (Gutman & Weissman 1973). T and B cells may leave the node by entering trabecular and medullary sinuses connected to the efferent lymphatic vessel. To analyse cellular events during the migration of lymphocytes from the HEV towards the nodal parenchyma and their recirculation towards the lymph, we examined the cell surface topography of T and B cells in HEV, in medullary sinuses and in efferent lymphatics. Fig. 1 shows the morphology of an HEV as observed with SEM. The perfusion technique clears the blood cells from the lumen of the HEV, leaving only those cells that are in contact with the wall of the venule. Many small groups (two to four) of these cells were seen. In the mesenteric node of T × BM mice and nude mice we found that about two-thirds of the cells exhibited microvilli, and one-third a smooth surface. On close inspection we found that many of the lymphocytes used their microvilli to contact endothelial cells (Fig. 2). These microvilli were frequently up to 1 μm in length. Many cells were located between the bulges of high endothelial cells and there it was sometimes difficult to establish the nature of their contact with the endothelial cells. Most attached cells were spherical but some exhibited a bipolar configuration accompanied by flat cytoplasmic extensions (Fig. 3). From *in vitro* studies it is known that such cells are in the process of active amoeboid migration (McFarland et al 1966). Most cells situated between endothelial cells were virtually depleted of microvilli. These cells were always situated in ridges or

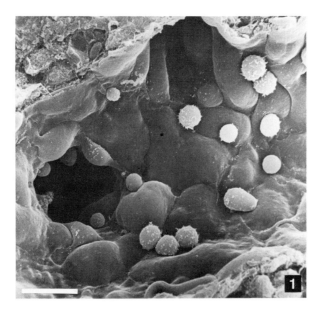

FIG. 1. Scanning electron micrograph of a high-endothelial venule in the mesenteric lymph node of a T × BM mouse. Scale marker: 10 μm.

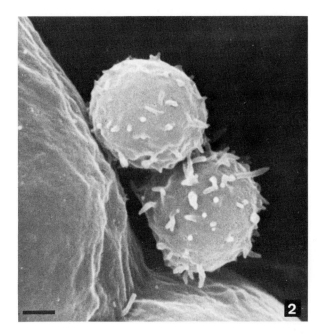

Fig. 2. B lymphocytes contact a high endothelial venule in the mesenteric lymph node of a T ×
BM mouse. Scale marker 1 μm. (From van Ewijk et al 1975.)

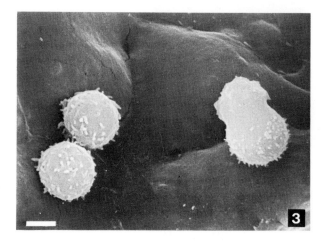

Fig. 3. B lymphocytes in contact with the wall of a high endothelial venule. The lymphocyte at the
right shows morphological signs of amoeboid migration across the endothelium. Scale marker
2 μm.

gaps between endothelial cells (Fig. 4). They were smaller in diameter than the villous cells. Sometimes they displayed a pseudopod projecting between endothelial cells. Apparently these cells are in the process of migration from the blood towards the nodal stroma. We found no evidence for emperipolesis of lymphoid cells by endothelial cells; migrating lymphoid cells were always located between endothelial cells. We were unable to locate a preferential site of migration for B cells in HEV. They left the venule in the paracortex as well as the outer cortex and in some instances in medullary cords.

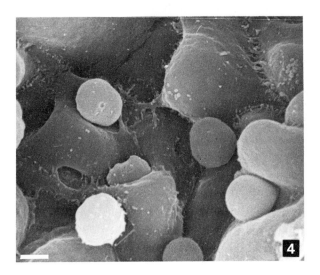

Fig. 4. B lymphocytes in the process of migration from blood towards the lymph node parenchyma. Note the route between the endothelial cells. Scale marker: 2 μm. (From van Ewijk et al 1975.)

B lymphocytes in the medullary sinus and the efferent lymphatic were always villous (Fig. 5). They used these microvilli to contact each other and other types of cells, such as macrophages and reticular cells.

T cells in the HEV of the mesenteric node of T and CRT mice were also villous. Some smooth-surfaced cells were observed here which, like B cells, were located in ridges between endothelial cells. T cells in medullary sinuses (Fig. 6) and in the efferent lymphatic were villous. Some T cells showed aggregation of microvilli over one pole of the cell; others also showed cytoplasmic extensions, obviously indicating cell movement.

FIG. 5. B lymphocytes in the efferent lymphatic duct of the mesenteric lymph node of a nude mouse. Scale marker: 2 μm.

FIG. 6. Lymphocytes in the medullary sinus of the mesenteric node of a CRT mouse. Lymphocytes (ly) exhibit short microvilli, whereas macrophages (m) demonstrate many cytoplasmic blebs. Scale marker: 5 μm.

Topography of lymphoid cells in specific microenvironments

SEM studies of the thymus of young AKR and (DBA/2 × C57BL/Rij) F1 mice revealed the presence of a meshwork of dendritic cells in which the lymphoid cells were interspersed. The dendritic pattern is caused by epithelial reticular cells. In immunoelectron-microscopic studies at Stanford University Dr I.L. Weissman and I have demonstrated the presence of antigens coded for by the major histocompatibility complex (MHC) on epithelial reticular cells in the thymus. Using monoclonal antibodies in combination with the immunoperoxidase method we were able to identify the presence of I-A and H-2K determinants on the cell surface of epithelial reticular cells. With light microscopy, we found a differential distribution pattern of H-2 antigens in the thymus. I-A determinants were expressed in a reticular pattern throughout the thymic cortex, while in the medulla the staining pattern was more confluent (Fig. 7). H-2K determinants were virtually absent from the cortex and only the medulla showed a confluent staining pattern. These observations support the contention by Zinkernagel et al (1978) that thymic epithelial cells govern the differentiation of T lymphocytes in terms of H-2 restriction. It appears that lymphoid subclasses are selected for their capacity to recognize particular H-2 markers during their differentiation in the thymus and that this 'learning' process is not influenced by the H-2 haplotype of the T cells themselves. The differential distribution pattern of I-A and H-2K antigens might indicate the existence of different microenvironments directing the

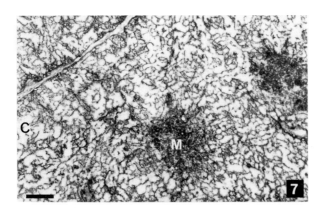

FIG. 7. Immunoperoxidase staining of a frozen section of the thymus incubated with monoclonal anti-Ia antibodies. Note the dendritic staining pattern in the cortex and the confluent staining pattern in the medulla (M). Scale marker: 100 μm.

maturation of functional different subclasses of T cells (R.V. Rouse et al 1979).

Lymphoid cells in the thymus were virtually devoid of microvilli. In some cases small microvilli were seen in contact with epithelial reticular cells.

T cells in the paracortical region of lymph nodes of T and CRT mice also showed a smooth surface. These cells were frequently distributed between reticular-type cells. The nature of these cells has not yet been identified, although interdigitating cells and reticular cells are possible candidates.

B lymphocytes, when studied in their specific homing areas, were smooth-surfaced. These cells, like T cells in the paracortex, were also situated between reticular elements.

So far these SEM studies of lymphocytes *in situ* have not revealed a clear distinction between the topography of the two major lymphoid subpopulations: microvilli are present on the cell surface of B as well as T lymphocytes, depending on the cell's *in situ* localization. One can, however, differentiate between homing and recirculating cells, by SEM. It appears that homing T and B cells are both virtually depleted of microvilli, whereas cells in recirculation pathways show many microvilli. Topographical studies of cells, collected from the thoracic duct of rats, enriched for either T cells or B cells have confirmed these observations (W. van Ewijk & P. Nieuwenhuis, unpublished data).

Topography of lymphoid cells in vitro

To investigate whether the conditions of preparation influenced the topography of lymphoid cells and could therefore explain the difference between our results and those of Polliack et al (1973, 1974), we decided to investigate the topography of lymphoid cells *in vitro* with a new method based on the use of a plastic film dish in which a Teflon film was mounted as a cell support. We seeded suspensions of lymphoid cells from different sources into the dish (Table 1). Cells were allowed to attach to the Teflon film for 30 min and prepared for SEM. We determined the diameter of the lymphoid cells and the number and size of the microvilli (see Table 1). Frequently, cells showed long microvilli at the base of the cell, oriented towards the substratum (Fig. 8).

From these *in vitro* studies, it is obvious that a reliable distinction between T and B lymphocytes based on cell surface topography is not possible. Although the cortisone-resistant cells in suspension as well as in the mesenteric node of CRT mice were slightly larger than thymocytes and B lymphocytes, and exhibited a large number of relatively short microvilli, the individual

TABLE 1

Scanning electron microscopy of lymphocytes seeded into a plastic film dish

Cell source	Diameter [a]	n	Mean number of microvilli/ cell	n	Mean length of microvilli	n
Thymus	3.9 ± 0.5	94	52 ± 48	146	0.47 ± 0.25	107
Thymus of dexamethasone-treated mice [b]	4.3 ± 0.7	190	148 ± 60	110	0.16 ± 0.08	225
Mesenteric node of T × BM mice	4.2 ± 0.5	122	48 ± 30	120	0.47 ± 0.13	175
Mesenteric node of nude mice	4.2 ± 0.5	214	68 ± 48	116	0.42 ± 0.26	155

[a]Values are expressed in μm ± standard error of the mean; n defines the number of cells counted.
[b]Mice injected (i.p.) two days previously with dexamethasone at a concentration of 30 mg/kg body weight.

variation between the cells of this subpopulation does not permit an accurate morphological definition. From the *in vitro* studies it is also obvious that lymphoid cells change their topography rapidly when made into a cell suspension. Thymocytes are smooth cells when observed *in vivo* but SEM studies of cell suspensions of thymus cells show that microvilli appear on many cells within 30 min.

It has been shown that experimental conditions such as the temperature (Lin et al 1973a) and molarity of the medium (Yahara & Kakimoto-Sameshima 1977) change the topography of lymphoid cells dramatically. Obviously, microvilli are very contractile cell organelles and their exposure is highly dependent on physiological conditions. Indeed, actin has been found in microvilli (Fagraeus et al 1974), providing direct evidence for the contractility of these organelles.

It is obvious that SEM studies on perfused lymphoid tissues provide insight into cell–cell interactions in lymphoid microenvironments and recirculation pathways; however for adequate submicroscopical characterization of lymphoid and non-lymphoid cell types immunoelectron microscopic studies detecting the presence of specific membrane markers are necessary. We are now developing labelling methods in which thin slices of lymphoid tissues are incubated with antisera conjugated to microspheres. This technique has enabled us to locate T and B cells in their specific microenvironments by SEM (W. van Ewijk, A.L. Rembaum & I.L. Weissman, unpublished observations).

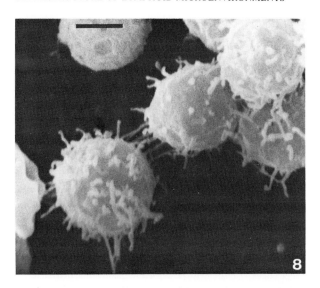

FIG. 8. Thymocytes *in vitro*, attached to a Teflon film. Lymphocytes show extended microvilli at the base of the cells. Scale marker: 2 μm.

The question arises of what functions microvilli on lymphoid cells might have. Based on the SEM studies of lymphocyte–endothelial interactions in HEV and on the SEM studies of lymphoid cells attaching to a substratum *in vitro* we propose that microvilli support the cells during the attachment of the cell to a substrate. The shear forces caused by the perfusion of fluids *in vivo* and *in vitro* were not sufficient to wash the cells from the lumen of HEV and to detach them from the plastic film. This indicates a tight bond between the cell and the substratum. Anderson & Anderson (1976), in transmission electron microscope studies of lymphocyte–endothelial wall interaction in HEV, have shown that microvilli fit into small pits in the membrane of endothelial cells. With SEM we could not demonstrate that these pits were pre-existing structures on the surface of HEV. However, these structures may develop *in vivo* during the initial attachment phase of lymphocytes and endothelium. Direct evidence for the existence of a specific receptor on the surface of lymphocytes for HEV comes from the work of Stamper & Woodruff (1976), Woodruff et al (1977), Woodruff & Kuttner (this volume), and Butcher & Weissman (this volume). From their *in vitro* experiments it appears that mature recirculating lymphoid cells specifically recognize the endothelium of HEV. Our studies indicate that microvilli are involved in lymphocyte–endothelial interactions. We speculate therefore that receptors for HEV are preferentially expressed on microvilli of recirculating cells. Since microvilli are also involved in lymphocyte–lymphocyte, lymphocyte–macrophage and lymphocyte–reticular cell interactions in lymph nodes and spleen (van Ewijk et al 1974, 1975), we propose that the general recognition of 'self' determinants on other cell types by recirculating lymphocytes is achieved through receptors preferentially expressed on microvilli. In HEV, the initial recognition of 'self' determinants on the endothelium by the recirculating lymphocyte could then initiate active amoeboid movement over the endothelium towards gaps between endothelial cells, followed by migration of the lymphocyte across the vascular lining.

ACKNOWLEDGEMENTS

I am indebted to Mr René Brons, Department of Cell Biology, Erasmus University, for technical assistance, to Dr Patricia Jones, Department of Genetics, Stanford University, for providing monoclonal anti H-2 reagents, and to Mrs Cary Meijerink and Rita Boucke for typing assistance. This work was done in part at the Department of Pathology, Stanford University, in cooperation with Dr I.L. Weissman, sponsored in part by Public Health Service International Research Fellowship No. 5 FO 5 TW 02564.

References

Anderson AO, Anderson ND 1976 Lymphocyte emigration from high endothelial venules in rat lymph nodes. Immunology 31:731-748

Butcher EC, Weissman IL 1980 Cellular, genetic and evolutionary aspects of lymphocyte inter-actions with high-endothelial venules. In this volume, p 265-281

Fagraeus A, Lidman K, Biberfeld G 1974 Reactions of human smooth muscle antibodies with human blood lymphocytes and lymphoid cell lines. Nature (Lond) 252:246-247

Gowans JL, Knight EJ, 1964 The route of recirculation of lymphocytes in the rat. Proc R Soc Lond B Biol Sci 159:257-282

Gutman GA, Weissman IL 1972 Lymphoid tissue architecture. Experimental analysis of the origin and distribution of T cells and B cells. Immunology 23:465-479

Gutman GA, Weissman IL 1973 Homing properties of thymus-independent follicular lympho-cytes. Transplantation 16:621-629

Lin PS, Wallach DF, Tsai S 1973a Temperature induced variations in the surface topology of cul-tured lymphocytes are revealed by scanning electron microscopy. Proc Natl. Acad Sci USA 70:2492-2496

Lin PS, Cooper AG, Wortis HH 1973b Scanning electron microscopy of human T cell and B cell rosettes. N Engl. J Med 289:548-551

McFarland W, Heilman H, Moorhead JF 1966 Functional anatomy of the lymphocyte in immu-nological reactions in vitro. J Exp Med 124:851-858

Nieuwenhuis P, Ford WL 1976 Comparative migration of B and T lymphocytes in the rat spleen and lymph nodes. Cell Immunol 23:254-267

Polliack A, Lampen N, Clarkson BD, Bentwich Z, Siegal FP, Kunkel HG 1973 Identification of human B and T lymphocytes by scanning electron microscopy. J Exp Med 138:607-624

Polliack A, Fu SM, Douglas SD, Bentwich Z, Lampen N, De Harven E 1974 Scanning electron-microscopy of human lymphocyte-sheep erythrocyte rosettes. J Exp Med 140:146-158

Rouse RV, van Ewijk W, Jones P, Weissman IL 1979 Expression of MHC antigen by mouse thymic dendritic cells. J Immunol 122:2508-2515

Stamper HB, Woodruff JJ 1976 Lymphocyte homing into lymph nodes: *in vitro* demonstration of the selective affinity of recirculating lymphocytes for high endothelial venules. J Exp Med 144:828-833

van Ewijk W, Hösli P 1975 A new method for comparative light and electronmicroscopic studies of individual cells, selected in the living state. J Microsc (Oxf) 105:19-31

van Ewijk W, Mulder MP 1976 A new preparation method for scanning electronmicroscopic studies of single selected cells, cultured on a plastic film. In: Johari O, Becker RP (eds) Scanning electron microscopy, Advances in biomedical applications of the scanning electron microscope. IIT Research Institute, Chicago, vol 5:131-134

van Ewijk W, Verzijden JHM, van der Kwast ThH, Luijck-Meijer SWM 1974 Reconstitution of the thymus dependent area in the spleen of lethally irradiated mice. A light and electronmicroscopic study of the T cell microenvironment. Cell Tissue Res 149:43-60

van Ewijk W, Brons NHC, Rozing J 1975 Scanning electron microscopy of homing and recirculating lymphocyte populations. Cell Immunol 19:245-261

van Ewijk W, Rozing J, Brons NHC, Klepper D 1977 Cellular events during the primary immune response in the spleen. A fluorescence, light and electronmicroscopic study in germfree mice. Cell Tissue Res 183:471-489

Veerman AJP, van Ewijk W 1975 White pulp compartments in the spleen of rats and mice. A light and electronmicroscopic study of lymphoid and non-lymphoid cell types in T and B areas. Cell Tissue Res 156:417-441

Weissman IL 1976 T cell maturation and the ontogeny of splenic lymphoid architecture. In: Battisto J, Streilein JW (eds) Immuno-aspects of the spleen. Elsevier/North-Holland, Amsterdam p 77-87

Woodruff JJ, Kuttner BJ 1980 Adherence of lymphocytes to the high endothelium of lymph nodes *in vitro*. In this volume, p 243-257

Woodruff JJ, Katz JM, Lucas LE, Stamper HB 1977 An in vitro model of lymphocyte homing. II Membrane and cytoplasmic events involved in lymphocyte adherence to specialized high endothelial venules of lymph nodes. J Immunol 119:1603-1610

Yahara I, Kakimoto-Sameshima F 1977 Ligand-independent cap formation: redistribution of surface receptors on mouse lymphocytes and thymocytes in hypertonic medium. Proc Natl Acad Sci USA 74:4511-4515

Zinkernagel RM, Callahan GN, Althage S, Cooper P, Klein PA, Klein J 1978 On the thymus in the differentiation of H-2 self-recognition by T cells: evidence for dual recognition? J Exp Med 147:882-896

Discussion

Owen: Eric Jenkinson and I (unpublished work) have been using the same antibodies as you to study Ia on early embryonic mouse thymus. We lightly trypsinize embryonic thymus from animals of 13 to 14 days' gestation, place the thymus fragments in culture and incubate overnight. We then carry out immunofluorescence tests on unfixed cells using monoclonal antibodies against the K and I-A subregions followed by a Fab anti-γ_2 antibody. By the 14-day stage there are cells which stain with the anti-Ia and anti-K antibodies.

These studies confirm that Ia- and K-positive cells are present in thymic embryogenesis when lymphopoiesis is beginning.

Gowans: Is there any information about the distribution of Ia in other tissues?

van Ewijk: Wiman et al (1978) give information on the Ia positivity of different cell types. They showed Ia in epithelia, particularly in the mammary gland and the gut. I only looked at Ia in the spleen. It appears that interdigitating cells in thymus-dependent areas are Ia-positive. The B cells also have Ia.

Gowans: As well as describing the epithelial localization (which might be relevant to the homing of lymphocytes into the gut and the mammary gland), Wiman et al reported that the amount of Ia on mammary epithelium varied with hormone levels: there was considerably more Ia on cells in the lactating than in the virgin gland. Have you observed changes in the amount of Ia on cells during development?

van Ewijk: Delovitch et al (1978) state that Ia antigens first appear on the surface of embryonic cells on day 11 after conception. Their expression is confined to the fetal liver, and precedes the expression of immunoglobulins. Our own results are preliminary. We plan to use a protein A-peroxidase conjugate as a label for looking at the distribution of Ia antigens in lymphoid organs during embryogenesis.

Weissman: J.Frelinger, J.Frelinger and L.Hood (personal communication) looked at immunoprecipitable Ia which they could then identify by peptide analysis and classify as donor or host type. They found that a lethally irradiated host restored with bone marrow had skin Ia of the bone marrow donor type rather than the host type. That fits with the idea that there are circulating precursors that give rise to Ia-positive cells, perhaps Langerhans cells (Rowden et al 1978). So the question of whether there is a definitive epithelial cell that is Ia-positive, other than in these studies with thymic epithelial cells in mice, is still open.

Richardson: Your hypothesis that microvilli are involved in slowing down the cell during its passage through the lumen of HEV may not be well supported by fluid mechanics, Dr van Ewijk. Howard Brenner's work on the rotation of a sphere near a surface in a shear flow shows that as the sphere gets closer to the surface, as would happen to the cell without microvilli, the rate of rotation goes down compared with the motion further away from the surface (Cox & Brenner 1971).

Nearly 20 years ago A.D.Bangham and B.A.Pethica suggested that one of the functions of microvilli is to reduce the effect of the radius of the membrane. If cells are trying to achieve adhesion in the face of fixed charges on the surfaces which tend to repel them, it would be easier for adhesion to occur if there were processes of very small radius.

Have you tried defluorinating the Teflon of the chamber you described? Sodium in solution can remove fluorine and change the adhesion characteristics of Teflon.

van Ewijk: We didn't try that.

Richardson: How long was your cell preparation in the chamber? Was it in any sense like a culture or were you essentially precipitating the cells?

van Ewijk: We kept the cells in the chamber for 30 minutes and then started introducing the fixation solution in the dish. So in principle there are cells in the thymocyte population which can develop microvilli within that period, since thymocytes *in vivo* have a smooth surface.

Ford: One of the most important things you have shown is that lymphocytes retract their microvilli when they begin to cross the endothelium. It seems probable that microvilli could impede lymphocytes while they are traversing the endothelium and basement membrane. It is much less obvious that the microvilli are essential for the initial adhesion step. The images of lymphocytes making contact with other cells by their microvilli are not convincing on this point because there is no other way in which a cell covered with spikes could make contact. The sort of evidence needed would come from testing the ability of lymphocytes that lack microvilli to adhere to high-endothelial cells. One possibility might be to somehow mark or isolate lymphocytes which had recently entered lymphoid tissues so that their adherence to HEV could be measured. Is there any evidence on these lines?

van Ewijk: Dr Butcher and I looked at the attachment of lymphoid cells to HEV *in vitro*. From both Dr Butcher's studies (this volume) and Dr Woodruff's studies (this volume) there is now evidence that recirculating cells have a specific receptor for HEV. Spleen cells and lymph node cells, but not thymocytes, stick to HEV when incubated on frozen sections of lymph nodes. In scanning electron microscopy we saw that all cells attached to HEV showed microvilli. I realize, however, that this is not direct evidence for the hypothesis that microvilli are involved in self-recognition. We will have to characterize the HEV receptor biochemically and demonstrate its presence on microvilli with immunoelectron-microscopical methods.

McConnell: Do the microvilli of polymorphonuclear cells attach to the plastic film? If not it would strengthen the idea that the microvillous-HEV interaction was relevant to lymphocyte traffic across the postcapillary venule.

van Ewijk: We did not do these experiments, but we do know that erythrocytes do not stick well to the plastic film.

McConnell: Do you know the cellular distribution of the Ia antigens detected by your monoclonal antibody? Does it detect macrophage subpopulations in mice? We have recently made a monoclonal anti-human Ia antibody which only reacts with about 60% of macrophages in human peripheral blood.

van Ewijk: The thymus doesn't seem to have a large number of Ia-positive macrophages. I haven't had time to study in detail the ultrastructure of Ia-positive cell populations in other lymphoid organs. Some of the few macrophages I have seen so far were positive in the thymus and some were negative. The patchy distribution of the Ia antigen on the cell surface of Ia-positive cells is complicating our observations. We find localized areas on the cell surface which are Ia-positive and also areas on the same cells which are negative. I don't know whether this is a preparation artifact caused by insufficient penetration of the label, or whether it is patch formation of Ia molecules on the surface of the epithelial cells and macrophages. Dr Owen observed patch formation of Ia molecules on thymic epithelial cells *in vitro*.

Vane: All your pictures show cells either with microvilli or without them. Do you ever catch them in a transitional state?

van Ewijk: I showed a picture (not reproduced here) of a migrating cell which has microvilli still on the luminal side of the cell surface while the front side oriented to the endothelium is smooth.

Vane: Your *in vitro* culture system would provide an almost ideal way of looking at the effects of drugs on the microvilli. Have you done any such studies?

van Ewijk: No.

Born: There is a possible *in vivo* experiment too. If the migrating cells behave like some other cells in putting out microvilli reversibly, they could perhaps be made to retract by an increase in intracellular cyclic AMP. This could possibly be done *in vivo* by administering a phosphodiesterase inhibitor together with an adenylate cyclase activator, both in quite low concentrations. Of course this might also do something to the endothelial cells where the migration occurs. If the cyclic AMP is raised in some other cells, the inhibition of their function persists for some time after the inhibitory agents have been removed.

van Ewijk: Perfusion of the vascular system with phosphate-buffered saline, followed by infusion of cells pretreated *in vitro* with reagents which, for example, affect the cytoskeleton, would be of interest in studies of the involvement of microvilli in the interaction of lymphocytes with HEV.

Humphrey: Olah et al (1968) described cells in the medulla of the rat thymus which contained Birbeck granules and would presumably correspond to interdigitating cells, and would contain Ia antigen. Dr George Janossy has shown me sections of human thymus stained for Ia in which there were very bright cells in the medulla as well as others in the cortex. Are the interdigitating cells and the reticular cells which you describe the same?

van Ewijk: No, absolutely not. From work published by Veerman (1974) we know that the interdigitating cells are in some way related to the mononuclear phagocyte system. The major Ia-positive cell of the thymus is definitely not the interdigitating cell but the epithelial reticular cell. At the ultrastructural level these cells are characterized by tonofilaments and desmosomes. From the work of Hoefsmit & Gerber (1975) we know that interdigitating cells form a minor cell population which preferentially occurs in the thymus medulla, not in the thymus cortex. Hoffmann-Frezer's group reported in 1978 that in peripheral lymphoid organs the Ia-positive cell types occur in thymus-dependent areas. They suggested that these cells were interdigitating cells. We also have some preliminary results indicating that interdigitating cells are Ia-positive.

References

Cox RG, Brenner H 1971 The rheology of a suspension of particles in a Newtonian fluid. Chem Eng Sci 26:65-93

Delovitch TL, Press JL, McDevitt HO 1978 Expression of murine Ia antigens during embryonic development. J Immunol 120:818-824

Hoefsmit EChM, Gerber GAM 1975 Epithelial cells and macrophages in the normal thymus. In: van Bekkum DW (ed) Biological activity in thymic hormones. Kooyker Scientific Publications, Rotterdam, p 63-68

Hoffmann-Frezer, Götze G, Rodt H, Thierfelder S 1978 Immunohistochemical localization of xenogeneic antibodies against Ia[k] lymphocytes on B cells and reticular cells. Immunogenetics 6:367-377

Olah I, Dunay C, Röhlich P, Toro I 1968 A special type of cell in the medulla of the rat thymus. Acta Biol Acad Sci Hung 19:98-113

Rowden G, Phillips TM, Delovitch TL 1978 Expression of Ia antigens by murine keratinizing epithelial Langerhans cells. Immunogenetics 7:465-478

Veerman AJP 1974 On the inderdigitating cells in the thymus dependent area of the rat spleen: a relation between the mononuclear phagocyte system and T lymphocytes. Cell Tissue Res 148:247-257

Wiman K, Curman B, Forsum U, Klareskog L, Malmnäs-Thernlund U, Rask L et al 1978 Occurrence of Ia antigens on tissues of non-lymphoid origin. Nature (Lond) 276:711-713

Structural and functional differentiation of microvascular endothelium

MAYA SIMIONESCU

Institute of Cellular Biology and Pathology, 8 Hasdeu Street, Bucharest 70646, Romania; and Section of Cell Biology, Yale University School of Medicine, New Haven, Conn., USA

Abstract Characteristic variations in endothelial components have been detected in precisely identified sequential segments of the microvasculature in rodents. Vesicle density is greater in capillary endothelium than in arterioles and venules. Transendothelial channels are especially frequent in the venular segment of capillaries. The endothelial junctions of arterioles and capillaries are of occluding type but about 30% of the junctions in postcapillary venules have gaps of about 6 nm and are characteristically sensitive to vasoactive amines.

The main pathway for transport of macromolecules ($\geq$ 2 nm) across the endothelium is provided by vesicles which shuttle between the two fronts (transcytosis) or fuse to form patent transendothelial channels. *In vitro* experiments have shown that endothelial cells have a constitutive endocytic capability, independent of the physical factors operating *in vivo*.

The distribution of electrostatic charges on the blood front of the capillary endothelium indicates that even more refined local differentiation exists: (a) the cell membrane is negatively charged; (b) the diaphragms of vesicles and those of transendothelial channels lack anionic sites; (c) fenestral diaphragms are heavily loaded with anionic sites. Experiments with proteolytic enzymes suggest that these acidic sites on the fenestral diaphragms are primarily glycosaminoglycans, most likely heparan sulphate. Capillary basement membrane is particularly rich in anionic sites.

The physical partition between the circulating blood and the surrounding tissue is represented by a continuous cell layer, the endothelium. In the microvasculature—the arterioles, capillaries and venules—endothelial cells constitute the main barrier mediating and controlling the exchange of substances between blood plasma and interstitial fluid. Although the endothelium is a simple squamous epithelium, the exchange of water and water-soluble molecules—as shown by the permeability and hydraulic conductivity coefficients—is two to three times higher than in other epithelia (Renkin 1977).

© *Excerpta Medica 1980*
Blood cells and vessel walls: functional interactions
(Ciba Foundation symposium 71) p 39-60

From physiological experiments with graded dextrans (Pappenheimer et al 1951, Grotte 1956) the existence has been postulated of water-filled channels or pores within the capillary wall through which massive transport of macromolecules from plasma to interstitial fluid takes place. Two sets of pores have been suggested: the small and large pores (Landis & Pappenheimer 1963). According to recent estimates, the more frequent small pores have a diameter of about 11 nm in their cylindrical version and a width of about 8 nm when they occur as open slits. Molecules up to 10 nm in diameter pass through small pores by a restricted diffusion which increases with the molecular dimensions of the injected probes (Grotte 1956, Crone & Christensen 1978). Large pores, which are less frequent, have a diameter of about 50–70 nm; molecules up to 30 nm in diameter permeate them freely.

The development of electron microscopy produced considerable interest in the identification of features which could account for the permeability properties of the capillary endothelium and may represent the structural equivalents of the two pore systems.

ULTRASTRUCTURE OF THE MICROVASCULAR ENDOTHELIUM

Electron microscope studies of small blood vessels have shown that in most vascular beds the endothelium consists of thin but relatively broad squamous epithelial cells ($\sim 25 \times 15 \times 0.2-0.5\ \mu$m), linked by simplified junctions and resting on a basement membrane which the cells produce. Endothelial cells possess the same organelles as any epithelial cells but, as a distinctive feature, they also have a large population ($\sim$ 10 000 to 15 000 per cell) of small, rather uniform ($\sim$ 70 nm diameter) *plasmalemmal vesicles* (Palade 1953). About 70% of these vesicles open on the blood or tissue front of the cell: this increases the exposed endothelial surface by a factor of about two (Fig. 1a). The remaining 30% of the plasmalemmal vesicles are located within the cytoplasm. Vesicles open onto the cell surface through a neck (10 to 40 nm diameter) which often has a thin diaphragm.

Some vesicles fuse and form patent *transendothelial channels* made up of either a single vesicle or a chain of vesicles opening on both cell fronts (Fig. 1a). Where they fuse with the plasmalemma and at the opening points, the channels usually have thin diaphragms and strictures of various sizes (up to 40 nm) which can impede the diffusion of molecules (N. Simionescu et al 1975). In addition to vesicles and channels, in visceral capillaries the endothelium is provided with round openings or *fenestrae* about 70 nm in diameter. These are usually closed by a single-layered diaphragm, the porosity and chemical nature of which are still unknown.

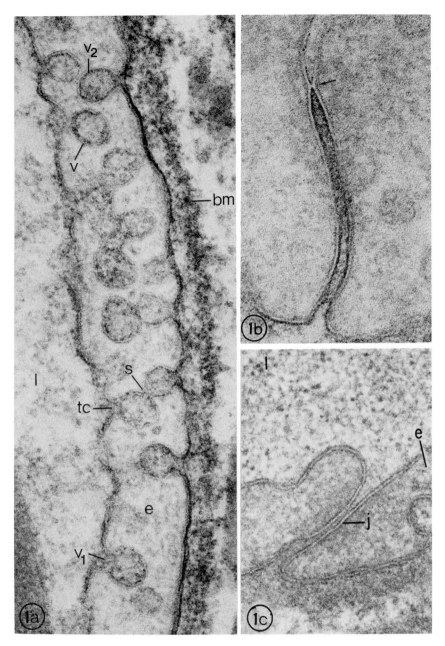

FIG. 1. Components of mouse microvascular endothelium potentially involved in permeability (diaphragm). (a) Continuous endothelium (e) of a capillary displays vesicles open on the blood front (v_1), on the tissue front (v_2) and located within the cytoplasm (v). A transendothelial channel (tc) is provided with a stricture (s) between the two fused vesicles. bm, basal lamina. (b) Endothelial junction of a capillary. Note the point of membrane fusion with elimination of the two outer leaflets of the opposed plasmalemmae (arrow). (c) Endothelial junction open to a gap of ~ 5 nm in a postcapillary venule. l, lumen, e, endothelial cell.

Parts of illustration taken from Simionescu & Simionescu 1977 (Figs 1a and b) and from M. Simionescu 1979 (Fig. 1c). (a) × 120 000; (b) × 175 000; (c) × 80 000

The occurrence of these intermediate forms suggests that vesicles are the key structure in endothelial organization.   They are highly modulating constituents which either appear as separate units or fuse as transendothelial channels, which in turn can be reduced in length to become fenestrae.

The endothelial cells are linked by occluding (tight) junctions (Fig. 1b) and in some segments of the vasculature by communicating (gap) junctions (M. Simionescu et al 1975).

SEGMENTAL DIFFERENTIATION OF ENDOTHELIAL STRUCTURES

In the work reported here mouse microvascular units were used, in which, as a result of their favourable geometry, sequential segments of the micro-vasculature—arterioles, capillaries and venules—could be identified reliably (bipolar microvascular fields in mouse diaphragm). The results show that the endothelial structures potentially involved in permeability vary charac-teristically along the microvasculature (Simionescu et al 1978a).

(a) The endothelium is particularly thin in the venular part of the capillaries ($\sim 0.17~\mu$m).

(b) There are more vesicles in true capillaries ($\sim 1000/\mu$m^3) than in arterioles ($\sim 190/\mu$m^3) or postcapillary venules ($\sim 645/\mu$m^3).

(c) Transendothelial channels, especially those made up of single vesicles, are more frequent in the venular portions of capillaries.

(d) As seen in both thin-sectioned and freeze-fractured specimens the endothelial cells in arterioles are linked by highly organized combinations of tight junctions with intercalated communicating (gap) junctions (Fig. 2a). Capillary endothelium contains occluding junctions only (Fig. 2b).   The endothelium of venules is characteristically provided with a very loose organi-zation of cell contacts with frequent discontinuities among rare and low profile junctional lines (Fig. 2c).   In sectioned specimens about 30% of venular junctions appear to be open to an extent of 6 nm (Simionescu et al 1978c) (Fig. 1c).   Communicating junctions are small and rare in muscular venules and absent from exchange vessels (capillaries and pericytic venules). The intercellular sealing is strong in arterioles (the site of 'peripheral resistance'), well represented in capillaries, and particularly loose in venules. Cell-to-cell communication via gap junctions is well represented in arterioles, occasionally present in muscular venules, and virtually absent from capillaries and postcapillary venules (M. Simionescu et al 1975).

$\longrightarrow$

FIG. 2. The appearance in freeze-fracture preparations of endothelial junctions in sequential segments of mouse microvasculature. (a) *Arteriole*: a tight junction appears on the E face (E) as a

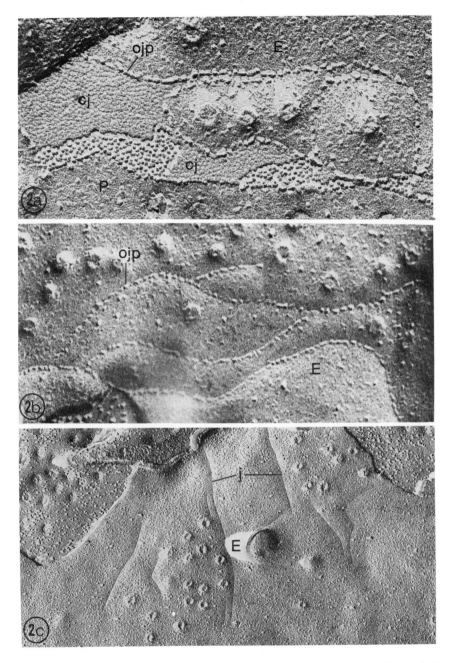

network of continuous interconnected grooves marked by particles (ojp). Communicating junctions (cj) are intercalated between the tight junctional strands. P,P face. (b) *Capillary*: a tight junction appears on the E face (E) as a maze of branching and continuous grooves marked by particles (ojp). (c) *Postcapillary venule*: the junction displays on the E face discontinuous, shallow grooves (j) devoid of particles. (From M. Simionescu et al 1975.) (a) × 140000; (b) × 120000; (c) × 60000.

TRANSENDOTHELIAL   PATHWAYS   OF   MACROMOLECULES: SEGMENTAL CHARACTERISTICS

Since none of the endothelial structures revealed by electron microscopy has the exact geometry and frequency postulated by the pore theory, investigators have tried to detect these pathways indirectly, using probe molecules of known size which are, or can be, rendered visible in the electron microscope. Some of these water-soluble macromolecular tracers can be detected as individual molecules or particles, e.g. dextran and glycogens ($\sim$ 15–30 nm diameter) and ferritin ($\sim$ 11 nm diameter). Other tracers can be visualized indirectly as electron-opaque products obtained in histochemical reactions for their peroxidase activity. Such tracers include horseradish peroxidase ($\sim$ 5 nm diameter) (Karnovsky 1967, Williams & Wissig 1975); myoglobin ($\sim$ 3.3 nm diameter) (Simionescu et al 1973), and haem peptides (2 nm diameter) (Simionescu et al 1975).

The continuous endothelium of skeletal mucle microvasculature is used here to show the type of information obtainable from tracer experiments.

As suggested by the kinetics of vesicle labelling by the tracer particles or the enzymic reaction product, probe molecules $\geq$ 2 nm diameter (e.g. haem peptides) cross the microvascular endothelium primarily via plasmalemmal vesicles (vesicular transport). The process occurs in successive, resolvable phases during which the label appears first in vesicles open on the blood front, then in vesicles located in the cytoplasm, and finally in vesicles associated with the tissue front, which discharge their contents into the adjacent pericapillary space (Fig. 3a-c). In the same specimens the reaction product also identified patent transendothelial channels made up of single vesicles or chains of fused vesicles open on both endothelial fronts (Fig. 3c). Channels are probably transient features with a very short lifespan: they can behave as either small or large pores, depending on the geometry and size of their opening and the porosity of their diaphragms.

In capillaries from the same specimens the endothelial junctions are not labelled by detectable amounts of the reaction product of the haem peptide (Fig. 3e). There is a permeable intercellular pathway but this seems to be restricted to the pericytic venules (Simionescu et al 1978b). From the earliest time-points examined, about 30% of the endothelial junctions of the venules were permeated by haem peptides (Fig. 3f) and to a lesser extent by horseradish peroxidase; they excluded haemoglobin (mol.diam. 6.4 $\times$ 5.5 $\times$ 5.0 nm) and native ferritin. These findings set the size-limit of molecules that can permeate the open venular junctions at about 6 nm (Simionescu et al 1978c).

$\rightarrow$

FIG. 3. Time-course experiments with haem peptides as tracer for studying the transendothelial pathways of such molecules. (a) *30 s after i.v. injection*: reaction product is present in the capillary lumen and vesicles open on the blood front. (b) *35 s after i.v. injection*: most of the

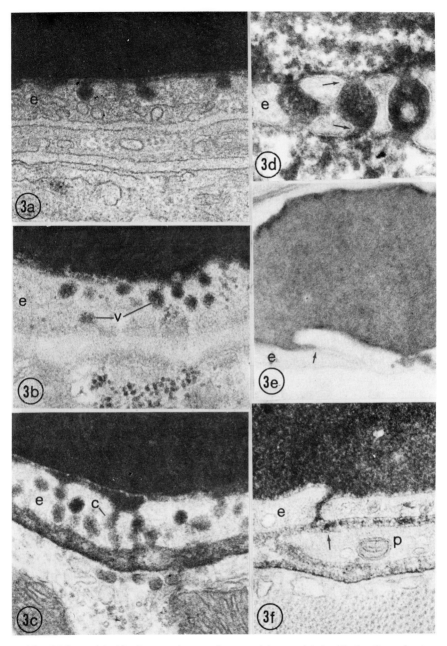

vesicles (v) located inside the cytoplasm and one vesicle associated with the tissue front are labelled. (c) *60 s after injection* there is extensive labelling of all plasmalemmal vesicles in all locations. Note one channel (c) formed by two fused vesicles. (d) Transendothelial channels marked by reaction product after interstitially injected tracer. Note the continuity of channel membrane with the plasmalemma (arrows) on both fronts of the cell. (e) Capillary endothelial junction which is not marked by haem peptide reaction product. (f) Open endothelial junction in a pericytic venule; a concentration gradient appears in the adjoining periendothelial space (arrow). e. endothelium, p, pericyte.

Parts of illustration taken from N. Simionescu et al 1975 (a-c, e); Simionescu et al 1976 (d); Simionescu et al 1978b (f). (a,b,c,) × 42 000; (d) × 180 000; (e,f) × 48 000

Tracer molecules (haem undecapeptide) interstitially injected into rat cremaster muscle can, to a certain extent, be transported through capillary endothelium by back-diffusion. In these experiments, vesicle labelling progresses from the tissue front to the blood front of the endothelium; channels were also found labelled (Fig. 3d) (Simionescu et al 1976).

The transport of substances across endothelium via vesicles or channels represents a special cellular process. Unlike other cells, including other epithelia, only negligible numbers of vesicles reach the lysosomes. Vesicle-shuttling between two endothelial fronts constitutes a short cut between endocytosis and exocytosis that bypasses the other intracellular compartments (e.g. the lysosomes). The vesicle-derived channels, as well as the fenestrae—most probably derived from the channels—may be considered the extreme and perhaps biologically most economical expression of the ability of the endothelial cell to efficiently couple endocytosis directly to exocytosis. This special process has been called *transcytosis* (N. Simionescu 1979). The energy requirements and the fractional contribution of each pathway involved in the transendothelial exchange of macromolecules remain to be established.

DIFFERENCES IN REACTIVITY TO HISTAMINE ALONG THE MICROVASCULATURE

The response of small vessels in well-defined microvascular fields (bipolar microvascular fields) from the mouse diaphragm was examined after histamine had been applied either intravenously or locally. Carbon black solution was injected into the circulation and used as a marker for leaky vessels; the whole diaphragm was screened for black vascular 'tattoo' marks. As shown in Fig. 4a, intramural deposits of carbon particles were found to be precisely limited to vessels of 12–25 μm diameter which in bipolar microvascular fields can be easily recognized as postcapillary venules. Arterioles and capillaries do not show deposits of carbon particles (Fig. 4a). Electron microscope examination of the tattooed areas of postcapillary venules shows characteristic focal separation of the endothelial cells at the junctions (Fig. 4b) (Majno et al 1961), with carbon particles accumulated against the basement membrane.

DIFFERENTIAL DISTRIBUTION OF CHARGES ALONG THE ENDOTHELIAL CELL

Experiments with isolated endothelial cells have shown that anionic probes such as native ferritin are taken up by vesicles more readily than cationic molecules (M. Simionescu & N. Simionescu 1978). As most plasma proteins are negatively charged, it becomes interesting to find out whether the

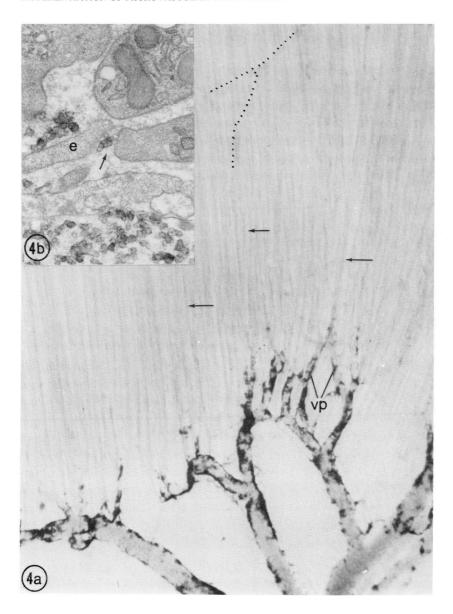

FIG. 4. Mouse diaphragm 30 min after i.v. injection of carbon black and topical application of histamine. Arterioles (dots) and capillaries (arrows) are not labelled by carbon whereas postcapillary venules (vp) show intramural deposits of carbon particles. (b) Thin section showing carbon particles accumulated in a largely open intercellular junction of a postcapillary venule. From Simionescu et al 1978c. (a) × 600; (b) × 65000.

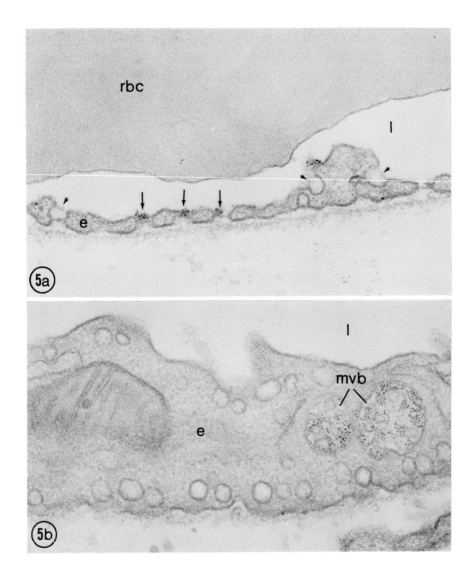

FIG. 5. Labelling of endothelial cell (mouse) by cationic ferritin. (a) Fenestral diaphragms are marked by particles of cationic ferritin (arrows) whereas the channel and vesicle diaphragms are not labelled (arrowheads). (b) 60 min after injection, particles of cationic ferritin appear in multivesicular bodies (mvb). (a) × 68 000; (b) × 68 000.

transport across the endothelium is facilitated by a particular distribution of charges along the endothelial cell surface. Cationic ferritin, pI 8.4, was injected into mice before or after the blood was washed out by perfusion with phosphate buffer saline. The fenestrated capillaries of the pancreas and jejunum were investigated, because each endothelial cell possesses vesicles, channels and fenestrae. From the earliest times examined, cationic ferritin was attached to the luminal membrane of the endothelium and to about 90% of the fenestral diaphragms (Fig. 5a). Over the same period only about 10% of the diaphragms associated with vesicles and transendothelial channels were labelled (Fig. 5a). Coated vesicles were consistently labelled by cationic ferritin. Sixty minutes after their injection ferritin particles had disappeared from all sites, except fenestral diaphragms, and began to appear intracellularly in multivesicular bodies (Fig. 5b). Fenestral diaphragms were labelled up to 6 h after injection but the labelling disappeared by 20 h.

To determine the chemical carrier (or carriers) of these particularly dense anionic sites of the fenestral diaphragms, we applied various enzymes before injecting cationic ferritin: neuraminidase, trypsin, hyaluronidase, chondroitinase A, B and C. None of them altered the binding pattern significantly. Recently, a crude preparation of heparitinase + heparinase (kindly provided by Dr J.E. Silbert, Veterans Administration, Boston, Mass. 02108) completely removed the acidic sites from the fenestral diaphragms without significantly changing the labelling pattern of the rest of the cell surface.

These findings suggest that sulphated glycosaminoglycans, and probably heparan sulphate in particular, are the major component of the luminal aspect of the fenestral diaphragms and carriers of anionic sites in these locations.

Studies with interstitially injected cationic ferritin, and also *in vitro* studies, have shown that the basement membrane of endothelial cells has acidic sites on both sides.

These experiments showed that very refined differentiations of the endothelial surface occur on a rather narrow domain of each cell. The results indicate that: (a) the affinity of fenestral diaphragms for cationic ferritin is higher than that of the plasmalemma;

(b) vesicle and channel diaphragms on the one hand, and fenestral diaphragms on the other, have different electrochemical natures;

(c) fenestral diaphragms are stable structures for at least 6 h;

(d) the plasmalemmal vesicles seem to be a system devised to favour the transport of anionic proteins.

CONCLUSIONS

The permeability characteristics of the microvascular endothelium can be associated with the existence of an unusually developed and highly modulating transport system made up of vesicles, channels and fenestrae. These features serve to carry solutes through the endothelium by a short cut between endocytosis and exocytosis (transcytosis). There are characteristic segmental differentiations in the endothelial structures involved in permeability, and in the pathways taken by probe macromolecules. Tracer molecules down to 2 nm diameter are transported in quanta by discrete vesicles, and by convective diffusion through channels and fenestrae. The intercellular passage of such molecules is restricted to a fraction of the endothelial junctions of the postcapillary venules. Fenestral diaphragms have a high density of anionic sites; the latter are almost absent from the vesicle and channel-associated diaphragms. Vesicles and channels appear to be differentiated in such a way that they discriminate in favour of anionic proteins.

ACKNOWLEDGEMENTS

This paper is based mostly on work carried out with Drs Nicolae Simionescu and George E. Palade at Yale University, Section of Cell Biology, New Haven, Connecticut, and supported by US Public Health Service Research Grant HL-170808.

References

Crone C, Christensen O 1978 Transcapillary transport of small solutes and water. In: Guyton AC, Young DB (eds) Cardiovascular physiology. University Park Press, Baltimore, vol 3:149-213
Grotte G 1956 Passage of dextran molecules across the blood-lymph barrier. Acta Chir Scand Suppl 211:1-84
Karnovsky MJ 1967 The ultrastructural basis of capillary permeability studied with peroxidase as a tracer. J Cell Biol 35:213-236
Landis EM, Pappenheimer JR 1963 Exchange of substances through capillary walls. In: Hamilton WF, Dow P (eds) Handbook of physiology, section 2. American Physiological Society, Washington DC, vol 2:961-1034
Majno G, Palade GE, Schoefl. GI 1961 Studies on inflammation. II. The site of action of histamine and serotonin along the vascular tree: a topographical study. Biochem Cytol 11:607-626
Palade GE 1953 Fine structure of blood capillaries. J Appl Phys 24:1424 (Abstr)
Pappenheimer JR, Renkin EM, Borrero LM 1951 Filtration, diffusion and molecular sieving through peripheral capillary membranes. A contribution to the pore theory of capillary permeability. Am J Physiol 167:13-46
Renkin EM 1977 Multiple pathways of capillary permeability. Circ Res 41:735-743
Simionescu M 1979 Transendothelial movement of large molecules in the microvasculature. In: Fishman AP, Renkin EM (eds) Pulmonary edema. American Physiological Society, Bethesda, Maryland, p 39-52

Simionescu M, Simionescu N 1978 Constitutive endocytosis of the endothelial cell. J Cell Biol 79:381 (abstr)

Simionescu M, Simionescu N, Palade GE 1975 Segmental differentiation of cell junctions in the vascular endothelium. The microvasculature. J Cell Biol 67:863-885

Simionescu N 1978 The microvascular endothelium: segmental differentiations; transcytosis; selective distribution of anionic sites. In: Weissman G et al (eds) Advances in inflammation research, Raven Press, New York, vol 1:61-70

Simionescu N, Simionescu M 1977 The cardiovascular system. In: Weiss L, Greep R (eds) Histology, 4th edn. McGraw-Hill, New York, p 373-431

Simionescu N, Simionescu M, Palade GE 1973 Permeability of muscle capillaries to exogenous myoglobin. J Cell Biol 57:424-452

Simionescu N, Simionescu M, Palade GE 1975 Permeability of muscle capillaries to small heme-peptides. Evidence for the existence of patent transendothelial channels. J Cell Biol 64:586-607

Simionescu N, Simionescu M, Palade GE 1976 Structural-functional correlates in the transendothelial exchange of water-soluble macromolecules. Thromb Res 8:257-269

Simionescu N, Simionescu M, Palade GE 1978a Structural basis of permeability in sequential segments of the microvasculature. I. Bipolar microvascular fields in the diaphragm. Microvasc Res 15:1-16

Simionescu N, Simionescu M, Palade GE 1978b Structural basis of permeability in sequential segments of the microvasculature. II. Pathways followed by microperoxidase across the endothelium. Microvasc Res 15:17-36

Simionescu N, Simionescu M, Palade GE 1978c Open junction in the endothelium of postcapillary venules of the diaphragm. J Cell Biol 79:27-44

Williams MC, Wissig SL 1975 The permeability of muscle capillaries to horseradish peroxidase. J Cell Biol 66;531-547

Discussion

Marcus: May I comment on your demonstration of transcytosis and a possible relationship of this process to the synthesis of prostacyclin and other prostaglandins. Drs Weksler, Jaffe and I (Marcus et al 1978) have been studying prostaglandin synthesis by cultured human endothelial cell monolayers from umbilical cord. When we added radioactive arachidonic acid to the endothelial cells, this fatty acid penetrated the cells quickly and was incorporated into the phospholipid fraction. In addition, the arachidonate was transformed into several prostaglandin products, but all of the prostaglandins were found in the cell supernatant. We next treated the endothelial cells with radioactive PGH_2, which is the precursor of prostacyclin (PGI_2) and other prostaglandins. Within a few seconds PGI_2 appeared in the cell supernatant along with other prostaglandins. No radioactivity was detectable in the endothelial cells themselves. We therefore postulated two mechanisms to explain the experimental results. First, the endoperoxide PGH_2 was converted to PGI_2 and other prostaglandins at the cell surface. The second was that the endoperoxide had penetrated the cell, the prostaglandins were synthesized, and the products were rapidly secreted into the surrounding medium. At present we do not know which mechanism was operative.

We noted that synthesis of prostacyclin from the endoperoxide was enzymic, whereas transformation of the endoperoxide into other prostaglandins could occur spontaneously. This phenomenon was also observed by Gorman (1979).

You mentioned that material could be seen to leave the endothelial cells within seconds. Your comments were thus reminiscent of what we have been observing in our prostaglandin-synthesizing system. It is interesting to know that endothelial cells are capable of rapidly secreting various biological substances.

Andrews: Is the process of transcytosis vectorial, with vesicles pinching off and migrating only from the lumen to the basal margin of the cell? If it is, presumably the migration of such relatively large amounts of surface membrane from the luminal face has to be accounted for.

Simionescu: What we have learnt so far is that molecules from the blood plasma are taken up by vesicles and transported to the tissue front, where the content is discharged into the interstitial space. The transport seems to be bidirectional. This process raises the important question of whether domains of the luminal front of the endothelial cell membrane move as vesicles and are subsequently incorporated into the plasmalemma of the other side of the cell, or whether the vesicles represent a separate population of membranes, which shuttle between the two fronts of the cell. We have no answer to this question yet.

Howard: What was the evidence that these were vesicles rather than channels with some kind of impermeability associated with them? In section, all one could see was holes, but they could be convoluted channels.

Simionescu: Bruns & Palade (1968) have reconstructed from serial sections a sector of an endothelial cell. They showed that the cell is provided with a large number of vesicles (about 70 nm in diameter) bound by a unit membrane. Vesicles occur either isolated in the cytoplasm or open on the cell surface; often vesicles fuse together. We have demonstrated (Simionescu et al 1975) that one vesicle or a chain of fused vesicles can open simultaneously on both endothelial fronts to form patent transendothelial channels. These channels are provided with diaphragms and strictures at fusion points, and they have been proposed as a candidate for the structural equivalent of the small pores.

Weiss: Where the endothelium is not full of vesicles, for example in the brain, is there a preferential distribution in relation to cell junctions or elsewhere?

Simionescu: I don't know in detail how the vesicles are distributed in the endothelium of brain capillaries. In muscle capillaries vesicles are particularly frequent in the thin peripheral zone of the endothelial cell, away from the intercellular junction (M. Simionescu et al 1974).

Hall: You said it was a two-way process and that material could move from the tissue side through the endothelium and into the blood. What was the limiting molecular size of such material?

Simionescu: So far, back-diffusion of the tracer molecules has been demonstrated with haem undecapeptide (microperoxidase) with a molecular diameter of about 2 nm (Simionescu et al 1976) and horseradish peroxidase of about 5 nm in diameter (Johansson 1977). Haem undecapeptide is about 1900 daltons, and horseradish peroxidase about 40 000 daltons.

Hall: Can a large molecule get right across?

Simionescu: The largest probe molecule so far tested was horseradish peroxidase.

Hall: But in practice very few large molecules can gain direct access to the blood and must go via the lymphatics.

Simionescu: In the experimental conditions mentioned, most of the haemoprotein molecules injected interstitially are removed by the lymphatics; only a small fraction can actually be detected crossing the capillary endothelium.

Gowans: Is it simply size that determines whether molecules in the extracellular fluid pass back into the bloodstream or into the lymphatics?

Hall: A lot of work has been done with labelled blood serum proteins, for example, and various classes of macromolecules seem to have very great difficulty in passing from tissue fluid directly into blood vessels. At least, that is what is said (Yoffey & Courtice 1970).

Morris: That is what happens.

There are two well-documented physiological facts that I am not sure are explained by your ultrastructural presentation, Dr Simionescu. First, the ease with which plasma proteins escape from the circulation into the tissue fluid is directly proportional to their mean effective diffusion diameters. The smaller plasma proteins escape more readily than the others and they appear in concentrations in the lymph which are in proportion to their diffusion diameters. Secondly, once they have gained the interstitial fluid they cannot normally return to the bloodstream directly.

There is good evidence that if plasma proteins are perfused through the lymph nodes they can be recovered quantitatively in the efferent lymph. The evidence that plasma proteins can return directly from the bloodstream across the vascular endothelium is not very strong. These two facts need to be explained by any ultrastructural analysis of the way in which materials are selectively transferred across the endothelium. I wasn't sure that your particular arrangement would accommodate those two well-documented physiological facts.

Simionescu: The experiments have shown that vesicles can transport molecules from 2 nm up to 30 nm in diameter. Both physiological and morphological experiments have shown that the rate of transport depends on molecular size, shape and charge. The time required for smaller tracers to traverse the endothelium is shorter than that for larger ones, i.e. for haem peptides about 40 s (Simionescu et al 1975), myoglobin about 60 s (Simionescu et al 1973), horseradish peroxidase about 2 min (Williams & Wissig 1975). Regarding the back-diffusion, as I previously pointed out our observations are limited to haem undecapeptide (about 2 nm) only.

Morris: My problem is, how is this differential transport accomplished in terms of the diffusion diameters of the molecules?

Born: An experimental approach introduced for a particular purpose (Begent & Born 1970) could perhaps be broadened to answer such questions. In principle one wants to apply molecules of different sizes, shapes and charges to restricted, defined sites on the outside of small vessels and determine how the molecules traverse the wall. For our purpose the technique was to microiontophorese ADP onto the outside of small venules. By several criteria this left the venule walls themselves entirely normal. In the lumen an adhering aggregate of platelets began to grow within a few seconds. The time from starting the iontophoretic current, which was very small, to the first appearance of adhering platelets turned out to be remarkably similar in different venules, with a minimum of about 12 seconds. We wondered whether most of this time could be due to diffusional delays within the venule walls, probably due mainly to the endothelial layer. To look into this (Begent et al 1972) we used two micropipettes on opposite sides of the vessel, one containing ADP and the other histamine, which has no effect on platelets but contracts the endothelial cells so that gaps form between them. During the application of histamine, there was a significant decrease in the time to the first appearance of the platelet aggregates, suggesting that the rate of diffusion of ADP from the outside of the vessel to the inside had increased. It seemed that histamine had diminished the diffusion barrier against this comparatively small, highly charged molecule. This was a reversible effect, because when the iontophoretic current for histamine was switched off the delay increased again as if the endothelial cells were relaxing and coming together again.

This kind of experiment could presumably be done with microapplications of larger molecules of the kind we have been talking about. It hinges, of course, on establishing some indication of the presence of the molecules inside the vessel, in the way that the platelet reaction indicated the presence of ADP. Kinetic analysis could perhaps show whether particular proteins move through intercellular junctions or in the vesicles.

Simionescu: There is a difference: your experiments were done on small venules, the junctions of which are known to be sensitive to histamine (Majno & Palade 1961). Moreover, in normal specimens about 30% of these venular junctions are open, with a gap of 3−6 nm (Simionescu et al 1978). In the same conditions, the junctions between the endothelial cells of capillaries are not permeable to molecules of 2 nm or more in diameter and do not open under the influence of histamine.

Born: The gaps are illustrated in the paper by Begent et al (1972).

Gowans: There is a large transport of protein molecules into the extra-vascular space in normal animals and then back into the lymph. Are these proteins carried by vesicles across the capillaries?

Simionescu: According to the tracer experiments reported, the answer is yes.

Morris: I think we can accept that these materials are transferred across the endothelium in vesicles. The difficulty I have is understanding how there comes to be this precise gradient of transfer which is related to the diffusion diameter of the molecule. This is very important when one is dealing with immunological phenomena and trying to establish the concentration of anti-body in tissue fluid. The content of antibody in the interstitial fluid has to be established by these transport mechanisms, and a molecule like IgM hardly makes it across the endothelium at all. The only alternative mechanism is for the cells which synthesize antibody to enter the tissue spaces and secrete the antibody at this site.

Born: Do those molecules get through the gaps quickly?

Morris: I don't know. IgM is present in the general tissue fluid in low con-centrations. This concentration in the capillary filtrate is very low. The con-centration in the lymph of course is much higher because water in the capil-lary filtrate is reabsorbed. Very little IgM makes it across the transport systems we have just heard described. People have suggested that the reason IgM doesn't get out is because the molecule has a very large diffusion diameter compared with IgG_1 or IgG_2.

Gowans: Professor Born made the point that histamine opens up gaps between endothelial cells. Does this allow large protein molecules to pass into the extravascular spaces?

Morris: During inflammatory responses the permeability of the capillaries is greatly increased, as it is when histamine is applied to blood capillaries—in these situations molecules such as IgM will certainly escape from the blood more readily.

Simionescu: Under the influence of histamine, the gaps formed between endothelial cells in postcapillary venules measure from 0.1 to 3 μm; through such spaces any large molecule easily gains access to the subendothelial space.

In normal conditions, besides driving forces such as the hydrostatic pressure, osmotic pressure and concentration gradients, transport of macromolecules across the endothelium depends on many other factors, such as the dimensions and shape of the molecule, charge, molecular environment (plasma, interstitial fluid), and temperature. Also important are the charge, porosity and chemistry of the vesicular and fenestral diaphragm, about which we don't have enough information.

Richardson: Have you some sort of model of how transcytosis occurs? Is there any locomotive capacity for these vesicles or do they, so to speak, take advantage of Brownian motion and then somehow recognize that they have got to the right side of the cell?

Simionescu: At the level of resolution so far attained with the electron microscope no special device has been observed for the movement of vesicles. The movement has been described as diffusion (Shea et al 1969) or Brownian movement (Green & Casley-Smith 1972). It seems to be mostly a random process over a short distance which does not require special energy. Energy may, however, be necessary for membrane fusion–fission. In general very little is known about the intimate mechanism and control of the vesicular transport.

Richardson: If everything was being put into a bag and sent across, so to speak, the filtration would have to be based on the molecule size at the entrance to the vesicle when it is loading. That would be the only mechanism with any control worth mentioning.

Gowans: What do you imagine controls the rate at which vesicles are formed to account for changing rates of transport across the vessel?

Richardson: There are several questions around this. One is whether there is even a possibility of size-independent transport into the vesicle loading up against the flow side. In those circumstances molecules might be brought in willy-nilly, irrespective of the particular size of the molecule. On the other hand there may be gates which, because of the size of the pores, are then available, for example in the curtain-like shrouds or fenestral diaphragms over the vesicle; these gates would provide some control over the rate at which molecules of different sizes enter. Then all the molecules have to go over in this little shuttle and they are all transported at the same rate. One can't attribute any variation to the shuttle in its movement across. The molecular-weight-based selection has to be done either at the entry or at the exit, not letting everything out, which doesn't sound very efficient. So the question is, what sort of control is exerted in the filling process and how? Is this going to be the major pathway, or is there, as Gustav Born suggested, a complementary pathway between the cells which can provide a more traditional pore-like filtration effect or control over the diffusion of different molecular sizes?

Morris: I don't believe that the ultrastructural explanation accords with the physiological facts as we know them. It is quite clear that proteins are indeed transported by these vesicles but there is a paradox that remains to be resolved. Differential permeability to proteins has to be explained in terms of differential transport by vesicles.

The other question which hasn't been touched on yet is the ultrastructural aspect of capillary permeability as it relates to cells.

Weissman: This moulding and changing of the membrane on a very small scale looks similar to phagocytosis and to virus budding. Those two events, as far as I know, involve a multivalent structure—a spheroid, let's say—as for a virus core binding on the underside of the plasma membrane to many transmembrane viral envelope protein receptor sites. Essentially the membrane must bind to all ligand sites on the spheroidal viral core, forming a unilamellar membrane envelope. The forming of the membrane around it is essentially due to contact points. One can make the same argument for antibody-coated red cells and macrophage phagocytosis. You seemed to say that there was a charge-dependence of vesicular transport of things like plasma proteins, Dr Simionescu. Are all of those molecules also transported by this same sort of process, by multivalent binding to endothelial cell surface receptors? Could one test whether that is so by having neutral substrates varying according to whether there was one charge or two charges or several charges per molecule? Does there have to be a multivalent charge interaction between the molecules to be transported? Or is diffusion everything, so that the molecules get across these vesicles without touching the membrane?

Simionescu: It appears that molecules gain access to vesicles opened to the blood front by simple diffusion. It is uptake in bulk of plasma and/or tracer, similar to fluid non-adsorptive pinocytosis. The fact that anionic tracers are taken up more readily by vesicles (Simionescu & Simionescu 1978a) and that, unlike the fenestral diaphragm, the vesicular diaphragm is devoid of anionic sites (Simionescu & Simionescu 1978b) suggests that plasmalemmal vesicles may represent a system devised to favour the transport of anionic proteins (like plasma proteins).

McConnell: There is very good immunological evidence that the vascular endothelium is not permeable to antigens of high molecular weight (10 000 or more). If antigen is present in a cannulated lymph node in sheep there is no systemic spread of antigen directly from the node to the blood and one fails to see priming of the whole animal for a secondary response. Even if small amounts crossed the vascular endothelium then priming for a secondary response would be observed, but that doesn't happen.

Born: But couldn't this be due simply to differences in rates? For example, fibrinogen is a long, thin molecule which diffuses into and through vessel walls, but very slowly. How is it with IgG?

Morris: This is the point about specifying diffusion diameters rather than molecular weights. While the fibrinogen molecule is of about the same order of molecular weight as the albumin molecule, it is much longer and has a much larger diffusion diameter than albumin. Albumin gets across much more easily than does fibrinogen.

IgM is the classic example of a large molecule which escapes from the bloodstream slowly. It is present in the capillary filtrate in very low concentrations.

Howard: Presumably the question of direction is simply a function of the net molecular concentration on each side of the barrier. In a way what immunologists take as dogma, which is the one-way traffic from tissue space to lymphatic, just reflects the fact that one's subcutaneous injections of proteins never manage to achieve intravascular protein concentrations.

Dr Simionescu showed molecules back-tracking down the transcytotic route, so immunologists who are concerned about the movement of proteins from tissue space in the vascular system must take account of that. Is it entirely an artifact? Dr Simionescu obviously achieved very high local concentrations. Is it simply a question of diffusion?

Hall: The late Professor Mayerson and his colleagues deliberately established high concentrations of proteins labelled with ^{131}I outside blood vessels in order to test this; there was no evidence of direct entry into the blood vessels (Patterson et al 1958).

Weiss: One interesting variation would be to do these experiments in blood vessels in which blood flow is stopped, to avoid the whole question of changes in concentration. Have you done that, Dr Simionescu?

Simionescu: No. In our experiments the tracer was injected while the blood was circulating; the circulation was arrested at given time intervals by fixing the tissue with glutaraldehyde *in situ*.

Weiss: Do you think that one major difference between the two sides of this endothelium could be that on one side there is still water and on the other a rather quickly moving fluid?

Simionescu: Yes, that might make a difference, which should be substantiated experimentally.

Butcher: In most of your electron micrographs of vesicles or transcytotic channels in endothelial cells there appeared to be a membranous barrier at the opening of the vesicles on the luminal aspect. That barrier could conceivably be a molecular sieve that determines the rate at which proteins of various sizes enter the vesicles. Is there always such a membranous diaphragm separating the lumen and the vesicle?

Simionescu: Yes. Usually the vesicle openings are provided with a single layered diaphragm, the porosity and nature of which are not known. Yet this diaphragm seems to be devoid of a phospholipid layer because it is readily permeated by water-soluble molecules such as tracer haemoproteins or carbohydrates. Together with the strictures occurring at the level of transendothelial channels, the diaphragms represent size-limiting structures which contribute to the control of the passage of macromolecules through the endothelium.

Butcher: This diaphragmatic barrier, then, seems a possible candidate for the morphological entity which determines what molecules enter the lymph from the blood.

References

Begent NA, Born GVR 1970 Growth rate *in vivo* of platelet thrombi produced by iontophoresis of ADP, as a function of mean blood flow velocity. Nature (Lond) 227:926-930

Begent NA, Born GVR, Sharp DE 1972 The initiation of platelet thrombi in normal venules and its acceleration by histamine. J Physiol (Lond) 223:229-242

Bruns RR, Palade GE 1968 Studies on blood capillaries. I. General organization of blood capillaries in muscle. J Cell Biol 37:244-276

Gorman RR 1979 Modulation of human platelet function by prostacyclin and thromboxane A$_2$. Fed Proc 38:83-88

Green HS, Casley-Smith JR 1972 Calculation of the passage of small vesicles across endothelial cells by brownian motion. J Theor Biol 35:103-111

Johansson BR 1977 The microvasculature in skeletal muscle. Thesis. University of Göteborg

Majno G, Palade GE 1961 Studies on inflammation. I. The effect of histamine and serotonin on vascular permeability: an electron microscopic study. J Biophys Biochem Cytol 11:571-605

Marcus AJ, Weksler BB, Jaffe EA 1978 Enzymatic conversion of prostaglandin endoperoxide H$_2$ and arachidonic acid to prostacyclin by cultured human endothelial cells. J Biol Chem 253:7138-7141

Patterson RM, Ballard CL, Wasserman K, Mayerson HS 1958 Lymphatic permeability to albumin. Am J Physiol 194:120-124

Shea SM, Karnovsky MJ, Bossert WH 1969 Vesicular transport across endothelium: simulation of a diffusion model. J Theor Biol 24:30-42

Simionescu M, Simionescu N 1978a Constitutive endocytosis of the endothelial cell. J Cell Biol 79:381a

Simionescu N, Simionescu M 1978b Differential distribution of anionic sites on the capillary endothelium. J Cell Biol 79:59a

Simionescu M, Simionescu N, Palade GE 1974 Morphometric data on the endothelium of blood capillaries. J Cell Biol 60:128-152

Simionescu N, Simionescu M, Palade GE 1973 Permeability of muscle capillaries to exogenous myoglobin. J Cell Biol 57:424-452

Simionescu N, Simionescu M, Palade GE 1975 Permeability of muscle capillaries to small heme-peptides. Evidence for the existence of patent transendothelial channels. J Cell Biol 64:586-607

Simionescu N, Simionescu M, Palade GE 1976 Structural functional correlates in the transen-dothelial exchange of water soluble macromolecules. Thromb Res 8:257-269

Simionescu N, Simionescu M, Palade GE 1978 Structural basis of permeability in sequential segments of the microvasculature. II. Pathways followed by microperoxidase across the endo-thelium. Microvasc Res 15:17-36

Williams MC, Wissig SL 1975 The permeability of muscle capillaries to horseradish peroxidase. J Cell Biol 66:531-555

Yoffey JM, Courtice FC 1970 Lymphatics, lymph and the lympho-myeloid complex. Academic Press, London

Haemodynamic and biochemical interactions in intravascular platelet aggregation

G.V.R. BORN

Department of Pharmacology, King's College, Strand, London WC2R 2LS

Abstract   The only undoubted function of platelets is their aggregation in injured vessel walls as haemostatic plugs. The aggregation of platelets as thrombi in atherosclerotic arteries may also be initiated by haemorrhage. The association of thrombocytopenia with petechial haemorrhages suggests that platelets are somehow required for the functional integrity of small vessels, but no mechanism has yet been established. Platelets adhere under some other conditions, e.g. when subendothelial tissues are exposed between contracting or deficient endothelial cells; but any functional significance of this is uncertain and made more so by the aggregation of platelets on the walls of *artificial* vessels wherever blood flow is non-laminar. Adhesion and aggregation of platelets are not prevented by unphysiologically high wall shear forces. Indeed, these processes may depend in some way on abnormal haemodynamic conditions.

This contribution is therefore mainly concerned with questions about how haemodynamic conditions in and around vascular leaks affect arriving platelets that are responsible for sealing the leaks, and about the chemical agents responsible for making the platelets adhesive. The effects of these agents are known mainly from *in vitro* experiments in which aggregation can be quantitatively correlated with biochemical effects by simple and reproducible methods. The relevance of *in vitro* observations on platelets to their reactions in haemostasis is uncertain. It is difficult to devise quantitative *in vivo* methods, mainly because of the rapidity with which platelets adhere and aggregate in a blood vessel after injury. When hypotheses for explaining *in vivo* platelet aggregation in biochemical terms are being considered the haemodynamic situation needs to be well imprinted on the mind. [Illustrated on film.]

When blood vessels are injured so that they bleed, circulating platelets adhere to the damaged vessel walls and aggregate, so diminishing or arresting the haemorrhage. This interaction between platelets and vessel walls therefore has an easily demonstrable physiological function. There is much clinical and experimental evidence that a deficiency or defect in circulating platelets is

© *Excerpta Medica 1980*
Blood cells and vessel walls: functional interactions
(Ciba Foundation symposium 71) p 61-77

associated with 'spontaneous' haemorrhages from small vessels.   This
suggests that platelets are somehow essential for the functional integrity of
these vessels, but no mechanism has yet been established.

Claims are made that agents released from platelets are able to damage
vessel walls, either acutely (Mustard et al 1977b) or by contributing to
atherogenesis (Ross & Glomset 1973). The evidence for these propositions is
indirect and circumstantial and no such effects have been incontrovertibly
established (see Walton 1975). On the other hand, there is conclusive evidence
that occlusive thrombi in arteries damaged by atherosclerosis contain platelets
as a major, if not the main, component. The formation of platelet thrombi
appears so similar to that of haemostatic plugs of platelets that analysis of the
mechanism of the latter is likely to provide an understanding of the former.

This contribution considers how the plugging mechanism depends on the
haemodynamic and biochemical environment in which platelets adhere to and
aggregate on vessel walls.

DOES INTERACTION BETWEEN PLATELETS AND ARTERIAL WALLS CONTRIBUTE
TO ATHEROGENESIS?

The old thrombogenic hypothesis of atherosclerosis (von Rokitansky 1841;
Duguid 1949) has reappeared in modern costume as claims that *platelets*
contribute to atherogenesis in three ways: first, through damaging arterial
endothelial cells by releasing injurious agents, presumably where circulating
platelets adhere (Mustard et al 1977a); secondly, through the release in such
situations of a factor responsible for smooth muscle proliferation in the
arterial wall (Ross & Glomset 1973); and thirdly through the formation of
persistent mural thrombi which are organized into intimal thickenings. Such
evidence as there is for these propositions fails to establish any of them as
relevant to atherosclerosis in animals or human beings (see Walton 1975).
Underlying all three claims is the assumption that some normal circulating
platelets settle on arterial walls for long enough to release some of their
contents.   There is no observational basis for this assumption in normal
arteries.   Therefore it is assumed further that arterial endothelium is
continuously subject to 'damage' or 'injury' of some kind as a precondition
for the adherence of platelets. There is no convincing evidence for this
generalization, especially not in human beings. The only finding that could
conceivably apply to human arteries is that guinea-pig aorta has a higher
replacement rate of endothelium around the openings of branches than
elsewhere (Payling-Wright & Born 1971). This is most simply explained by
assuming that endothelial turnover depends, *inter alia,* on haemodynamic

effects due to non-laminar blood flow over such areas. But this should be thought of more correctly as a quasi-physiological effect and, even there, platelets are rarely if ever seen adhering to the walls. The turnover rate of endothelium is increased in experimental hypertension (Payling-Wright 1972). This is compatible with hypertension as a 'risk factor' for coronary heart disease. It seems more likely that this is due to an accelerating effect on plasma lipoprotein accumulation through interendothelial gaps (Stehbens 1965; Caro 1977) than to an increase in the indiscriminate or even selective deposition of platelets on arterial walls. Indeed, there are other ways in which hypertension could accelerate atherosclerosis (see Caro 1977).

INTERACTION BETWEEN PLATELETS AND VESSEL WALLS IN THROMBOGENESIS

Both the gross and the histological appearance of arterial thrombi establish that the central mass consists mainly of aggregated platelets. What, therefore, is the mechanism responsible for rapid and extensive platelet aggregation in an artery as an apparently random event in time (see Born 1979)? Close serial sectioning of obstructed coronary arteries established some time ago that the platelet thrombus responsible is invariably associated with recent haemorrhage into an underlying atherosclerotic plaque (Friedman 1970, Constantinides 1966). The haemorrhages occur through fissures or fractures in the plaque; and the sudden appearance of such a fissure or fracture may well be the random, individually unpredictable event affecting coronary arteries that has to be assumed to occur to account for the clinical onset of acute coronary thrombosis (Born 1979).

How does haemorrhage into a ruptured plaque start off platelet thrombogenesis? This can be regarded as part of the general question of how platelets are caused to aggregate through haemorrhage, and most effectively through haemorrhage from arteries. Until recently this question was commonly answered by assuming that the process depends on the adhesion of platelets to collagen which is exposed where damaged vessel walls are denuded of endothelium. Adhering platelets then release other agents, including thromboxane A_2 and ADP, which in turn are responsible for the adhesion of more platelets as growing aggregates. This explanation is unlikely to be correct, for the following reasons. First, haemostatic and thrombotic aggregates of platelets grow without delay and very rapidly. For example, when an arteriole 200 μm in diameter is cut into laterally, the rate of accession of platelets to the haemostatic plug is of the order of 10^4/s (Born & Richardson 1979). In contrast, although the adhesion of platelets to collagen itself is almost instantaneous, the subsequent aggregation of platelets, even under optimal conditions for their reactivity, begins only after a delay or lag period of at least 15 to 30 s (Wilner et al 1969). Secondly, platelets tend to aggregate as mural thrombi when anti-coagulated blood flows through the plastic vessels of artificial organs such as oxygenators or dialysers (Richardson et al 1976) that contain no collagen or anything else capable of activating platelets similarly. This implies that there are conditions under which platelets are activated in the blood by something other than collagen or other constituents of the walls of living vessels.

The plaque on which a thrombus grows has usually narrowed the arterial lumen. At constant blood pressure the flow of blood is faster through the constriction than elsewhere in the artery. Therefore, high flow and wall shear rates are no hindrance to the aggregation of platelets as thrombi (Born 1977). Indeed, the question arises of whether the activation of platelets which precedes their aggregation depends in some way on such abnormal haemo-dynamic conditions.

Measurements of the haemodynamic forces required to activate platelets directly (Hellums & Brown 1977) indicate that the blood flow over athero-sclerotic lesions *in vivo* is unable to do this (Colantuoni et al 1977). Therefore, the activation must be indirect. Now it has been known for many years that platelets can be activated by at least one agent, namely ADP, derived from the red cells which outnumber and surround the platelets in the blood. Indeed, the discovery of the activation of platelets by ADP which is highly specific among nucleotides and related substances began with the demonstration that the adhesion of platelets in columns of small glass beads depended on the presence of red cells and varied in proportion to their con-centration; the agent was identified as ADP (Gaarder et al 1961).

This paper is therefore mainly concerned with questions, most of them un-answered, about the effects of the fluid dynamics in and around a vascular leak on the interactions of platelets with the walls of the vessel and with the red cells, and about the chemical agents responsible for making the platelets adhere and aggregate.

Most of what is known about these agents and their effects on platelets has come from *in vitro* experiments (for a recent review see Ciba Foundation 1975) in which adhesion and aggregation can be correlated with biochemical changes by comparatively easy and highly reproducible methods (Born 1962). The relevance of *in vitro* observations to platelet aggregation as seen *in vivo* during haemostasis is still uncertain. This is probably because it is much more difficult to devise satisfactory quantitative methods for investigating the process *in vivo* than *in vitro;* and the main cause of the difficulty is the astonishing rapidity with which platelets adhere and aggregate in a blood vessel after it has been injured, in spite of extremely rapid blood flow which would be expected to counteract these processes. The haemodynamic situation needs to be well imprinted on the mind, preferably by micro-cinematographic observation of the haemostatic process, when hypotheses that attempt to explain *in vivo* platelet aggregation in biochemical terms are being considered.

HAEMODYNAMIC CONDITIONS DURING ARTERIOLAR HAEMOSTASIS

The effectiveness of platelet aggregation in plugging a leak is at least as effective in arterioles as in venules. As the haemodynamic situation should be more unfavourable to the formation of aggregates in arterioles than in venules, an explanation of arteriolar haemostasis is likely to account in principle also for that in venules. For that reason, the following considerations are limited to arterioles.

When an arteriole is cut, platelets are seen to adhere with great rapidity to the damaged vessel wall, while the red cells continue to rush by. The high flow velocity in relation to the small size of the vessels implies the presence in the fluid of strong mechanical forces acting normally and tangentially on and near the vessel walls. The cut causes peripheral resistance to the flow to diminish suddenly; and if the inflow pressure remains constant the mean flow velocity increases. Thus the fluid-mechanical forces on platelets adhering and aggregating on the vessel wall become greater still. With increasing size the platelet aggregates tend to constrict the cut, causing a further, although usually temporary, increase in flow velocity.

In spite of wall shear stresses of 10^5 to 10^6 μN/cm^2 which are one or two orders of magnitude greater than anywhere in the normal circulation, platelets succeed in aggregating into haemostatically effective plugs. The blood-flow velocities that would be experienced by platelets closest to the vessel wall and therefore with the highest probability of colliding with the sites of damage can be calculated (Schmid-Schönbein et al 1976). Human platelets have a major diameter of about 1.5 μm. In an arteriole of medium size the flow velocity of plasma and of any cells in it at a distance of 1 μm from the wall is of the order of $10-100$ μm/ms. Therefore, a platelet flowing within a distance no greater than its own diameter would pass an injury site 100 μm long in at most 10 ms. In the absence of other influences, this would seem to be the time available for such a platelet to adhere to the damaged wall.

ACTIVATION TIME OF PLATELETS

The time just calculated as available to circulating platelets 'at risk' for adhering to a wall lesion has to be compared with what is known about the time required for platelets to be activated into a condition in which their collision with such a lesion would very probably result in adhesion. That a process of activation is an essential prerequisite for adhesion and aggregation is inferred from the non-reactivity of normal circulating platelets.

As activation is indicated by adhesiveness, the change must involve one or more constituents of the outer surface of platelets. There is evidence that the essence is the exposure of surface receptors for fibrinogen, which has long been known to be an essential and specific plasma co-factor for platelet aggregation (Born & Cross 1964, Cross 1964). The activation time of platelets may then be defined as the interval between the encounter of platelets with an activating agent such as ADP and their ability to react with plasma fibrinogen.

Until recently, it seemed reasonable to suppose that the activation of platelets was accompanied by the gross morphological changes which are quantifiable morphometrically or photometrically *in vitro* (Born et al 1978). A prominent component of this change is the extrusion of long thin spikes; and it seemed reasonable to assume that the ability of platelets to approach closely enough for adhesion is like that of other cells with fixed surface charges, facilitated by the extrusion of pseudopodia. The first question is, therefore, whether anything approaching the gross morphological changes demonstrable *in vitro* invariably precedes platelet adhesion and aggregation *in vivo*. However, recent electron-microscopical observations indicate that platelets can adhere to vessel walls and to each other without any obvious deviations from their normal morphology (Born 1977). Whether this is really so can be conclusively established only by quantitative statistical electron-microscopical methods of the kind that have established, for example, the reversibility of the shape-changing effects of ADP (Born et al 1978). If, therefore, activation is accompanied by changes in morphology these changes are apparently beyond the limits of resolution in time and amplitude of the *in vitro* photometric techniques so far used. Minimal morphological changes should occur in much less time than the time constant determined for the gross changes in shape.

At all events, activation as defined above is likely to precede the gross morphological changes which, at 37 °C, have a time constant of the order of 1 s (Born et al 1978). This is some two orders of magnitude greater than the time available for adhesion under the conditions likely to occur *in vivo*. The discrepancy is so great that the assumptions underlying the calculations have to be reconsidered on the basis of additional experimental evidence.

PLATELET ACTIVATION *IN VIVO*

That circulating platelets can be activated to adhere in much less time than that required by the gross changes in shape is indicated by direct experimental observations. An arteriole can be irradiated by a laser in such a way that damage is limited to a few square micrometres of endothelium (Arfors et al 1976). The site of damage is covered almost immediately with platelets that must have been activated in small fractions of a second.

Very similar events follow the application of the activating agent ADP by micro-iontophoresis to the outside of an arteriole or venule under conditions in which appropriate controls indicate that there is no evidence at all of damage to the endothelial layer (Begent & Born 1970). Platelet aggregates grow in the vessel exactly opposite the tip of the micropipette, while the blood continues to flow rapidly and without noticeable disturbance over the site. This is explained most simply by assuming that sufficient ADP diffuses between the endothelial cells into the blood to reach platelets passing close to the wall and that this ADP activates them in a few milliseconds.

An extension of this technique has provided a basis for calculating an average activation time for circulating platelets. It was found that the size of platelet aggregates produced by the iontophoretic application of ADP increases exponentially. The rate constant of this increase depended on the mean blood flow velocity, determined in the same vessels at the same time (Begent & Born 1970). The shape of the experimentally determined curve was simulated closely by a theoretical curve (Richardson 1973), which was derived on the single assumption that platelets require an activation time of about 100 ms to 200 ms. This time is still one order of magnitude greater than that indicated by the earlier theoretical considerations, so either this experimental derivation overestimates the true activation time or the earlier considerations failed to take something into account that would allow flowing platelets more than a few milliseconds for activation. More time would, for example, be available if the blood flow near the vessel wall were non-laminar, so that platelets caught up in vortices, however small, might be exposed to localized activating conditions for longer than they would otherwise be. When branching vessels of the microcirculation are observed microscopically, platelets can often be seen trapped in vortices for variable times of up to several seconds. Such delays may occur in the immediate vicinity of major vessel wall lesions, whether caused by disease such as the sudden rupture of an atheromatous plaque (Constantinides 1966, Friedman & Van den Bovenkamp 1966) or by traumatic injury such as a puncture or transection. However, there is no evidence of even the smallest disturbances in the flow of blood in a normal vessel in which platelets are caused to adhere by ionto-phoretically applied ADP. Moreover, it seems most unlikely that any endothelial unevenness produced by laser injury would give rise to flow disturbances large enough to delay the passage of platelets.

PHYSICAL VERSUS CHEMICAL ACTIVATION OF PLATELETS

The major question is, therefore, the nature of the stimulus or stimuli which activate platelets so very rapidly under conditions in which they function haemostatically. One type of stimulus is primarily physical, i.e. the effect of fluid-mechanical forces on the platelets. It has been established that platelets are activated by shear stresses greater than about 1030 μN/cm^2 and that the time required for activation varies inversely with the applied shear stress (Colantuoni et al 1977). But even with shear stresses as great as those calculated for arteriolar lesions only a few platelets are activated in the time during which they are passing by.

Fluid-mechanical effects are, however, apparently able to activate platelets *indirectly* by acting on the red cells that surround and outnumber them in the blood. Clear evidence for this was provided by experiments in which blood was made to flow through branching channels in extracorporeal shunts (Mustard et al 1962). Chambers made of different plastic materials were introduced into a shunt through which heparinized blood flowed from a carotid artery to a jugular vein of an anaesthetized pig. Deposits of platelets formed consistently on the shoulders of a bifurcation in the flow chamber but nowhere else in the channels. Clearly, therefore, this deposition did not depend on the properties of the materials from which the chambers were made. Furthermore, when the chambers were perfused not with blood but with platelet-rich plasma no deposit formed, showing that red cells were essential for the increased reactivity of the platelets that resulted in their mural deposition. The augmenting effect of red cells on the deposition of platelets can also be demonstrated with blood flowing through chambers of other geo-metrical conformations or other types of wall surface. For example, the endo-thelium can be removed by the introduction of a balloon catheter into rabbit aortas, exposing a subendothelial surface composed mainly of connective tissue; such a subendothelial surface can also be exposed to blood flowing in annular chambers of different diameters, so providing a variety of shear rates at the blood–surface interface (Turitto & Baumgartner 1975). When blood was perfused over such a surface, platelets soon covered almost all of it and there were numerous platelet aggregates or thrombi on the adhering layer. When platelet-rich plasma was perfused instead of blood, few platelets were deposited.

The increased deposition of platelets from flowing blood associated with the presence of the red cells could be caused by physical or chemical mechanisms or, of course, by both acting synergistically. A physical mechanism would depend essentially on an increase in the diffusivity of platelets caused by the flow behaviour of the erythrocytes. Indeed, the diffusivity of platelets in flowing blood has been estimated to be two orders of magnitude greater than that predicted for platelets diffusing in plasma (Turitto et al 1972, Turitto & Baumgartner 1975). This is consistent with the enhanced radial fluctuations of erythrocytes and latex microspheres (2 μm in diameter) in flowing suspensions of red-cell ghosts (Goldsmith 1972).

There is evidence also of a chemical mechanism for the increased adhesiveness of platelets in the presence of red cells, i.e. through release of their ADP. The concentrations of ADP required are small (less than 10^{-6} M) (Milton et al 1979) so its direct demonstration in plasma involves two important considerations: first, red cells are such a large reservoir of ATP and ADP that the slightest damage to them swamps the plasma with ADP; and, secondly, because the outer surfaces of the cells as well as the plasma contain enzymes that catalyse the rapid breakdown of ADP (Bolton & Emmons 1967, Haslam & Mills 1967, Parker 1970). Therefore, the release of ADP into plasma has been inferred indirectly by demonstrating that the effect of red cells on platelets is prevented by the addition of enzyme systems capable of utilizing ADP specifically. Thus, in the presence of the pyruvate kinase system, which removes ADP by enzymic phosphorylation to ATP, the difference in the adhesiveness to glass of platelets from whole blood or from platelet-rich plasma is abolished and so is the increase in platelet adhesiveness caused when red cells are added to platelet-rich plasma (Harrison & Mitchell 1966). Similar results are obtained with added apyrase, which catalyses the hydrolysis of ADP to AMP. Indirect evidence of this kind is analogous to the conclusion that the abolition by atropine of, say, a secretion indicates that it is mediated physiologically by acetylcholine which, unless its destruction is prevented by an anticholinesterase, is too rapidly destroyed to be demonstrated directly.

It has recently become possible to demonstrate the appearance of free ADP in blood directly in concentrations sufficiently high to activate platelets (Schmid-Schönbein et al 1979). In specially designed apparatus whole blood or resuspended cells are exposed to controlled, different shear stresses for known time periods. The apparatus is designed to cover the range of the variables presumed to be relevant *in vivo*. The experiments show that ADP appears in the plasma in concentrations above those required for platelet activation (0.1 to 1.0 μM) but in direct proportion to free haemoglobin, indicating that platelet activation can result from small degrees of haemolysis due to haemodynamic stresses such as occur during haemorrhage, whether external or through a plaque fissure. It seems, moreover, that the appearance of free ADP is rapid enough to account for *in vivo* aggregation. This process is much faster than the release of ADP from the platelets themselves or than the release of thromboxane A_2 produced by them, which in any case induces aggregation via ADP (B. Samuelsson and A. Marcus, personal communications).

Other experiments (Born et al 1976) provide further evidence that the indirect activation of platelets via erythrocytes is mediated chemically rather than physically. When a polyethylene tube of 200 μm internal diameter is perfused with heparinized blood at 37 °C, a small puncture is sealed off within two minutes or so by a haemostatic plug of platelets, just as in a living blood vessel. This 'bleeding time' is prolonged when the blood contains chlorpromazine in low concentrations which have no effect on platelets but which increase the resistance of erythrocytes to hypotonic haemolysis (Seeman 1972). Under these conditions, therefore, the activation of platelets is inhibited by a membrane-stabilizing drug acting on the accompanying red cells by diminishing either their physical collision-mediated effect, which seems improbable, or their release of platelet-activating agent, presumably ADP.

Recent discoveries about prostaglandins and related substances suggest another way in which fluid-mechanical forces could initiate chemical changes resulting in the haemostatic aggregation of platelets. The first step in the formation of thromboxane A_2 by platelets, which it causes to aggregate (Hamberg et al 1975, Svensson et al 1976), is the release of arachidonic acid from phospholipids in the cell membrane. This release is catalysed by the enzyme phospholipase A_2, which is normally inactive in platelets. How the enzyme is activated physiologically is not known, but perhaps activation is initiated by small distortions of the outer membrane of platelets passing through a field of fluid-mechanical forces greater than those in the normal circulation, for example during arteriolar haemorrhage. Such fluid-mechanical activation of platelets would not involve the red cells, which apparently do not contain the thromboxane-forming system.

SIGNIFICANCE OF THE HAEMOSTATIC REACTIVITY OF PLATELETS

Platelets can be activated by an extraordinary variety of naturally occurring agents, including ADP, collagen, thrombin, arachidonic acid, thromboxane A_2, some prostaglandins, and several neurotransmitters including, at least in the dog, even acetylcholine. Human platelets, those of other mammals and the functionally homologous thrombocytes of birds (Belamarich & Simoneit 1973) are activated by 5-hydroxytryptamine (serotonin) which is also accumulated in the platelets by mechanisms similar to those in tryptaminergic neurons. Because of these similarities, platelets are used increasingly as accessible models for nerve cells, with promising results.

The question remains why such extraordinary pharmacological reactivity should be associated with this type of circulating cell, the only certain function of which is in haemostasis. Local haemostasis must now be one of the few homeostatic mechanisms of the organism, if not the only one, in which neither the central nervous system nor the endocrine system is directly involved. The ability of platelets to be activated by so many different agents may, therefore, be an evolutionary adaptation to the comparative isolation in which they perform their literally vital function.

References

Arfors KE, Cockburn JS, Gross JF 1976 Measurement of growth rate of laser-induced intravascular platelet aggregation, and the influence of blood flow velocity. Microvasc Res 11:79-87

Begent NA, Born GVR 1970 Growth rate *in vivo* of platelet thrombi produced by iontophoresis of ADP, as a function of mean blood flow velocity. Nature (Lond) 227:926-930

Belamarich FA, Simoneit LW 1973 Aggregation of duck thrombocytes by 5-hydroxytryptamine. Microvasc Res 6:229-234

Bolton CH, Emmons PR 1967 Adenosine diphosphate breakdown by the plasma of different species and by human whole blood and white cells. Thromb Diath Haemorrh 18:779-787

Born GVR 1962 Aggregation of blood platelets by adenosine diphosphate and its reversal. Nature (Lond) 194:927-929

Born GVR 1977 Fluid-mechanical and biochemical interactions in haemostasis. Br Med Bull 33:193-197

Born GVR 1979 Arterial thrombosis and its prevention (Proc VIII World Congr Cardiol, Tokyo 1978). Excerpta Medica, Amsterdam, ICS 470, p 81-91

Born GVR, Cross MJ 1964 Effects of inorganic ions and of plasma proteins on the aggregation of blood platelets by adenosine diphosphate. J Physiol (Lond) 170:397-414

Born GVR, Richardson PD 1979 Activation time of blood platelets. Proc R Soc Lond B Biol Sci (submitted)

Born GVR, Bergquist D, Arfors KE 1976 Evidence for inhibition of platelet activation in blood by a drug effect on erythrocytes. Nature (Lond) 259:233-235

Born GVR, Dearnley R, Foulks JG, Sharp DE 1978 Quantification of the morphological reaction of platelets to aggregating agents and of its reversal by aggregation inhibitors. J Physiol (Lond) 280:193-212

Caro CG 1977 Mechanical factors in atherogenesis. In: Cardiovascular flow dynamics and measurement. University Park Press, Baltimore.

Ciba Foundation 1975 Biochemistry and pharmacology of platelets. Excerpta Medica, Amsterdam (Ciba Found Symp 35)

Colantuoni G, Hellums JD, Moake JL, Alfrey CP Jr 1977 The response of human platelets to shear stress at short exposure times. Trans Am Soc Artif Intern Organs 23:626-630

Constantinides P 1966 Plaque fissures in human coronary thrombosis J Atheroscler Res 6:1-17

Cross MJ 1964 Effect of fibrinogen on the aggregation of platelets by adenosine diphosphate Thromb Diath Haemorrh 12:524-527

Duguid JB 1949 Pathogenesis of atherosclerosis. Lancet 2:925

Friedman H 1970 Pathogenesis of coronary thrombosis, intramural and intraluminal hemorrhage. In: Halonen LA (ed) Thrombosis and coronary heart disease. Karger, Basel, vol 4:3

Friedman M, Van den Bovenkamp GJ 1966 The pathogenesis of a coronary thrombosis. Am J Pathol 48:19-44

Gaarder A, Jonsen J, Laland S, Hellem A, Owren PA 1961 Adenosine diphosphate in red cells as a factor in the adhesiveness of human blood platelets. Nature (Lond) 192:531-533

Goldsmith HL 1972 The flow of model particles and blood cells and its relation to thrombogenesis. In: Spaet TH (ed) Progress in hemostasis and thrombosis. Grune & Stratton, New York, vol 1:90

Hamberg M, Svensson J, Samuelsson B 1975 Thromboxanes: a new group of biologically active compounds derived from prostaglandin endoperoxides. Proc Natl Acad Sci USA 72:2994-2998

Harrison MJG, Mitchell JRA 1966 The influence of red blood cells on platelet adhesiveness. Lancet 2:1163-1164

Haslam RJ, Mills DCB 1967 The adenylate kinase of human plasma, erythrocytes and platelets in relation to the aggregation of adenosine diphosphate in plasma. Biochem J 103:773-784

Hellums JD, Brown CH 1977 Blood cell damage by mechanical forces. In: Hwang NHC, Norman NA (eds) University Park Press, Baltimore, ch 20

Milton JG, Yung W, Glushak C, Frojmovic MM 1979 Kinetics of ADP-induced human platelet shape change: apparent positive cooperativity. Can J Physiol Pharmacol, in press

Mustard JF, Murphy EA, Rowsell HC, Downie HG 1962 Factors influencing thrombus formation. Am J Med 33:621-647

Mustard JF, Moore S, Packham MA, Kinlough Rathbone RL 1977a Platelets, thrombosis and atherosclerosis. Proc Biochem Pharmacol 13:312-325

Mustard JF, Packham MA, Kinlough Rathbone RL 1977b Platelets, thrombosis and atherosclerosis. Adv Exp Med Biol 104:127-144

Parker JC 1970 Metabolism of external adenine nucleotides by human red blood cells. Am J Physiol 218:1568

Payling-Wright HP 1972 Mitosis patterns in aortic endothelium. Atherosclerosis 15:93-95

Payling-Wright HP, Born GVR 1971 Possible effect of blood flow on the turnover rate of vascular endothelial cells. In: Hartert HH, Copley AL (eds) Theoretical and clinical hemorheology. Springer-Verlag, Berlin, p 220-226

Richardson PD 1973 Effect of blood flow velocity on growth rate of platelet thrombi. Nature (Lond) 245:103-104

Richardson PD, Galetti P, Born GVR 1976 Regional administration of drugs to control thrombosis in artificial organs. Trans Am Soc Artif Intern Organs 22:22-29

Ross R, Glomset JA 1973 Atherosclerosis and the arterial smooth muscle cell. Proliferation of smooth muscle is a key event in the genesis of the lesions of atherosclerosis. Science (Wash DC) 180:1332-1339

Schmid-Schönbein H, Rieger H, Fischer T 1976 Blood vessels: problems arising at the borders of natural and artificial blood vessels. In: Effert S, Meyer-Erkelenz J (eds) Springer, Berlin, p 57-63

Schmid-Schönbein H, Born GVR, Richardson PD, Cusack J, Rieger H, Forst R, Rohling-Winckel I, Blasberg P, Wehmeier A 1979 Rheology of thrombotic processes in flow: the interaction of erythrocytes and thrombocytes subjected to high flow forces. Biorheology, in press

Seeman P 1972 The membrane action of tranquilisers. Pharmacol Rev 24:583-655

Stehbens WE, 1965 Endothelial cell mitosis and permeability. J Exp Physiol (Cogn Med Sci 50:90-92

Svensson J, Hamberg M, Samuelsson B 1976 On the formation and effects of thromboxane A_2 in human platelets. Acta Physiol Scand 98:285-294

Turitto VT, Baumgartner HR 1975 Platelet interaction with subendothelium in a perfusion system: physical role of red blood cells. Microvasc Res 5:167-179

Turitto VT, Benis AM, Leonard EF 1972 Platelet diffusion in flowing blood. Ind Eng Chem Fund 11:216-223

von Rokitansky (1841-46) Handbuch der pathologischen Anatomie. Braumüller & Seidel, Vienna

Walton KW 1975 Pathogenetic mechanisms in atherosclerosis. Am J Cardiol 35:542-558

Wilner GD, Nossel HL, LeRoy EC 1969 Aggregation of platelets by collagen. J Clin Invest 47-2616-2621

Discussion

Gowans: If platelet emboli are responsible for transient ischaemic attacks can patients at risk be identified if you monitor the blood for platelet aggregates?

Born: Yes. The history of this began when people with transient visual disturbances were seen to have small emboli shooting through their retinal vessels (Pickering 1968); it was reasonable to suggest that the emboli caused the visual disturbances. Since then, evidence has accumulated that transient disturbances of consciousness are also due to passing emboli that produce transient ischaemic attacks. Fields and his colleagues in Texas have seen over 800 patients and are using them to test the clinical effectiveness of drugs capable of inhibiting platelet aggregation *in vitro,* for example aspirin (Fields et al 1977). This investigation is still going on, and so are similar ones elsewhere.

Gowans: And such patients are at risk?

Born: Yes.

Richardson: We have been developing a technique to try to detect aggregates in blood passing through blood vessels. At present we are still using plastic tubes for this. We have a pulsed ultrasound system where we put the echo through a fast Fourier transform in comparison with the incident signal and are able to interpret whether the particle is a gas bubble or an aggregate. We think now that we can watch embolized aggregates in the bloodstream. We can't yet apply that to the vessels in the patient but the technology is evolving (Abts et al 1978).

Vane: One of the features of prostacyclin is that it disaggregates platelets as well as preventing their aggregation. There are several animal models where this has been shown, but Ryszard Gryglewski has done some relevant work on human beings.

Gryglewski: We have estimated the circulating platelet aggregates in human blood. Wu & Hoak (1974) were first to describe the existence of circulating platelet aggregates and we reproduced their findings. Healthy volunteers had from 0 to 30% of platelets circulating in the form of aggregates. In healthy human beings who volunteered to be infused with prostacyclin one of the most striking effects was that prostacyclin completely disaggregated all these circulating platelets in all subjects, even though the suppression of ADP-induced platelet aggregability in platelet-rich plasma was not very much affected at a low dosage of prostacyclin (5 ng kg^{-1} min^{-1}).

Weiss: In a good deal of this discussion in relation to red blood cell haemo-lysis, platelets and prostacyclin, we seem to have been talking about a model of the spleen. The spleen normally holds about a third or more of the platelets of the body in reserve. There is always a certain amount of haemolysis in the spleen. Small platelet aggregates are present and there are even platelets indi-vidually mixed amongst red cells. How does the spleen manage to sequester 35% or more of the body's platelets and then allow them to be released? Is it possible that the spleen's normal business may be to induce haemolysis, there-by causing the release of ADP? When a spleen stores platelets and releases them, moreover, it keeps them in pretty good shape. In most cases when platelets touch on collagen or aggregates they degranulate and the changes seem to be irreversible. But in some of these aggregations that are due to red cell haemolysis, or in some that are carried out without too much perturba-tion, is platelet morphology maintained? Are those platelets perhaps detained transiently and may they be reused?

Born: There is evidence that rapid initial aggregation is wholly reversible and that the half-life of the platelets in the circulation is about the same as that of platelets which have not undergone such aggregation. The disintegration of red cells in the spleen might indeed be a mechanism for entrapping platelets there. It is often asked whether haemolytic diseases such as sickle-cell anaemia, paroxysmal haemoglobinuria or malaria should not, on our hypothesis, be associated with arterial thrombotic episodes. This does not necessarily follow, because the haemolytic situations are quite different. In haemolytic diseases, red cells release their adenine nucleotides into the general plasma circulation in which they are enzymically dephosphorylated. What one has to postulate is that the haemolysis that activates platelets through ADP occurs strictly at the point of highest shear stress at the very edge of the cut.

Weiss: Regarding the two instances of haemolytic disease that you mentioned, there are animals with moderate spectrin deficiency where haemolysis is well controlled and there may be increased platelet storage in the spleen.

Youdim: Professor Born, you mentioned that chlorpromazine inhibits aggregation of platelets *in vitro*. We have reported similar results with this drug and other neuroleptics (Oppenheim et al 1979). Recent studies in my laboratory have indicated that the action of chlorpromazine *in vivo* might be quite different (Oppenheim et al 1978, Heifetz et al 1979). When platelet function was used as a peripheral model (bioassay) for examining the efficacy of drugs such as chlorpromazine, which act in the CNS as anti-psychotic agents, the platelet aggregation response to 5-hydroxytryptamine (serotonin) in patients receiving chlorpromazine is opposite to its *in vitro* action. That is, the aggregation response to 5-HT is enhanced, resulting in an irreversible aggregation similar to that observed with ADP (Fig. 1). The aggregation response is dependent on the concentration of 5-HT. One cannot fail to see that the aggregation response to 5-HT has become supersensitive, due either

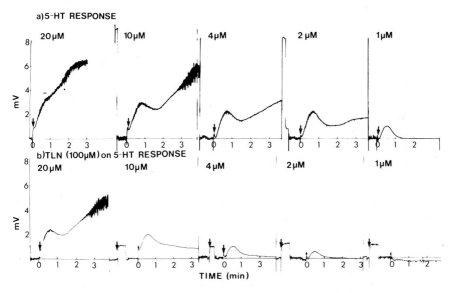

FIG. 1. (Youdim). (a) Platelet aggregation response to 5-hydroxytryptamine (5-HT) in a schizophrenic patient treated with chlorpromazine. This patient is considered to be a clinical responder. (b) In the same patient, the platelet aggregation responses to 5-HT after preincubation for 3 min with a 5-HT analogue, tryptoline (TLN, tetrahydro-β-carboline). TLN inhibits the aggregation response. Preincubation with chlorpromazine will also inhibit the irreversible platelet aggregation response to 5-HT that is obtained in chlorpromazine responders. (See Oppenheim et al 1978, 1979.)

to unmasking of other receptors or to synthesis of new ones. Therefore, one must be very careful about the *in vitro* action of the membrane-stabilizing compounds such as chlorpromazine. This supersensitivity of platelets to 5-HT could be an undesirable side-effect of chlorpromazine treatment.

Patients with true iron-deficiency anaemia also show the same functional changes in 5-hydroxytryptamine-induced aggregation (Woods et al 1977). Using a platelet resuspension technique, we have shown that the alteration in platelet aggregation responses in chlorpromazine-treated and iron-deficient patients was due to a platelet defect (Oppenheim et al 1979). I am not aware of any report which suggests that schizophrenic patients on longer-term neuroleptic therapy have an increased incidence of thrombosis. However, recent preliminary studies in the Nuffield Department of Medicine at Oxford have indicated that patients with deep vein thromboses have an increased incidence of iron-deficiency anaemia. If this can be confirmed, together with reports that in subjects with iron-deficiency anaemia, 5-HT as well as noradrenaline catabolism is diminished due to the reduction in enzymic activity of monoamine oxidase, one can envisage a role for 5-hydroxytryptamine.

Born: The evidence that I am aware of indicates that 5-HT is irrelevant to the initial adhesion of platelets to vessel walls. Furthermore, only the second phase of aggregation associated with the release reaction is inhibited by chlorpromazine at rather higher concentrations than the ones effective against haemolysis. The second phase of aggregation comes on later than the extremely fast effect that has to be explained.

Howard: Presumably the concentration of ADP declines extremely rapidly downstream after haemolysis. Is it possible to have a measure of instantaneous ADP concentration in a natural haemodynamic lesion?

Born: Probably, yes. The red cells constitute a very large reservoir of all the adenine nucleotides. We have recently determined the concentration of free ADP directly. This can so far be done within a few, probably two or three, seconds. That is very fast compared to the rate of breakdown of ADP by the enzymes present on the cells and in the plasma, where the half-life is a few minutes.

Howard: But the other consideration is the flow rate, isn't it?

Born: Yes. One has to postulate that a small proportion of red cells, presumably the most fragile, haemolyse as they arrive. The concentration of ADP at the exact site would be difficult to determine. The same plasma enzymes dephosphorylate ATP, ADP, and AMP. The dephosphorylation of ATP to ADP is much faster than that of ADP. As red cells contain much more ATP and ADP, ATP released by haemolysis represents a reservoir for free ADP in the plasma.

Marcus: I think chlorpromazine is also an antioxidant. One might speculate that another explanation for the bleeding phenomenon is that chlorpromazine blocked conversion of arachidonic acid to prostaglandins. Therefore, one saw an effect on the bleeding time comparable to that observed with aspirin. One way to determine whether this might be true is to stimulate platelets with chlorpromazine and observe whether an oxygen burst occurs (Bressler et al 1979). If chlorpromazine blocks the oxygen burst, it may be acting in a manner comparable to aspirin.

Born: I would like to know the time course of such an antioxidant action of chlorpromazine. It would have to be a very rapid effect. And as I said, what evidence there is makes it unlikely that endogenous 5-HT is involved in the rapid phase of haemostatic platelet aggregation.

Marcus: I was referring to the bleeding time experiments carried out with chlorpromazine. Thus, the lengthening of the bleeding time might be related to an inhibitory effect of chlorpromazine on thromboxane synthesis in the platelet.

References

Abts LR, Beyer RT, Galletti PM, Richardson PD, Karon D, Massimino R, Karlson KE 1978 Computerized discrimination of microemboli in extracorporeal circuits. Am J Surg 135:535-538

Bressler NM, Broekman MJ, Marcus AJ 1979 Concurrent studies of oxygen consumption and aggregation in stimulated human platelets. Blood 53:167-178

Fields WS, Lewak NA, Frankowski RF, Hardy RJ 1977 Controlled trial of aspirin in cerebral ischaemia. Stroke 8:301-315

Heifetz A, Oppenheim B, Glantz B, Youdim MBH 1979 Chlorpromazine therapy and platelet aggregation in response to serotonin. Isr J Med Sci, in press

Oppenheim B, Youdim MBH, Goldstein S, Heifetz A 1978 Human platelets as a model for the study of the pharmacological activity of tryptolines and neuroleptics. Isr J Med Sci 14:1096-1097

Oppenheim B, Hefez A, Youdim MBH 1979 Serotonin receptor site in human platelets from control and chlorpromazine treated subjects. In: Burgen A, Heldman E (eds), Neuroactive compounds and their cell receptors. Elsevier, Amsterdam, in press

Pickering GW 1968 High blood pressure. Churchill, London

Woods HF, Youdim MBH, Boullin D, Callender S 1977 Monoamine metabolism and platelet function in iron-deficiency anaemia. In: Iron metabolism. Excerpta Medica, Amsterdam (Ciba Found Symp 51) p 227-248

Wu KK, Hoak JC 1974 A new method for the quantitative detection of platelet aggregates in patients with arterial insufficiency. Lancet 2:924-926

Prostacyclin

J.R. VANE and S. MONCADA

Wellcome Research Laboratories, Langley Court, Beckenham, Kent BR3 3BS

Abstract   Prostacyclin (PGI$_2$), generated by the vascular wall, is a strong vasodilator and a potent inhibitor of platelet aggregation. It can be continually released by the lungs into the arterial circulation and, therefore, can also be a circulating hormone. Thus, platelets in the bloodstream are subjected constantly to prostacyclin stimulation and it is via this mechanism, together with PGI$_2$ formation in the vascular endothelium, that platelet aggregability *in vivo* is controlled. Conditions which decrease prostacyclin availability, such as atheroma deposition, will tend to allow platelet thrombi to form.

The prostacyclin/thromboxane A$_2$ (TXA$_2$) ratio is important in the control of thrombus formation; manipulation of this ratio by a selective inhibitor of thromboxane formation might be a useful anti-thrombotic therapy. Arachidonic acid (C20:4ω6) (obtained from dietary linoleic acid and from farm animal meats) is the precursor of PGI$_2$ and TXA$_2$ and it is the balance between these potent substances with opposing activities which determines the thrombotic state. Eicosapentaenoic acid (C20:5ω3) obtained from dietary marine animals can also lead to a potent prostacyclin-like compound (PGI$_3$), but TXA$_3$ formed by platelets has no aggregating activity. Thus, the balance is shifted towards the anti-aggregating state. This could be the basis of the low incidence of thrombotic diseases and the tendency to bleed in Eskimos and also for consideration of dietary change or supplement as a prophylactic measure against thrombotic diseases.

Prostacyclin has interesting potential for clinical application in conditions where enhanced platelet aggregation is involved or in increasing the biocompatibility of extra-corporeal circulation systems.

During the summer of 1975 we isolated and partially characterized an enzyme in platelet microsomes that is responsible for the generation of thromboxane A$_2$ (TXA$_2$) from prostaglandin endoperoxide. This enzyme was different from cyclo-oxygenase in its behaviour and sensitivity to inhibitory drugs. We called it thromboxane synthetase. Three months later, we searched for

© *Excerpta Medica 1980*
Blood cells and vessel walls: functional interactions
(Ciba Foundation symposium 71) p 79-97

thromboxane synthetase in other tissues, including blood vessels (see Moncada & Vane 1978, 1979).

There were two questions in our minds. First, does thromboxane generation in the vasculature synergize with platelet TXA_2 and help in haemostatic plug formation, especially through the immediate vasoconstriction which follows when a small vessel is cut? Secondly, Morrison & Baldini (1969) had suggested that platelets and vascular tissue shared some antigenic properties, and we wondered whether the two structures shared some enzymic proteins.

The project, in which Gryglewski and Bunting collaborated, quickly demonstrated (to our disappointment) that TXA_2 was not formed by aortic microsomes, as measured by enhancement of the contracting activity of endoperoxide on rabbit aorta in the cascade bioassay. Indeed, the endoperoxide precursor was consumed enzymically into an unknown product not identifiable by the standard bioassay tissues—rat stomach strip, chicken rectum, rat colon, and rabbit aorta. The substance resulting from this enzymic activity was labile and relaxed the coeliac and mesenteric arteries of the rabbit. We began to refer to it as prostaglandin X (PGX).

The second most exciting moment came when we realized that the vessel wall might be synthesizing the biological opposite of TXA_2, as a defence mechanism against platelet aggregation. We tested PGX as an inhibitor of platelet aggregation and found that it was many times more active than PGE_1 or PGD_2.

Later work further characterized PGX; it was a potent relaxant of vascular strips *in vitro*, a strong vasodilator *in vitro*, the most potent inhibitor of platelet aggregation yet discovered, and anti-thrombotic. Furthermore, it was the major metabolite of arachidonic acid in vascular tissue. PGX was the unstable intermediate in the formation of 6-oxo-$PGF_{1\alpha}$, a compound described at about that time by Pace-Asciak (1976) as a product of prostaglandin endoperoxides in the rat stomach. In the work which led to the elucidation of the structure of PGX (Johnson *et al* 1976) we collaborated with scientists from the Upjohn Company, Kalamazoo. PGX was then renamed prostacyclin, with the abbreviation PGI_2. For detailed accounts of the formation and activity of prostacyclin, see Vane & Bergstrom (1979).

PROSTACYCLIN: BIOSYNTHESIS AND BIOLOGICAL ACTIVITY

Biosynthesis

In vessels, Moncada *et al* (1977) showed that the enzyme which metabolizes prostaglandin endoperoxides to prostacyclin (prostacyclin synthetase) is most highly concentrated in the intima and progressively decreases in activity towards the adventitia. Measurement of prostacyclin formation by cultured cells from vessel walls of rabbits or human beings also shows that endothelial cells are the most active producers of prostacyclin. Moreover, this production persists after numerous sub-cultures *in vitro* (Christofinis *et al* 1979). Initially, we demonstrated that vessel microsomes could utilize prostaglandin endoperoxides, but not arachidonic acid, to synthesize prostacyclin. Later, we showed that fresh vascular tissue could utilize both precursors, although it was far more effective in utilizing prostaglandin endoperoxides. We also demonstrated that vessel microsomes, fresh vascular rings or endothelial cells treated with indomethacin could, when incubated with platelets, generate an anti-aggregating activity, the release of which was abolished by 15-hydroperoxy arachidonic acid (15-HPAA), a selective inhibitor of prostacyclin formation.

From all these results we concluded that the vessel wall can form prostacyclin from its own endogenous precursors, but that it can also synthesize prostacyclin from endoperoxides released by the platelets, thus suggesting a biochemical cooperation between platelets and vessel wall when they are in intimate contact.

This hypothesis has proved to be controversial. Needleman et al (1978) demonstrated that, although arachidonic acid was rapidly converted to prostacyclin by perfused rabbit hearts and kidneys, PGH_2 was not readily used. The authors concluded that some degree of vascular damage is necessary for the endoperoxide to be utilized by prostacyclin synthetase. However, incubation of platelet-rich plasma with fresh indomethacin-treated arterial tissue leads to an increase in platelet cyclic AMP (cAMP) which parallels the inhibition of the aggregation and can be abolished by previous treatment of the vascular tissue with tranylcypromine, an inhibitor of prostacyclin formation. Additionally, Tansik et al (1978) showed that lysed aortic smooth muscle cells could be fed prostaglandin endoperoxides by lysed human platelets. Further, Marcus et al (1978) showed that undisturbed endothelial cell monolayers readily utilize PGH_2, transforming it into prostacyclin.

The fact that prostacyclin inhibits platelet aggregation (platelet–platelet interaction) at much lower concentrations than those needed to inhibit adhesion (platelet–collagen interaction) suggests that, indeed, prostacyclin allows platelets to stick to vascular tissue and interact with it, at the same time suppressing thrombus formation. Clearly, a platelet adhering to a site where prostacyclin synthesis is possible could feed the enzyme with endoperoxide, thereby producing prostacyclin and preventing other platelets from aggregating. Indeed, after electrical damage to the carotid artery of the rabbit, prostacyclin infused intravenously inhibits thrombus formation while allowing a layer of platelets to form at the site of damage (Ubatuba et al 1979). Thus, prostacyclin has an important role as a defensive hormone in that it protects against thrombus formation inside normal vessels. Moreover, when there is endothelial damage or when a vessel is cut, platelets in contact with subendothelial layers (with pro-aggregating activity due to the presence of collagenous structures and a weaker ability to form prostacyclin) will start adhering. The balance between the pro- and anti-aggregating forces, perhaps depending on the degree of damage, will control the degree of clumping, thrombus formation, etc. (see Moncada & Vane 1978, 1979).

It is also possible that other formed elements of blood, such as the white cells, which produce endoperoxides and TXA_2, can interact with the vessel wall to enhance prostacyclin formation, as do the platelets. Thus, by modulation of white cell behaviour, prostacyclin may reduce white cell migration during the inflammatory response.

Prostacyclin inhibits platelet aggregation by stimulating adenyl cyclase (adenylate cyclase, EC 4.6.1.1), raising cAMP levels in the platelets. In this respect, prostacyclin is many times more potent than either PGE_1 or PGD_2. 6-Oxo-$PGF_{1\alpha}$ has weaker anti-aggregating activity and is almost devoid of activity on platelet cAMP.

In contrast to prostacyclin, prostaglandin endoperoxides and TXA_2 reduce or inhibit cAMP formation in platelets. These opposing effects suggest that a balance between TXA_2 and prostacyclin formation regulates platelet cAMP *in vivo* and, therefore, platelet aggregability. This proposition has been reinforced by the recent finding that prostacyclin is a circulating hormone. Unlike other prostaglandins, such as PGE_2 and $PGF_{2\alpha}$, prostacyclin is not inactivated on passage through the pulmonary circulation. Indeed, the lung constantly releases small amounts of prostacyclin into the circulation. The concentrations of prostacyclin are higher in arterial than in venous blood because about 50% is inactivated in one circulation (20−30 s) through peripheral tissues (Dusting et al 1978). This short half-life in the circulation contrasts with the chemical half-life of 3 min at blood temperature and pH. Thus, platelet aggregability *in vivo* is modulated by circulating prostacyclin which will reinforce the actions of locally-produced prostacyclin throughout the vasculature.

Circulating platelets, therefore, could be constantly stimulated by prostacyclin, and they might have higher cAMP levels and be less aggregable than has ever been revealed by *in vitro* measurements. Such measurements are made only after a 10−30 min delay during which the blood is processed. In this period, the unstable prostacyclin, and also platelet cAMP levels, will decay.   This concept helps to explain the known differences in reactivity between platelets *in vitro* and *in vivo*.

These observations also throw light on the control of cAMP levels in platelets. Although platelet-aggregating agents, including the prostaglandin endoperoxide PGH_2, reduce cAMP levels *in vitro*, some authors have found that strong pro-aggregating agents, such as prostaglandin endoperoxides or TXA_2, do not affect 'basal' levels of cAMP; they only find activity of these compounds in antagonizing PGE_1 or PGI_2-induced increases of cAMP levels. Clearly, those 'basal' levels might be low due to the decay of platelet cAMP during the preparative procedure. Thus, it may be impossible to measure the small further decrease necessary to cause aggregation, whereas the change becomes measurable when the cAMP levels are increased with PGE_1 or prostacyclin.

The discovery of prostacyclin and the elucidation of its biological activities have enhanced our insight into several physiological and pathophysiological mechanisms such as thrombus and haemostatic plug formation. Further, it became possible to explain the mechanism of anti-thrombotic activity of phosphodiesterase inhibitors such as dipyridamole. Prostacyclin has also provided an endogenous compound which could well give a new therapeutic approach to thrombo-embolic disorders and other conditions where platelet aggregation is enhanced. Any function of prostacyclin in preventing the development of atherosclerosis has still to be assessed.

PROSTACYCLIN, TXA_2, THROMBOSIS AND HAEMOSTASIS

Prostacyclin and TXA_2 represent, in biological terms, opposite poles of the same general homeostatic mechanism for regulation of platelet aggregability *in vivo*.   Manipulation of this control mechanism will, therefore, affect thrombus and haemostatic plug formation. Selective inhibition of the formation of TXA_2 should lead to an increased bleeding time and inhibition of thrombus formation, whereas inhibition or loss of prostacyclin formation should be propitious for intravascular thrombus formation. The amount of control exerted by this system can be tested, for selective inhibitors of each pathway have been described (see Moncada & Vane 1978, 1979).

There is ample evidence showing that aspirin inhibits platelet cyclo-oxygenase *in vivo* and *in vitro*. Whereas the analgesic and anti-inflammatory dose in humans is about 1.5 g per day, a single tablet of aspirin (0.325 g) inhibits the cyclo-oxygenase of platelets by 90% (Burch *et al* 1978). Moreover, this effect is irreversible, due to acetylation by aspirin of the active site of the enzyme. Platelets are unable to synthesize new protein and do not replace the cyclo-oxygenase; therefore, the inhibition is long-lasting and will only be overcome by entry of new platelets into the circulation after the blockade of synthesis in megakaryocytes has worn off. Vessel wall cyclo-oxygenase, at least *in vitro*, is much less sensitive to aspirin than is that of platelets, and there is also the suggestion that the endothelial cells recover from aspirin inhibition by regeneration of their cyclo-oxygenase.

Recent studies in rabbits (Moncada & Korbut 1978) suggest that low doses of aspirin reduce TXA_2 formation to a greater extent than prostacyclin formation. These experiments also showed that inhibition of TXA_2 formation is longer-lasting than that of prostacyclin. Indeed, infusions of arachidonic acid into rabbits or cats lead to an anti-thrombotic effect and to an increase in bleeding time, which can be potentiated by low doses of aspirin and blocked by large doses that inhibit prostacyclin and TXA_2 formation.

The anti-thrombotic activity of dipyridamole depends mainly on its inhibition of phosphodiesterase, which amplifies the increase in cAMP induced by circulating prostacyclin (Moncada & Korbut 1978). Dipyridamole is most effective when there is a favourable PGI_2/TXA_2 ratio, after a small dose of aspirin or more than 24 h after a high dose. A selective inhibitor of thromboxane formation together with a phosphodiesterase inhibitor should now be tested for anti-thrombotic efficacy, since theoretically this provides an advantage over aspirin in leaving endoperoxides from platelets available for the vessel wall or other cells to synthesize prostacyclin.

These results also suggest that, when aspirin is used, a small daily dose or large doses at weekly intervals, alone or in combination with a phosphodiesterase inhibitor such as dipyridamole, would be a useful therapeutic combination. Clearly, it is important not to use too high a dose of aspirin, for that will neutralize the whole system, including prostacyclin formation.

Until the discovery of prostacyclin the use of aspirin as an anti-thrombotic agent, based on its effect on thromboxane formation, appeared simple. Now, however, the situation needs further clarification, especially with respect to the optimal dose of aspirin. Aspirin in high doses (200 mg/kg) increases thrombus formation in a model of venous thrombosis in the rabbit, and *in vitro* treatment of endothelial cells with aspirin enhances thrombin-induced platelet adherence to them. Interestingly, recent studies in humans show that single small doses of aspirin (0.3 g) increase cutaneous bleeding time while large doses are devoid of such an effect (O'Grady & Moncada 1978).

THERAPEUTIC POTENTIAL OF PROSTACYCLIN

We have suggested that prostacyclin or a chemical analogue could be used as a 'hormone replacement' therapy for conditions in which platelet aggregation is involved, such as acute myocardial infarction or crescendo angina and other conditions where there is excessive platelet aggregation in the circulation (see Vane & Bergstrom 1979). Moreover, we have suggested the use of prostacyclin in extracorporeal circulations such as cardiopulmonary bypass, renal dialysis and charcoal haemoperfusion. In these three systems, there is a substantial fall in platelet count, accompanied by the formation of microaggregates which, when returning to the patient, can cause cerebral and renal impairment. Side effects, especially osteoporosis, are also produced by prolonged use of heparin.

Several anti-platelet drugs have been tested, some with moderate success. We and others (see Vane & Bergstrom 1979) have used prostacyclin in all three systems in dogs and demonstrated that it has a strong heparin-sparing effect. Moreover, there is *no* formation of microaggregates or platelet loss. Although the long-term effects of this substance have yet to be studied, evaluation in humans has started and its clinical possibilities seem bright.

Vasodilatation

Prostacyclin relaxes arterial strips *in vitro* and causes vasodilatation and hypotension in animals and man. Dusting et al (1977) showed that prostacyclin is the unstable metabolite of arachidonic acid responsible for the paradoxical coronary relaxation described by Kulkarni et al (1976); endoperoxides produce a biphasic effect consisting of an initial sharp contraction followed by a longer-lasting relaxation. When prostacyclin formation is prevented by 15-HPAA, the relaxation is abolished (as in the coeliac and mesenteric arteries of the rabbit), unmasking a pure contraction which is due to the direct effect of endoperoxides.

In the anaesthetized dog, prostacyclin is hypotensive and a potent inducer of coronary vasodilatation when given as bolus injections directly into the left circumflex artery. These latter effects are enhanced by indomethacin or meclofenamate.

Given intravenously to the anaesthetized rabbit or rat prostacyclin causes a fall in blood pressure and is four to eight times more potent than PGE_2, and at least 100 times more potent than 6-oxo-$PGF_{1\alpha}$. Since it is not inactivated by the pulmonary circulation, prostacyclin is equipotent as a vasodilator when given either intra-arterially or intravenously in the rat, rabbit or dog. In this respect, prostacyclin is different from PGE_2 which, because of pulmonary metabolism, is much less active when given intravenously.

Many authors have suggested a vasodilator role for locally generated PGE_2 in the vascular wall and others have suggested that PGE_1 is released. There is little evidence that PGE_1 is a naturally occurring prostaglandin in the cardio-vascular system of mammals, and prostacyclin is the major product of arachidonic acid metabolism in all vascular tissues so far studied. Indeed, prostacyclin is responsible for the vasodilator properties of arachidonic acid in several vascular beds (see Vane & Bergstrom 1979), such as those of the heart, kidney, mesenteric and skeletal muscle. Moreover, it may mediate the release of renin from the renal cortex. Indeed, arachidonic acid, prosta-glandin endoperoxides and prostacyclin all stimulate renin release from slices of rabbit renal cortex, but PGE_2 does not (see Vane & Bergstrom 1979). Infusion of arachidonic acid or prostacyclin, but not PGE_2, into the renal artery in dogs increases renin release several-fold. Since 6-oxo-$PGF_{1\alpha}$ has been identified in incubates of PGG_2 with renal cortical microsomes, prosta-cyclin may be the obligatory endogenous mediator of renin secretion by the kidney.

Protection of gastrointestinal mucosa

6-Oxo-$PGF_{1\alpha}$ was first isolated as the major product of endoperoxide metabolism in homogenates of rat stomach (Pace-Asciak 1976). We later showed that prostacyclin is the major prostaglandin product of the gastric mucosa of several species and, furthermore, that it is a potent vasodilator of rat stomach mucosa *in vivo*, where it also reduces acid secretion induced by pentagastrin (see Moncada & Vane 1979). Similar results have been obtained in the canine stomach. Thus, prostacyclin release may be involved in functional hyperaemia of the mucosa during acid secretion and may also act as a natural brake on the secretion, a role previously ascribed to PGE_2. Inhibition of prostacyclin production by prostaglandin synthetase inhibitors, such as aspirin and indomethacin, could explain why this general group of substances tends to cause gastrointestinal irritation. These findings are already leading to the development of 'anti-ulcer' compounds based on the prostacyclin molecule which are potent and more selective than the previous generation of compounds based on PGE_2.

FATTY ACIDS AND THROMBOSIS

Before the discovery of prostacyclin, it was suggested that the use of dietary dihomo-γ-linolenic acid, the precursor of the E_1 series of prostaglandins, could be an approach to the prevention of thrombosis, for PGG_1 and TXA_1 are not pro-aggregating and PGE_1 is anti-aggregating. Other reports tend to agree with this proposal but there is some controversy, for rabbits fed with dihomo-γ-linolenic acid have an increased tissue content of this acid without change in platelet responsiveness, at least to ADP. The main criticism of all this work, including that on human platelets, is that the conclusions are based on studies performed *in vitro* in which platelets are studied as isolated cells without contact with vessel walls (see Moncada & Vane 1978, 1979).

It is now evident that the use of dihomo-γ-linolenic acid to direct the synthetic machinery of the platelets is not the most rational approach to the prevention of thrombosis. This is because the endoperoxides PGG_1 and PGH_1 are not substrates for the formation of prostacyclin or because their precursor might adversely affect the prostacyclin protective mechanism.

Human beings have the enzymes needed to elongate and desaturate linoleic acid (C18:2ω6) to arachidonic acid (C20:4ω6). We also obtain preformed arachidonic acid from the meat of land animals. Eicosapentaenoic acid (C20:5ω3) comes from marine animals and is the precursor of a different series of prostaglandins. It can be used to generate an anti-aggregating agent in the vessel wall which is probably a prostacyclin (Dyerberg et al 1978). However, if a thromboxane is formed by platelets, it does not induce their aggregation (Raz et al 1977). The consumption of this fatty acid, then, could afford a dietary protection against intravascular thrombosis, for it would swing the balance towards the anti-thrombotic side of the system. Indeed, the low incidence of myocardial infarction in Eskimos and their increased tendency to bleed is probably due to the high eicosapentaenoic acid and low linoleic and arachidonic acid content of their marine diet and, therefore, of their tissues (Dyerberg et al 1978). It has been known for some years that this is associated with a lowered blood cholesterol.

In a well-controlled cross-over study von Lossonczy et al (1978) measured the effects of the addition of fish (to give 8 g daily of ω3 fats) or cheese to a lacto-ovo-vegetarian diet in men and women. Serum cholesterol and triglycerides were lowered and high density lipoprotein was increased by the fish diet as compared with the cheese diet. There was also a strong decrease in the very low density lipoprotein. Each of these changes is thought to lower the risk of arterial disease. Interestingly, low density lipoprotein inhibits prostacyclin formation by human endothelial cells in culture (Nordy et al 1978).

All these results, taken together with the findings that the balance of the prostacyclin/thromboxane system can also be changed by consumption of fish oils, lend great importance to future research into the effects of *specific* rather than *general* polyunsaturated fats in the diet. The 'polyunsaturated' fats as a group may well have less relevance to the prevention of cardio-vascular disease than those that can lead to prostacyclin, especially eicosapentaenoic acid.

Cod liver oil contains 10% eicosapentaenoic acid, but to equal an Eskimo's intake would mean drinking 700 calories (2940 kJ) worth of oil a day. Indeed, the fats of most common fish contain 8–12% eicosapentaenoic acid. The fats of the more exotic sea foods, such as scallops, oysters and red caviare, contain more than 20%!

α-Linoleic acid (C18:3ω3) is the vegetable oil which, if elongated and desaturated, would lead to eicosapentaenoic acid. Whether this occurs in human beings appears to be uncertain, but Sanders et al (1977) made an important comparison of the fatty acid composition of plasma choline phospho-glycerides and red cell lipids in vegans and omnivores. The only striking differences were in the ω3 series: the vegans had only 12–15% of the levels of eicosapentaenoic acid found in omnivores. Clearly, it is important to establish whether humans convert sufficient linolenic acid to eicosapentaenoic acid, for the latter is found especially as a constituent of brain cell lipids.

What bearing does all this have on the continuing debate about dietary polyunsaturates? Margarines contain anything from 9 to 48% of linoleic acid, depending on the brand. The rest of the fats are C16:0 (7–18%), C18:0 (1.5–14%), C18:1 (8–53%) and C18:3 (0.1–5%) (Weihrauch et al 1977). Of these fatty acids, only C18:2 can act (when in the cis form) as a precursor (when elongated) for prostacyclin. However, a substantial and variable proportion of the linoleic acid in margarine is in the trans form (Weihrauch et al 1977). Certainly, choosing a margarine with a high content of C18:2ω6 in the cis form needs more information than is provided on the label, and no margarine has the high content of C20:5ω3 which may be preferable.

Thus, as providers of precursors for prostacyclin and other prostaglandins, many of the 'polyunsaturated' fats in margarine are unimportant and could even have a deleterious effect if the non-precursor fats are peroxidized, for lipid peroxides inhibit prostacyclin production. Polyunsaturated fatty acids are easily auto-oxidized or enzymically converted to the corresponding linear hydroperoxy fatty acids. Peroxides of fatty acids, including arachidonic, dihomo-γ-linolenic, α-linolenic and γ-linolenic acids, inhibit prostacyclin synthetase when incubated with this enzyme at low concentrations (1–2 μM). Interestingly, in the absence of vitamin E, which is a natural anti-oxidant, a polyunsaturated fatty-acid-rich diet fed to pigs causes endothelial damage and subsequent thrombosis (Nafstad 1974).

The role of lipid peroxides in the development of atherosclerosis has been debated for the last 25 years, since Glavind et al (1952) described the presence of lipid peroxides in human atherosclerotic aortas. They found the peroxide content in diseased arteries to be directly proportional to the severity of the atherosclerosis. Subsequent investigations suggested that Glavind's findings were artifactual, ascribing the formation of lipid peroxides to their formation during the preparative procedure. Despite this, the presence of conjugated diene hydroperoxides in lipids of human atheroma has again been described and lipid peroxides have been found in atherosclerotic rabbit aortas (Iwakami 1965) subjected to an extraction procedure which avoids lipid peroxidation *in vitro*.

More recent papers favour the suggestion that lipid peroxides are present in atherosclerotic plaques (Brooks et al 1971, Harland et al 1971). Whether these peroxides act by inhibiting prostacyclin formation, and as a consequence reduce the vessel wall defence mechanism, is not finally determined but deserves further research.

References

Brooks CJW, Steel G, Gilbert JD, Harland WA 1971 Lipids of human atheroma, part 4. Characteristics of a new group of polar sterol esters from human atherosclerotic plaques. Atherosclerosis 13:223-237

Burch JW, Stanford N, Majerus PW 1978 Inhibition of platelet prostaglandin synthetase by oral aspirin. J Clin Invest 61:314-319

Christofinis GJ, Moncada S, MacCormick C, Bunting S, Vane JR 1979 Prostacyclin (PGI$_2$) release by rabbit aorta and human umbilical vein endothelial cells after prolonged subculture. In: Vane JR, Bergstrom S (eds) Prostacyclin. Raven Press, New York, in press

Dusting GJ, Moncada S, Vane JR 1977 Prostacyclin (PGX) is the endogenous metabolite responsible for relaxation of coronary arteries induced by arachidonic acid. Prostaglandins 13:3-16

Dusting GJ, Moncada S, Vane JR 1978 Recirculation of prostacyclin (PGI$_2$) in the dog. Br J Pharmacol 64:315-320

Dyerberg J, Bang HO, Stoffersen E, Moncada S, Vane JR 1978 Polyunsaturated fatty acids, atherosclerosis and thrombosis. Lancet 2:117-119

Glavind J, Hartmann S, Clemmesen J, Jessen KE, Dam H 1952 Studies on the role of lipid peroxides in human pathology II. The presence of peroxidized lipids in the atherosclerotic aorta. Acta Pathol Microbiol Scand 30:1-6

Harland WA, Gilbert JD, Steel G, Brooks CJW 1971 Lipids of human atheroma, part 5. The occurrence of a new group of polar sterol esters in various stages of human atherosclerosis. Atherosclerosis 13:239-246

Iwakami M 1965 Peroxides as a factor of atherosclerosis. Nagoya J Med Sci 28:50-66

Johnson RA, Morton DR, Kinner JH, Gorman RR, McGuire JR, Sun FF et al 1976 The chemical structure of prostaglandin X (prostacyclin). Prostaglandins 12:915-928

Kulkarni PS, Roberts R, Needleman P 1976 Paradoxical endogenous synthesis of a coronary dilating substance from arachidonate. Prostaglandins 12:337-353

Marcus AJ, Weksler BB, Jaffe EA 1978 Enzymatic conversion of prostaglandin endoperoxide H$_2$ and arachidonic acid to prostacyclin by cultured human endothelial cells. J Biol Chem 253: 7138-7141

Moncada S, Korbut R 1978 Dipyridamole and other phosphodiesterase inhibitors act as anti-thrombotic agents through potentiating endogenous prostacyclin. Lancet 1:1286-1289

Moncada S, Vane JR 1978 Unstable metabolites of arachidonic acid and their role in haemostasis and thrombosis. Br Med Bull 34:129-136

Moncada S, Vane JR 1979 Prostacyclin in perspective. In: Vane JR, Bergstrom S (eds) Prostacyclin. Raven Press, New York, in press

Moncada S, Herman AG, Higgs EA, Vane JR 1977 Differential formation of prostacyclin (PGX or PGI$_2$) by layers of the arterial wall. An explanation for the anti-thrombotic properties of vascular endothelium. Thromb Res 11:323-344

Morrison FS, Baldini MG 1969 Antigenic relationship between blood platelets and vascular endothelium. Blood 33:46-57

Nafstad I 1974 Endothelial damage and platelet thrombosis associated with PUFA-rich, vitamin E deficient diet fed to pig. Thromb Res 5:25-28

Needleman P, Bronson SD, Wyche A, Sivakoff M, Nicolaou KC 1978 Cardiac and renal prostaglandin I$_2$. Biosynthesis and biological effects in isolated perfused rabbit tissues. J Clin Invest 61:839-849

Nordoy A, Svensson B, Wiebe D, Hoak JC 1978 Lipoproteins and inhibitory effect of human endothelial cells on platelet function. Circ Res 43:527-534

O'Grady J, Moncada S 1978 Aspirin: a paradoxical effect on bleeding time. Lancet 2:780P

Pace-Asciak C 1976 Isolation, structure and biosynthesis of 6-keto prostaglandin F$_{1\alpha}$ in the rat stomach. J Am Chem Soc 98:2348-2349

Raz A, Minkes MS, Needleman P 1977 Endoperoxides and thromboxanes. Structural determinants for platelet aggregation and vasoconstriction. Biochim Biophys Acta 488:305-311

Sanders TAB, Ellis FR, Dickerson JWT 1977 Polyunsaturated fatty acids and the brain. Lancet 1:751-752

Tansik RL, Namm DH, White HL 1978 Synthesis of prostaglandin 6-keto F$_{1\alpha}$ by cultured aortic smooth muscle cells and stimulation of its formation in a coupled system with platelet lysates. Prostaglandins 15:399-408

Ubatuba FB, Moncada S, Vane JR 1979 The effect of prostacyclin (PGI$_2$) on platelet behaviour, thrombus formation *in vivo* and bleeding time. Thromb Haemostasis, in press

Vane JR, Bergstrom S (eds) 1979 Prostacyclin. Raven Press, New York, in press

von Lossonczy TO, Ruiter A, Bronsgeest-Schoute HC, van Gent CM, Hermus RJJ 1978 The effect of a fish diet on serum lipids in healthy human subjects. Am J Clin Nutr 31:1340-1346

Weihrauch JL, Brignoli CA, Reeves JB, Iverson JL 1977 Fatty acid composition of margarines, processed fats and oils. Food Technol 31:80-85

Discussion

Gryglewski: Prostacyclin has been successfully used in the treatment of thrombotic complications of arteriosclerosis obliterans (Szczeklik et al 1978). There is no doubt that prostacyclin acts as a local hormone at the site of its generation in arteries and prevents adhesion and aggregation of platelets. Is prostacyclin a circulating hormone? We have demonstrated that the arterial blood of anaesthetized cats contains more prostacyclin-like activity than mixed venous blood (Gryglewski et al 1978b), and therefore we put forward a hypothesis that the lung is an endocrine gland that secretes prostacyclin. To demonstrate this phenomenon we used a new bioassay technique but this may give some contamination of our results. In order to measure the dis-aggregatory action of prostacyclin in venous and arterial blood, we used collagen strips superfused with heparinized blood of cats (Gryglewski et al 1978a). The strips gained in weight because platelet thrombi were deposited on the surface. Prostacyclin caused a dose-dependent loss in weight, due to its disaggregatory activity. When exogenous prostacyclin was infused intra-venously we found that disaggregation on the venous side was much less than on the arterial side. However, one could argue that superficial layers of platelet clots could be brought with venous blood into the lung and stimulate the generation of endogenous prostacyclin. It is difficult to refute this argument.

When blood-superfused strips of bovine coronary artery were used to bioassay prostacyclin in cat blood, angiotensin II released prostacyclin from the lungs and from the kidney. Using the multiple ion detection technique we were able to confirm that prostacyclin is continuously generated by the perfused cat lung (1-3 ng/ml of 6-oxo-PGF$_{1\alpha}$ is spontaneously released to the effluent) and is stimulated by angiotensin II. At the moment we know that *in vivo* in anaesthetized cats angiotensin II stimulates the generation of prosta-cyclin, and that not only the lungs but also the kidney contribute to this release.

Born: Do ADP and thromboxane A$_2$ act independently of each other, and perhaps additively, or does one of them act through the other? Evidence from several people, including Dr Marcus, suggests that thromboxane acts via ADP, so ADP remains the final common agent.

Marcus: Our working hypothesis now is that the final common pathway of irreversible platelet aggregation depends on the release of ADP from stimulated platelets. In our opinion thromboxane A$_2$ is another agent that can induce ADP release from the platelet itself, possibly through its capacity for mobilizing calcium from storage sites in the platelet into the platelet cyto-plasm itself. This calcium mobilization may also be closely correlated with a decrease in platelet cyclic AMP (Gorman et al 1979).

Vane: But aspirin, which prevents thromboxane formation, only prevents the second phase of platelet aggregation, so the first phase must be due to something else.

Born: An important observation by Dr Marcus is that there are diseases in which the patients bleed abnormally because there is no release of ADP from their platelets. These platelets are not aggregated by thromboxane A$_2$.

Marcus: There are probably three pathways responsible for the platelet aggregation response (Packham et al 1977). In one of these pathways platelet arachidonic acid is transformed to thromboxane A$_2$. In the second, ADP is released from intracellular granules. The third is an as yet unknown pathway that may represent calcium translocation from storage vesicles into the cytoplasm. Agents such as thrombin or collagen probably aggregate by all three mechanisms. At present, however, I believe that the final common pathway is ADP release (Bressler et al 1979, Marcus 1978).

Davies: There seemed to be two phenomena—one which concerned the aggregation of platelets outside the vascular system in the extravasated mass and the other which concerned the build-up of platelets in the ruptured vessel. When locally important factors are involved, being inside or outside the principal traffic flow could make a big difference.

Vane: I think the platelets have a positive feedback mechanism. Once they start aggregating they release both ADP and thromboxanes, each of which can cause further aggregation. It is not until this positive feedback is prevented by a substance such as prostacyclin that the platelet plug will stop being formed.

Davies: So wherever they aggregate they create a local environment which initially is overriding.

Gryglewski: During experimental atherosclerosis in rabbits the generation of prostacyclin by blood vessels is suppressed by more than 60% (Dembińska-Kieć et al 1977). One would expect thrombosis to occur in all arteries but it occurs only in certain regions of the arterial tree. The explanation may be that prostacyclin deficiency would be amplified by the local release of ADP from erythrocytes when they are smashed in 'rheologically dangerous areas' because of the shear phenomenon.

Morris: How are breaches in vascular endothelium stopped up and repaired?

Marcus: There is an unknown platelet component that seems to 'nourish' capillary endothelium. Such a component may also serve to maintain the integrity of lymphocyte channels. We know that abrupt reductions in platelet count in patients result in increases in vascular fragility, as evidenced by the appearance of petechiae and ecchymoses. Infusions of freshly prepared platelets are capable of arresting such bleeding. On the other hand, there are patients whose platelet counts approach low levels in a very gradual manner over long periods of time. These individuals frequently develop the capacity to maintain capillary fragility, even in the absence of platelets. As I have already speculated, it is possible that platelets pick up this unknown substance at a site of synthesis in a manner comparable to that of serotonin uptake from the intestine. Perhaps, in the setting of prolonged thrombocytopenia, this unknown substance manages to gain entry into the bloodstream and operate in the absence of platelets to maintain reasonable capillary integrity.

Gowans: You mentioned an individual who has no platelets but is apparently healthy. Is his bleeding time normal?

Marcus: This patient's bleeding time is several minutes over normal. We have not followed the bleeding times serially since it is probably beyond ethical standards to measure bleeding times repeatedly in such a subject. I don't know where this substance is synthesized, or where it is circulating to. Although we know where serotonin is synthesized, we really do not know where the platelets are transporting it to.

Gowans: I am not sure of the evidence for the existence of this 'nourishing' platelet-derived material in the recirculation but your views suggest that platelets do not normally maintain the integrity of capillaries by mechanically sealing holes in them.

Marcus: If platelets have nothing to do with sealing holes in blood vessels, it is hard to explain why the bleeding time is so prolonged in thrombocytopenia. One would also need a new explanation for 'primary haemostasis' (Marcus 1978).

Weiss: Craig Kitchens and I did an experiment in rabbits in which the platelet count was reduced with either busulphan or anti-platelet antibody. In consequence, purpura occurred. We examined the blood vessels in the tongue and found that the endothelium, which is usually very thick, became markedly attenuated and actually developed perforations. We were able to rationalize the purpura as due to vascular damage. Then, using prednisone, Kitchens was able to reduce the platelet level yet prevent the endothelial changes and the purpura. So here prednisone substituted for platelets. When platelets are responsible for endothelial maintenance but are not present, something like prednisone may be able to substitute for them.

Marcus: I think the steroid effect must be regarded as relatively temporary. When patients are given large doses of steroids for prolonged periods of time, there is an increase in vascular fragility. For example, we know from clinical experience that when thrombocytopenic patients require much more than 20 mg prednisone daily for the maintenance of vascular integrity, they will eventually experience difficulty because of a paradoxical effect of steroids on vascular fragility. This may also explain the petechiae and ecchymoses seen in patients with Cushing's syndrome.

Born: John Humphrey first showed that anti-platelet antibodies tend to react also with endothelial cells. So it seems that there is antigenic overlap, which is not too surprising. How much more is known about this now?

Humphrey: The anti-platelet antibody would certainly react with platelets. It didn't react well with other formed elements in the blood, but was it anti-endothelial? It is hard to say that there aren't any bits of material left from platelets on the endothelial surface. I have certainly heard it stated that platelets are necessary to maintain the integrity of vascular endothelium. For example if you have to perfuse a kidney for any length of time it is much better if you perfuse it with some platelets present. This suggests that something is being released from platelets. It could be something pharmacological that maintains the integrity of the vascular endothelium, but this is a field I don't know much about.

Weiss: In the experiments that I referred to, while we were able to depress the platelets with an anti-platelet antibody, the possibility that an anti-endothelial antibody was present was controlled by using a second, independent, means of depressing the platelets. That is, we used busulphan which caused platelet depletion and endothelial changes similar to those present after anti-platelet antibody.

McConnell: What is the half-life of the effect of prostacyclin on inhibiting platelet aggregation *in vitro?* If you wash the platelets after they have been incubated with prostacyclin, how long do they remain refractory to aggregation?

Vane: The increase in cAMP lasts much longer than the life of prostacyclin itself. Stimulation of adenylate cyclase leads to an inhibitory effect on platelets which lasts for 30–60 min.

McConnell: Did you say that prostacyclin does not affect adherence but does affect function?

Vane: It affects adherence in far higher concentrations than are needed to prevent aggregation and the release reaction.

McConnell: Complement-coated zymosan particles, particularly those carrying C3, will bind to platelets and secondarily induce them to exocytose. In that situation would prostacyclin inhibit adherence or would it inhibit release?

Vane: We have only looked at the one situation.

Born: Prostacyclin inhibits release but not adherence. Adherence of platelets to collagen is difficult to inhibit because it has very specific binding sites for components of the platelet surface.

Youdim: You gave the impression that arachidonic acid can go either to PGI$_2$ in the endothelium or to thromboxane in the platelet, Dr Vane. One gets the impression that neither of these are formed in the other site. Do you have evidence for that?

Vane: All the evidence is that the endothelial cells only produce prostacyclin; they can't produce thromboxane A$_2$. The platelets can't produce prostacyclin but only produce thromboxane A$_2$ plus some hydroxy fatty acids and a very small amount of PGE$_2$.

Youdim: Do you know why there is this regulatory system?

Vane: Presumably it is because the different isomerase enzymes which take the endoperoxides through to either thromboxane A_2 or to prostacyclin are only present in the one tissue or the other. Morrison et al (1978) have perfused isolated kidneys which normally produce prostaglandin E_2 and prostacyclin. When they damage the kidneys by making them hydronephrotic the kidneys start producing thromboxane A_2. So there is a possibility that all the enzymes are present but that the pathways are being controlled in some way as yet undiscovered.

Gowans: You have identified your materials by incubating segments of blood vessels. Am I correct in assuming that the evidence that it is the endothelium which is synthesizing the materials comes from the incubation of scrapings from the internal surface of the vessels?

Vane: That is part of the evidence. There is also evidence from our laboratory, from Aaron Marcus's laboratory and from several others that endothelial cells in culture will make prostacyclin. There is some dispute as to whether smooth muscle cells in culture can make it too.

Gowans: There is no dispute that pure endothelium can make prostacyclin?

Vane: No.

Zigmond: What was the time course of the rise in cyclic AMP?

Vane: It is very rapid: decay takes about 15–30 min.

Howard: In the thromboxane–prostacyclin system ADP generation seems to be endogenous. That doesn't seem to be a good initiator for the system. There is a major theoretical advantage to Professor Born's version, where the ADP is exogenous. Did I misunderstand something or is there still ground for a preference here?

Vane: I think that the initiating event is the contact between platelets and subendothelial structures such as collagen. This causes adhesion of platelets, degranulation, and liberation of endoperoxide, thromboxanes and ADP. All these call in other platelets and this is a positive feedback mechanism. Gustav Born says that exogenous ADP coming from damaged red cells might be another initiating event. Whether Gustav regards those two as being complementary to each other or separate events I am not quite sure.

Born: I suspect that they will turn out to be complementary. The haemostatic mechanism, of which thrombosis is presumably an aberration, is so important that there are probably back-up systems in it. I come back to the rapidity of the events; it is that which most strongly suggests the requirement of an initiating agent from outside the platelets themselves. On the other hand, what appeals to me most about your system is the extraordinary potency of the agents. Adenosine produces exactly the same biochemical effect as prostacyclin but is less active by some three orders of magnitude.

Gowans: I suppose from the point of view of the animal it is an unfortunate consequence if the evils of intravascular haemolysis are compounded by the initiation of platelet aggregation but I imagine that the dominant consideration from your own standpoint will be the provision of a mechanism for ensuring the rapid formation of a haemostatic plug. It is the large shear forces which haemolyse the red cells as they encounter the hole in the vessel and thus release the initiating factors.

Born: That is the crux. It is *arterial* bleeding which is most dangerous and needs the fastest homeostatic reaction. In venous bleeding it seems that platelets have comparatively little effect through aggregation as haemostatic plugs but a much more important effect by accelerating clotting, which is indeed their only other known physiological function. Down the evolutionary scale in cold-blooded animals with lower blood pressures, this is the predominant function of cells corresponding to mammalian platelets.

Vane: Does adenosine disaggregate platelet clumps?

Born: Yes. In this, too, it acts just like prostacyclin and, incidentally, also like prostaglandin E.

Richardson: We have a prototype oxygenator which is a capillary device. So that blood can be distributed efficiently through several thousand capillaries this has a shallow dish-shaped manifold, and the entrance port was put tangentially at one side of this device. The inlet blood line is first run with heparinized blood and then washed with heparinized saline. We found that quite a bit of haemolysis was caused, which we identified with bad flow conditions around the inlet port. This was ultimately cured by making a different design of port. Downstream of the port causing the damage there was a deposit of platelet-dominated thrombi with a veneer of red cells on them. These pile up on the capillary entrances in the region nearest to the inflow of the blood. The blood swirls around in the manifold and after some time in the manifold, apparently, the effect is much less. So after the damage to the red cells there seems to be a strong local enhancement of platelet plug formation (Friedman et al 1971).

Howard: Is that in a circuit?

Richardson: It was placed in an extracorporeal circuit at the time, of course. It was run on partial bypass with a sheep for about 3 h.

Howard: So it is not clear that one pass is sufficient for the whole aggregate to form?

Richardson: No, but Gustav Born showed some slides from a similar device where you can watch how the accumulation of this affected the pressure drop across the device. It takes time for the aggregates to grow.

Born: There is a whole range of observations to which we have not referred at all.   It concerns the absolute requirement for von Willebrand factor in the adhesion of platelets to collagen and probably also to other subendothelial structures.   Patients with von Willebrand's disease bleed because they are deficient in this protein, which is closely related to anti-haemophilic globulin (Factor VIII).   Von Willebrand factor is presumably essential for the adhesion of platelets to vessel walls.

References

Bressler NM, Broekman MJ, Marcus AJ 1979 Concurrent studies of oxygen consumption and aggregation in stimulated human platelets. Blood 53:167-178

Dembińska-Kieć A, Gryglewska T, Zmuda A, Gryglewski RJ 1977 The generation of prostacyclin by arteries and by the coronary vascular bed is reduced in experimental atherosclerosis in rabbits. Prostaglandins 14:1025-1034

Friedman LI, Richardson PD, Galletti PM 1971 Observations of acute thrombogenesis in membrane oxygenators. Trans Am Soc Artif Intern Organs 17:369-375

Gorman RR, Hamilton RD, Hopkins NK 1979 Stimulation of human foreskin fibroblast adenosine $3':5'$-cyclic monophosphate levels by prostacyclin (prostaglandin I_2). J. Biol Chem 254:1671-1676

Gryglewski RJ, Korbut R, Ocetkiewicz A, Splawiński J, Wojtaszek B, Swies J 1978a Lungs as a generator of prostacyclin—hypothesis on physiological significance. Naunyn-Schmiedeberg's Arch Pharmakol 304:45-50

Gryglewski RJ, Korbut R, Ocetkiewicz A, Stachura J 1978b In vivo method for quantification of anti-platelet potency of drugs. Naunyn-Schmiedeberg's Arch Pharmakol 302:25-30

Marcus AJ 1978 The role of lipids in platelet function: with particular reference to the arachidonic acid pathway. J Lipid Res 19:793-826

Morrison AR, Nishikawa K, Needleman P 1978 Thromboxane A_2 biosynthesis in the ureter obstructed isolated perfused kidney of the rabbit. J Pharmacol Exp Ther 205:1-8

Packham MA, Guccione MA, Greenberg JP, Kinlough-Rathbone RL, Mustard JF 1977 Release of ^{14}C-serotonin during initial platelet changes induced by thrombin, collagen, or A23187. Blood 50:915-926

Szceklik A, Gryglewski RJ, Nizankowski R, Musial J, Pietoń R, Mruk J 1978 Circulatory and anti-platelet effects of intravenous prostacyclin in healthy men. Pharmacol Res Commun 10:545-556

General discussion I

Gowans: Two main topics have been suggested for this general discussion: Ia antigens and the permeability of fine blood vessels to cells. But first I would like to ask John Vane and Gustav Born to list for us the pharmacological agents that endothelium is known to make.

ENDOTHELIAL PRODUCTS

Vane: Several enzymes in the endothelial cells metabolize vasoactive hormones. For instance, bradykininase inactivates bradykinin. The same enzyme moonlights as 'converting enzyme' to make angiotensin II from angiotensin I. There is a system for removing 5-hydroxytryptamine (serotonin) and inactivating it once it is inside the cell. Similarly, there are other enzymes and uptake mechanisms for removing ADP and some of the prostaglandins. Since the lungs contain about half of the total endothelial cells of the body, the lungs are where most of these changes are seen. These include the generation of prostacyclin. So the endothelial cells have important metabolic functions, especially in the lining of the pulmonary circulation.

Gowans: Does the endothelium secrete any materials into the circulation?

Vane: The only secretory product I know of is prostacyclin.

Born: Endothelium also secretes von Willebrand factor.

Owen: Do we know definitively which cells make the basement membrane?

Simionescu: Howard et al (1976) have shown that cultured endothelial cells derived from calf synthesize collagen that resembles basement membrane collagen.

Weiss: The endothelial cells are covered with proteoglycans. They stain deeply with ruthenium red. These cells probably produce protein–carbohydrate complexes which are washed by the fluid and contribute to some extent to the blood levels of these materials. To what degree do endothelial cells detach and occasionally circulate in the blood?

Vane: They certainly do that after endotoxins.

Morris: There is a population of cells lining the sinusoids of the liver that detach and migrate out of the liver by way of the hepatic lymph (Smith et al 1970).

Weiss: Aren't those just macrophages that slip in and pass out again?

Morris: Florey would have disputed the proposition that phagocytic cells lining the hepatic sinusoids are just macrophages. Certainly some of the endothelial components lining the hepatic sinusoids are phagocytic.

Weiss: Did Volkman & Gowans (1965) say that these cells originate in the marrow?

Gowans: It has been shown in radiation chimeras that Kupffer cells can be derived from bone marrow.

Born: The turnover rate of endothelium, as far as it is known, is very low. Helen Payling Wright measured the turnover rate of the guinea-pig aortic endothelium, on the idea that it might be faster where the blood flow was non-laminar. We found that it was indeed faster by a factor of about two; but the absolute turnover rates were very slow. Almost nothing is known about the turnover rates of endothelium elsewhere.

Vane: The turnover of endothelial cells may be low but the turnover of the enzymes within them can be quite fast. Aspirin acetylates the cyclo-oxygenase enzymes which form prostaglandins and inactivate them irreversibly. When this happens in platelets they can't form thromboxanes for the rest of their lives. When it happens in endothelial cells, recent evidence suggests that the enzyme is regenerated within eight hours or so (see Moncada & Vane 1979).

Gowans: Dr Howard, would you tell us what Ia is and why it may be interesting for us at this conference?

Ia ANTIGENS

Howard: Ia was referred to earlier in the particular context of its location in some cells in the thymus. Behind that there lies a great wealth of immunological history, not all of it very clear. The Ia antigens are a series of polymorphic antigens specified by genes in the major histocompatibility complex (MHC). They were identified originally by people who were making antisera to what they hoped were going to be the T cell receptors (David et al 1973, Götze et al 1973). It was hoped that animals that differed in their MHC-linked immune response *(Ir)* genes could be mutually immunized to produce antibodies against specific T cell receptors. What came out was a series of antisera which identified lymphocyte membrane molecules which were called the Ia antigens (Shreffler et al 1974). There are basically three developments from those original findings.

First, there is the analysis of these things as molecules (Crumpton & Snary 1979). There appear to be at least two and probably more loci which specify the polypeptide chains of Ia antigens, and the molecules have a reasonably consistent structure. That is, they consist of two chains, one with a molecular weight of about 33 000 and the other of about 28 000–29 000, which are not disulphide-bridged but which are nevertheless very tightly attached to each other. The cells which carried Ia antigens in greatest abundance turned out to be in the periphery and were the B (marrow-derived) lymphocytes.

At about the same time as these antigens were being identified and their distribution on cell surfaces was becoming known, experiments were being done on the genetics of interactions between lymphocytes in the induction of immune responses. Those experiments led to the idea that antigenic differences at loci which were probably the same as those specifying Ia antigens themselves prevented T and B cells, for example, from interacting. This work ultimately led us to suggest that the exogenous material which we administer to an animal and which we consider to be an antigen is in some way recognized by the immune system in association with an Ia antigen on a cell membrane. That rather complicated idea has led to what amounts to a re-definition of antigens. Antigens can be thought of as being operationally a complex between Ia antigen on the surface and the thing which one thinks one is giving as an antigen. There are now several theories (e.g. Benacerraf 1978) which deal with the possible activity of the Ia molecules in the membrane trapping exogenous antigenic material. These theories are the most extreme version of the role of Ia antigens in immune responses—that they actually do in some specific way interact with antigen. The parallel implication of this is that the T cell receptor that recognizes the antigen has in some way to accommodate to the presence of the Ia antigen itself. In other words the recognition of exogenous antigen alone is not sufficient. That has led to the polemic about the precise nature of the T cell receptor. Is it two things, one recognizing Ia antigen and the other recognizing exogenous antigen? Or is it a complex receptor which recognizes both simultaneously? That thread is still being developed.

The third thread about the immune system is that the specificity of the T cell repertoire may be governed by the MHC antigenicity of the thymus (Jerne 1971). In this context, the antigenicity means the polymorphic determinants of Ia. That is not the whole story but it is certainly part of it. That is, the family of polymorphic determinants which the thymus can present to differentiating lymphocytes ultimately governs the recognition potential of lymphocytes when they reach the periphery. This interacts with what I said about the peripheral recognition process of exogenous antigen by lymphocytes being dominated by the presence of the Ia antigens themselves. That leads to a formal necessity to find Ia antigens in the thymus. If one can show that the Ia genetics of an animal operates through its thymus to determine the T cell repertoire, then it seems to be convenient to find Ia in the thymus. That is what Dr van Ewijk described—namely the actual identification, in the thymus, of the Ia antigen in a cell which was apparently not itself a lymphocyte. On this argument, that is the cell that would be responsible for diverting the development of the T cell repertoire in a particular direction. It is a selective agent.

A fourth theme is that until it was shown that MHC antigens in the thymus could influence the repertoire there really was no necessary role for Ia except in the presentation of antigen peripherally, that is in cells that are directly concerned with the presentation of antigen to the immune system. Ia antigens are found in abundance in B cells but cells of the macrophage-like series also carry Ia antigens. This is compatible with their role as antigen presenters to T cells. The coexistence of Ia on B cells and on macrophages was a limited role, sufficient to deal with that family of interactions.

The thymus epithelial cell takes the Ia into other tissues. This is where a whole new debate is just beginning about the localization of these antigens that were previously thought to have purely domestic functions within the immune system. The reticular cells of the thymus have one foot in the immune system and one foot out. The reticular cell is not itself a member of the lymphomyeloid series (as far as is known) but it may be governing the development of the repertoire. Now Ia is being found all over the place, including tissues which might appear to have no direct role in the interactions of lymphocytes with antigen.

Until recently it would have been something of a surprise if these molecules were found in places where antigen presentation is not obvious. The kidneys of certain strains of rat, for example, contain very large amounts of material which behave like Ia antigen (Davies & Butcher 1978). This material has not been demonstrated clearly to be associated with non-macrophages but the amount suggests that it isn't going to be macrophages. The epithelial linings of several tissues are turning out to contain large quantities of material which is antigenically Ia. The problem is that we as immunologists don't have any immunological justification for this material at the moment. It would be nice to think of one.

The chemical nature of Ia antigen has been best defined as a glycoprotein. However Parish and colleagues have produced evidence that is compelling but so far not confirmed that certain Ia specificities may take the form of very simple carbohydrate moieties like disaccharides (Parish et al 1976, McKenzie et al 1977). It is just possible that some of the non-lymphoid material, Ia-like material, which has been discovered will turn out to be the 'glyco' side of the antigenicity rather than the protein side. That may reconcile these two lines of thinking.

Weissman: The usual way to demonstrate whether a cell is derived from precursors in the haematolymphoid system is to make a chimera; that is, irradiate a mouse, inject bone marrow cells and then test whether the Ia antigen in a particular tissue has the allotypic specificity of the bone marrow donor or the irradiated host. I mentioned earlier (p 34) that this type of analysis revealed, surprisingly, that the epithelial cells had a low proportion of Langerhans cells that accounted for 'epidermal' Ia. Jonathan Howard might agree that there are reasonable immunological explanations of why *those* cells are there.

R. Rouse and I (unpublished work) have demonstrated that the thymic cortical Ia-positive presumptive epithelial cells are from the radio-resistant population, and not the bone marrow donor. One still would have to characterize those Ia-positive skin cells in Peterson's system. Unfortunately, Peterson is looking across species differences so he might not be able to recognize allotypic specificity. One really needs the allotypic marker to be sure that what one is seeing is not a cell that migrated in from bone marrow and has just changed its shape.

Gowans: One interest in the context of this conference is the speculation that Ia molecules might be involved in the homing of lymphocytes into particular sites, for example mammary gland and the gut. In this context, has Ia been identified on endothelial cells?

Howard: Ia has never been convincingly demonstrated on the endothelial surface, as far as I know. That may well be subject to re-evaluation now that we have new reagents.

Ford: Because functional interactions between T cells and other cells are often restricted by components determined by the major histocompatibility complex it has frequently been suggested that the migration of lymphocytes across vascular endothelium might be similarly restricted. To the contrary, two groups have recently shown that in rats the migration of lymphocytes from the blood is the same in allogeneic as in syngeneic recipients (Rolstad 1979, P. Nieuwenhuis, personal communication). This conclusion was supported by careful measurement of cell numbers in the blood early after i.v. injection. After lymphocytes have entered allogeneic lymphoid tissues their migration and survival are complicated by host versus graft and graft versus host reactions. Interpretation of the data is also complicated by strain differences between lymphocytes with respect to the rate at which they leave the blood, and there may be strain differences in vascular endothelium as well (Rolstad 1979).

Born: What does a difference in efficiency mean? If even a small proportion of the sialic acid on glycoproteins is removed, their half-life in the circulation is greatly decreased. If recognition is not very specific, could it be something rather simple like that?

Ford: The mechanism underlying the observation that strain A lymphocytes will consistently migrate into lymph nodes faster than strain B lymphocytes whether they are injected into strain A or strain B recipients is a matter for speculation. With respect to the role of sialic acid, Judith Woodruff (Woodruff et al 1977) and ourselves (Ford et al 1976) agree that neuraminidase-treated lymphocytes are perfectly capable of adhering to high-endothelial venules, but the liver removes them from the blood with remarkable effectiveness. The low numbers in the lymph nodes and spleen after i.v. injection simply reflect rapid removal from the blood.

PERMEABILITY OF SMALL VESSELS TO CELLS

Morris: It is worth reopening the general question of permeability of the blood vasculature in relation to formed elements of the blood rather than restricting the discussion to the structural and physiological aspects of blood vessels in relation to their permeability to molecules such as the plasma proteins. Probably not many people here would understand how difficult it was back in 1956, when I first arrived in Oxford, to accept that someone along the corridor could show that the lymphocytes were going round and round in the body, escaping from the blood vessels, and reappearing in the lymph. That discovery had such important implications for the physiology of the blood vascular system and for understanding the permeability of blood capillaries that I remained sceptical of it until I was invited to come here and talk about it. If lymphocytes were indeed recirculating continually we had to account in some way for the permeability of the blood vasculature to these very large elements. Clearly this permeability had to be dissociated entirely from the permeability of the blood vasculature to plasma proteins. There is no precise correlation between the numbers of lymphocytes escaping from a capillary bed and the amount of protein that is getting out. We had to explain why these cells could escape selectively and independently.

It is my belief that at the time immunology needed this discovery, because it represented for the first time a physiological basis on which the immune response could be set. Jim Gowans' experiments provided the first physiological evidence that enabled people to talk about the immune response and the immunologically competent cell in a physiological way.

When one thinks about the permeability of fine blood vessels to lymphocytes certain facts must be taken into account. The first is that there is no strict correlation between the permeability of blood vessels to proteins and to lymphocytes. The second is that the vascular element which is said to be uniquely responsible for the transmission of lymphocytes, that is the high-endothelial venule (HEV), is not the only type of blood vessel that transmits lymphocytes selectively. Blood vessels are permeable to lymphocytes throughout most tissues. Rather more lymphocytes escape through HEV in certain species, such as rats, mice and rabbits, than in others. There are species such as the sheep in which HEV are not striking features and yet lymphocytes recirculate on a very large scale.

Certain non-lymphoid tissues (such as the liver and the gut) have a very much larger traffic of cells through them than other tissues. Some of these have a high degree of permeability to plasma protein while others have not.

Another thing that is important to understand is that the permeability of the vascular bed to lymphocytes is not immutably fixed. It can be transformed rapidly by events that occur in the tissue, and these events are not necessarily antigenically derived. A classic example is the creation of a draining nephrosis by partial obstruction of the ureter. This immediately increases the traffic of lymphocytes through the vasculature of the kidney by some 10-fold. As many cells are then seen in peripheral lymph coming from the kidney as in lymph coming from the lymph nodes. There is evidence that lymphocytes circulate preferentially through the blood vasculature in many situations in the body, irrespective of any antigenic stimulus. The imposition of antigen on the system of course changes the character of lymphocyte recirculation in very obvious ways.

In discussing permeability we also need to understand what it is about the interaction between the lymphocyte and the endothelial cell that enables the lymphocyte to escape while the escape of plasma and of other blood cells is restricted.

That physical mode of egress has been a question for debate. Early on it seemed incredible that these vast numbers of cells could be escaping without compromising the integrity of the vasculature. One could imagine most easily that lymphocytes crossed the endothelial membrane through endothelial junctions. An early paper by Marchesi & Gowans (1964) described an intracellular pathway for transmitting lymphocytes across the endothelium. That provided a very acceptable explanation for why the plasma proteins weren't getting out at the same time. Gutta Schoefl and others have looked again at the anatomical aspects of this migration. With the help of Roger Miles, a specialist in geometric probability, Dr Schoefl analysed the probability that the lymphocytes were going between cells or through cells. In fact she found that it is very unlikely that many, if any, of them are going through cells (Schoefl 1972).

Those are the aspects of permeability that we ought to discuss before we go on to look at the immunological implications. The discovery that lymphocytes recirculate from blood to lymph was as significant in terms of its contribution to the physiology of the blood vascular system as it was for immunology, even though immunology claims the discovery. That needs to be recorded in this commemorative session.

Gowans: Several papers have now suggested that lymphocytes migrate between the endothelial cells of high-walled venules. The scanning pictures we saw earlier were said to show that they went between the cells. However, the experiments of Schoefl (1972) which you quoted were performed on Peyer's patches, not on lymph nodes. Did she subsequently look at lymph nodes?

Morris: She certainly has looked at lymph nodes. Her analytical work was done on HEV in Peyer's patches. The mode of egress could be different in other places, but high endothelium is not essential for the transmission of lymphocytes selectively from the blood vascular compartment to the tissue spaces. They go across low endothelial vessels quite nicely and selectively.

Weissman: They may go across any vessel but what is the likelihood that they will end up on one endothelial cell compared to another? Judith Woodruff and Eugene Butcher are both going to talk about that later.

What is the evidence that there is no mass transport of plasma constituents across the high endothelial venule along with the lymphocyte? Has anybody asked the permeability question or is it just an assumption?

Morris: The level of plasma proteins in the lymph from the hind leg of the sheep is around 20% of the level in the plasma whereas the level of protein in ovarian lymph during the luteal phase of the oestrous cycle is about the same as in plasma—there are very few lymphocytes in lymph from either of these sources. Large numbers of lymphocytes may be escaping from the blood into the lymph, say in the kidney developing a hydronephrosis, yet the level of protein in the lymph is relatively low. So there is no correlation between the number of lymphocytes in the lymph and its protein concentration. That to me is a good indication that the lymphocytes and plasma proteins do not escape through by the same pathway.

Weissman: Has there been a direct marker experiment?

Morris: The venules that Dr Simionescu described earlier, through which lymphocytes pass, are fairly leaky. Dr Schoefl postulated that the escape of plasma was restricted by virtue of the fact that most of the migration across the endothelial barrier related to the activity of the endothelial cells. In fact the lymphocytes were thought to remain fairly spherical and not become greatly deformed on their way out. It was the endothelial cells which actually wrapped around the lymphocytes and more or less 'squeezed' these cells out. That is an interesting proposition.

Weiss: There have been several examples of lymphocytes going through cells rather than between them. In the bone marrow, for example, they definitely go across the cells, a step away from the junctions. There seems to be a selective place in the endothelium of the venous sinuses of the bone marrow where transmural cell passage occurs. In the transcytosis that Dr Simionescu described (this volume) there are passageways right across cells and these seem to occur without undue loss of fluid components of the blood. There are also different ways of going between cells. Lymphocytes may well go between cells in the postcapillary venules but they certainly don't go directly between cells. They deeply dent the cells on either side so it is a special type of passage, perhaps different from the kind that granulocytes or monocytes would undertake in inflammation. In delayed hypersensitivity, where lymphocytes go from the blood into a local site, there is some evidence that they go across cells in the venules rather than between them. So lymphocytes and other cells can cross blood vessels, usually in a venular part of the vasculature. They can go between cells in different ways and they can also go through cells in different ways.

Gowans: In Dr Schoefl's pictures was the head of the lymphocyte ever seen on one side of the endothelium while the tail was on the other?

Morris: Yes (Schoefl 1972, Fig. 5). The junctions in these HEV are so tortuous that seeing a cell migrating at some distance from an interendothelial junction is not in the least relevant to whether the progression is through a cell or through a junction. The only way to identify that is to make serial sections of a particular lymphocyte all the way through the junction. Something like 150 sections may be needed to be able to say that the migratory cell is in the junction. Every time that has been done it has been shown to be in the junction. I still don't believe that anybody has demonstrated satisfactorily that lymphocytes migrate by passing through the cytoplasm of the endothelial cell.

A leakage of protein accompanies the passage of lymphocytes across the endothelium. Dr Schoefl makes the point specifically that the lymphocytes are not just burrowing through the endothelium—the movements of both lymphocyte and endothelial cells are important.

Gowans: If you are correct and large numbers of lymphocytes pass through the intercellular junctions, I suppose the insurance against a simultaneous leakage of molecules is that the junction is always sealed ahead of and behind the migrating cell. But is there any evidence for leakage in these vessels?

Morris: There is some evidence that afferent lymph going to a peripheral node has a lower plasma protein concentration than lymph coming out the other side, so it may be that there is a degree of protein loss associated with the egress of lymphocytes. However, differences in the protein content of afferent and efferent lymph could be explained simply in terms of a difference between the capillary permeability of the blood vessels in general, in the two situations.

Born: Is it possible to inflame these vessels in the way one can inflame other small venules for the purpose of investigating the emigration of polymorphs through intercellular junctions? When polymorphs migrate through these junctions, does this hinder or promote the egress of lymphocytes?

Gowans: That was part of our original study (Marchesi & Gowans 1964). When lymph nodes were inflamed polymorphs and macrophages could readily be seen passing through intercellular junctions in the high endothelial vessels, with the migrating cells occupying the whole length of junctions— the path was not tortuous. The appearance was quite different from that of migrating lymphocytes. It was this difference, together with the evidence from serial sections, which led us to believe that lymphocytes did not migrate between the endothelial cells.

Morris: The cellular effluent from the lymph node where that traffic is occurring consists almost entirely of polymorphs. Lymphocytes in an inflamed area are very much in the minority. So it could be said that the progression of the polymorph in this intercellular pathway is in fact inhibiting the lymphocytes from escaping out. That is just another example of the selective permeability that can occur in a vascular bed which a few minutes before had been transmitting only one class of cell. Again, as we know, there seems to be a preferential, selective permeability in the somatic tissues if we presume that the macrophage cells characteristic of peripheral lymph are derived from blood monocytes. We can say that there is a selective permeability in that class of cell over and above lymphocytes in peripheral somatic tissues. This is another interesting point that bears on the general questions of homing.

Vane: Has the interaction between lymphocytes and living endothelial cells been studied in culture?

Woodruff: De Bono (1976) prepared monolayer cultures of endothelium obtained from large blood vessels and found that lymphocytes which adhered to the cells then migrated between them.

Vane: Were you saying that it isn't the lymphocyte that burrows but the endothelial cell that opens, Professor Morris?

Morris: Dr Schoefl's interpretation of the mode of egress of lymphocytes through high-endothelial venules in the Peyer's patches of the rat is that the movements of both the lymphocytic and the endothelial cell are important. She makes the point that the lymphocytes are mostly rounded up during their passage out of the bloodstream. As a consequence she sees the endothelial cell as being more easily deformed than the lymphocytes.

Gowans: That is also what Dr van Ewijk showed us in his scanning pictures —the lymphocytes were rounded up and did not have the appearance of cells in motion.

van Ewijk: Many cells attached to the wall of HEV were round cells. Some of the cells showed a 'handmirror' configuration, indicating amoeboid movement. These cells are therefore migrating over the endothelium. Where the cells are actively emigrating from the bloodstream towards the parenchyma, we found clefts between the endothelial cells. In other sites in the high endothelial venules the lymphocytes are just sitting on the endothelium as round cells. So maybe various parts of the endothelium have a different intercellular spacing, and as such determine the site of emigration from the bloodstream.

Ford: Dr Weissman raised the question of the permeability of HEV to colloidal material. Anderson & Anderson (1975) perfused suspensions of horseradish peroxidase or thorotrast through the vasculature of lymph nodes. These colloidal markers traversed the high endothelium, often in association with lymphocytes, but they were effectively stopped by the basement membrane. If the basement membrane is the physiological barrier to large molecules then presumably they can return to the blood from the sub-endothelial space. In the context of permitting lymphocytes to pass but stopping proteins it may not matter whether lymphocytes go through or between endothelial cells because the selective turnstile is the elaborate basement membrane.

Gowans: How does the lymphocyte get through the basement membrane?

Ford: By insinuating itself through pockets in the basement membrane. Under physiological conditions more lymphocytes may be seen within the basement membrane than between endothelial cells. Traversing the basement membrane may place the greatest demands on the metabolic energy of the migrating lymphocyte.

Gowans: Does the basement membrane have holes in it or is it a continuous layer when looked at *en face*? What are pockets?

Ford: Pockets are conceived of as potential gaps that can be prised open by lymphocytes. A possible analogy is a rather stiff turnstile.

Weiss: Many basement membranes are crossed by cells relatively easily. In older work there are pictures of venules so situated in a lymph node that the lymphocytes crossed on one aspect of the vessel and not on the other. On the aspect where lymphocytes crossed the endothelium was high, while on the other aspect the endothelium was low. Perhaps the high endothelium is a consequence of lymphocyte crossing and represents the interaction of lymphocyte and endothelium?

Gowans: The lymphocyte-depleted animal has been alleged to have rather atrophic postcapillary venules which are restored when the animal's lymphocytes are restored (Goldschneider & McGregor 1968).

Woodruff: The high endothelium of lymph nodes did not change its appearance in rats subject to prolonged thoracic duct drainage (Anderson & Anderson 1976).

Owen: When lymphoid stem cells enter the thymus, particularly the early embryonic thymus, the cells have to migrate out of capillaries because the thymus isn't vascularized; they must cross mesenchyme and go through a basement membrane into the thymus, where they differentiate. It has been proposed, particularly by Le Douarin and her group, that because these stem cells enter at a particular time in ontogeny, there is a chemotactic influence which emanates from the thymus at this time (Le Douarin 1978). Presumably this has to reach the cells which are originally in the capillaries. We should perhaps be concentrating not on the endothelial cells but on other cells deeper in the tissue which may be liberating substances which may then pass through the endothelial cells.

Gowans: This is a central topic which we shall be discussing again later. The signal for migration does not have to be produced by the endothelial cell but if it is produced elsewhere, how and where is it displayed in order to engage the attention of circulating lymphocytes?

References

Anderson AD, Anderson ND 1976 Lymphocyte emigration from high endothelial venules in rat lymph nodes. Immunology 31:731

Anderson ND, Anderson AD 1975 Studies on the structure and permeability of the microvasculature in normal rat lymph nodes. Am J Pathol 80:387-418

Benacerraf B 1978 A hypothesis to relate the specificity of T lymphocytes and the activity of I-region-specific Ir genes in macrophages and B lymphocytes. J Immunol 120:1809-1812

Crumpton MJ, Snary D 1979 Cell surface antigens with special reference to histocompatibility antigens. Int Rev Biochem 23:127-176

David CS, Frelinger JA, Shreffler DC 1973 New lymphocyte antigen system (Lna) controlled by the Ir region of the mouse H-2 complex. Proc Natl Acad Sci USA 70:2509-2514

Davies HffS, Butcher GW 1978 Kidney alloantigens determined by two regions of the rat major histocompatibility complex. Immunogenetics 6:171

de Bono D 1976 Endothelial-lymphocyte interactions in vitro. I. Adherence on nonallergised lymphocytes. Cell Immunol 26:78-88

Ford WL, Sedgley M, Sparshott SM, Smith ME 1976 The migration of lymphocytes across specialized vascular endothelium. II. The contrasting consequences of treating lymphocytes with trypsin or neuraminidase. Cell Tissue Kinet 9:351-361

Götze D, Reisfeld RA, Klein J 1973 Serologic evidence for antigens controlled by the Ir region in mice. J Exp Med 138:1003-1008

Goldschneider I, McGregor DD 1968 Migration of lymphocytes and thymocytes in the rat. I. The route of migration from blood to spleen and lymph nodes. J Exp Med 127:155

Howard BV, Macarak EJ, Gunson D, Kefalides NA 1976 Characterization of the collagen synthesized by endothelial cells in culture. Proc Natl Acad Sci USA 73:2361-2364

Jerne NK 1971 The somatic generation of immune recognition. Eur J Immunol 1:1-9

Le Douarin NM 1978 Ontogeny of hematopoietic organs studied in avian embryo interspecific chimeras. In: Differentiation of normal and neoplastic hematopoietic cells. Cold Spring Harbor Laboratory, New York, p 5-31

Marchesi VT, Gowans JL 1964 The migration of lymphocytes through the endothelium of venules in lymph nodes: an electron microscope study. Proc R Soc Lond B Biol Sci 159:283-290

McKenzie IFC, Clarke A, Parish CR 1977 Ia antigenic specificities are oligosaccharide in nature: hapten inhibition studies. J Exp Med 145:1039

Moncada S, Vane JR 1979 Mode of action of aspirin-like drugs. In: Stollerman GH (ed) Advances in internal medicine, Year Book Medical Publishers, Memphis, vol 24:1-22

Parish CR, Chilcott AB, McKenzie IFC 1976 Low molecular weight Ia antigens in normal mouse serum. I. Detection and production of a xenogeneic antiserum. Immunogenetics 3:113-128

Rolstad B 1979 The influence of strong transplantation antigens (Ag-B) on lymphocyte migration in vivo. Cell Immunol, in press

Schoefl GI 1972 The migration of lymphocytes across the vascular endothelium in lymphoid tissue. A reexamination. J Exp Med 136:568-584

Shreffler DC, David CS, Götze D, Klein J, McDevitt HO, Sachs DJ 1974 Genetic nomenclature for new lymphocyte antigens controlled by the I region of the H-2 complex. Immunogenetics 1:189-190

Smith JB, McIntosh GH, Morris B 1970 The traffic of cells through tissues: a study of peripheral lymph in sheep. J Anat 107:87-100

Volkman A, Gowans JL 1965 The production of macrophages in the rat. Br J Exp Pathol 46:50-61

Woodruff JJ, Katz IM, Lucas LE, Stamper HB 1977 An in vitro model of lymphocyte homing. Membrane and cytoplasmic events involved in lymphocyte adherence to specialized high-endothelial venules of lymph nodes. J Immunol 119:1603-1610

The function and pathways of lymphocyte recirculation

J.L. GOWANS* and H.W. STEER †

MRC Cellular Immunology Unit, Sir William Dunn School of Pathology and † Nuffield Department of Surgery, Radcliffe Infirmary, Oxford, UK

Abstract The early work on lymphocyte recirculation assumed that all recirculating lymphocytes composed a common pool and that the composition of this pool could be inferred from studies on thoracic duct lymph. These propositions are examined in the light of more recent evidence, particularly from experiments on the traffic of lymphocytes through the lamina propria of the small intestine.

There has been little speculation on the functional significance of lymphocyte recirculation apart from the suggestion that it increases the efficiency of regional immune responses by allowing antigen-induced selection of precursors from a pool larger than that accommodated by the regional nodes alone. In addition, the mounting of a local immunity is dependent on a peripheral recirculation through the tissues, notably in the case of the secretory immune system of the intestine.

The discovery that lymphocytes recirculate from blood to lymph (Gowans 1957, 1959, Gowans & Knight 1964) explained the apparently high turnover of lymphocytes in the blood and eliminated from consideration two of the current theories about the fate of lymphocytes—that they either died in large numbers each day, or that they emigrated in large numbers into the bone marrow to become haemopoietic stem cells—but did not in itself provide any new clues about function. A dominant theme in the older literature of descriptive haematology, that small lymphocytes were not 'end' cells, but that they discharged their function by transforming into other cell types, was given some limited credence by the demonstration that small lymphocytes differentiate and divide in response to antigen, and initiate immune responses (Gowans et al 1962). Thus, an immunological significance has been sought for the phenomenon of recirculation. This problem remains unsolved, but a

---

* *Present address:* Medical Research Council, 20 Park Crescent, London W1N 4 AL
† *Present address:* John Radcliffe Hospital, Headington, Oxford

© *Excerpta Medica 1980*
Blood cells and vessel walls: functional interactions
(Ciba Foundation symposium 71) p 113-126

plausible suggestion is that the movement of lymphocytes through lymphoid tissue increases the efficiency of regional immune responses by permitting selection of precursor T and B cells from the total recirculating pool. The most elegant demonstration of such a process is the specific removal from the recirculating pool of T cells which initiate graft-versus-host reactions (Ford & Atkins 1971). The flux of cells within lymph nodes would certainly favour both lymphocyte–lymphocyte and lymphocyte–macrophage contacts, and from what is now known about T/B collaboration, and the activity of suppressor T cells, could mediate both inductive and regulatory interactions. Our understanding of cellular mechanisms in immunity has been considerably enhanced by studies *in vitro* employing suspensions of lymphoid cells, but the relative inefficiency of *in vitro* systems may reflect their failure to maintain lymphoid architecture and to provide a pool of recirculating lymphocytes. More physiological approaches are now required, but the technical difficulties are considerable.

The early work on lymphocyte recirculation led to the view that the repeated passage of lymphocytes from blood to lymph through all the lymph nodes resulted in a complete mixing of the recirculating pool, so that a sample from the thoracic duct lymph would reflect the composition of cells in efferent lymph generally. Additionally, such evidence as there was suggested that few cells recirculate from blood to lymph through non-lymphoid tissues; consequently, the input of cells into the nodes from afferent lymph was thought to be relatively small, with the exception of the lymph from the Peyer's patches (Baker 1932). These views will no doubt be modified as new knowledge about the heterogeneity of lymphocytes is applied to the problem, but some modifications are already necessary. The first is the possibility that subpopulations of cells recirculate preferentially through particular lymph nodes. The extent to which this occurs is not clear, but biases of this kind have been reported (Scollay et al 1976a, Cahill et al 1977). The second is the dramatic change in the cellular composition of the efferent lymph from antigenically stimulated lymph nodes, which has been studied in detail by Morris and his colleagues (Hall & Morris 1963, Morris 1972). Similarly, the presence of large lymphocytes in intestinal lymph (and thus in thoracic duct lymph) reflects immunological events in the gut. Lastly, it is necessary to reassess the extent to which lymphocytes recirculate through non-lymphoid tissues under both normal and pathological conditions. The possibility that lymphocytes recirculate normally through the lamina propria of the small intestine will be considered here.

The defence of the gut against microorganisms and bacterial toxins is mediated by a local immunity, which has its basis in the secretion of IgA antibody by plasma cells located in the lamina propria. The plasma cells in the lamina propria arise from precursors which enter the blood in the thoracic duct lymph (Gowans & Knight 1964, Guy-Grand et al 1974, Pierce & Gowans 1975). Among these lymph-borne precursors are large lymphocytes, which carry internal and surface IgA (Williams & Gowans 1975) and which have a predilection for migrating from the blood into the tissues of adult, fetal (Parrott & Ferguson 1974) and neonatal (Halstead & Hall 1972) small intestine. Since the fetal and neonatal gut do not contain microorganisms, or their products, it appears that the signal for the entry of large lymphocytes into the lamina propria must depend upon some intrinsic gut property and is not due to the antigen which originally provoked their formation in the intestinal lymphoid tissue. The nature of this signal is not known, nor do the fine blood vessels in the intestine possess any anatomical peculiarities analogous to those of the high-endothelial venules in lymph nodes.

The idea that the migration of plasma cell precursors into the lamina propria is antigen-independent runs counter to the observation that the local application of antigen to the gut gives a restricted local immunity (Ogra & Karzon 1969), as indicated by a higher local concentration of specific IgA-secreting cells and higher concentrations of locally secreted antibody. This phenomenon can be readily demonstrated in a rat primed parenterally with cholera toxoid and then challenged with toxoid into the lumen of only one of a pair of isolated Thiry-Vella intestinal loops. The concentration of the specific antibody-producing cells which develop in the lamina propria is considerably higher in the challenged than in the unchallenged loop (Husband & Gowans 1978). The problem is to determine how antigen deposited locally leads to this local accumulation of plasma cells.

When a challenging dose of cholera toxoid is injected into the lumen of the small intestine or into an isolated intestinal loop, antitoxin-containing cells (ACC) appear in the thoracic duct lymph, reaching maximum numbers about four days after challenge and then declining (Pierce & Gowans 1975, Husband & Gowans 1978). This cellular response in the efferent lymph draining the gut provides the blood-borne precursors of the plasma cells which later accumulate in the lamina propria. The influence of local antigen on the accumulation of these plasma cells is illustrated by an experiment in which rats carried a pair of Thiry-Vella intestinal loops (Fig. 1). The thoracic ducts were cannulated immediately after intestinal challenge with cholera toxoid into one of the loops and lymph was drained to the exterior throughout the experiment. Thoracic duct lymphocytes (TDL) were collected between Days 4 and 5 after challenge and injected back into the same animal. Fig. 1 shows the striking difference in the density of ACC which developed in the challenged as

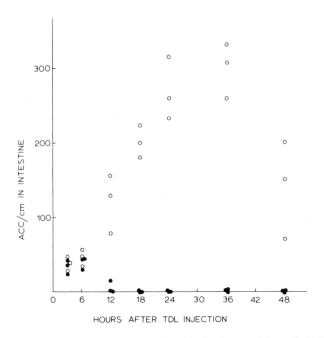

FIG. 1. Comparison of the density of antitoxin-containing cells (ACC) in the proximal (•) and distal (○) loop of rats at various times after an injection of TDL. Rats were primed with 100 μg of purified cholera toxoid in Freund's complete adjuvant intraperitoneally and two Thiry-Vella loops of small intestine were prepared 10 days later. Four days later still the thoracic ducts were cannulated and 1 mg toxoid was injected immediately into the lumen of the distal loop. TDL were collected from Day 3 to Day 4 after loop challenge, washed and injected back intravenously into each rat. At each time interval three rats were killed and the density of ACC in sections of each loop estimated by immunofluorescence. The thoracic ducts continued to drain throughout the experiment. (From Husband & Gowans 1978.)

compared with the non-challenged loops. This experiment provides clear evidence that the magnitude of a local intestinal immune response is determined by locally applied antigen, and runs counter to earlier studies suggesting that the migration of plasma cell precursors into the lamina propria occurred independently of antigen. This contradiction would be resolved if antigen could be shown to lead to a local proliferation of the precursors after they had emigrated from the blood, and, indeed, the injection of tritiated thymidine into rats during the evolution of the response illustrated in Fig. 1 showed that a significant proportion of the plasma cells became labelled, under conditions where precursors were no longer available to the gut, since they were being drained away continuously from the thoracic duct fistula (Husband & Gowans 1978).

The interpretation of the previous experiments rests upon the assumption that the precursors of intestinal plasma cells are large dividing lymphocytes, which already contain IgA antibody, and which enter the blood by way of the thoracic duct. These are the cells which have been shown to migrate preferentially into the small intestine, and which increase dramatically in numbers after local intestinal challenge with specific antigen. An experiment was done to determine whether such antitoxin-containing cells recirculate from blood to lymph or whether, having entered the intestine, they are arrested and live out the rest of their lifespan locally (Husband & Gowans 1978). Thoracic duct lymphocytes, collected from rats at the height of the intestinal ACC response, were transfused into the blood of normal rats with thoracic duct fistulae. ACC were readily detected in the lymph from the recipients, showing that recirculating blood-borne precursors could give rise to ACC in thoracic duct lymph. The obvious but, as it turned out, the wrong interpretation of the finding was that the large lymphocytes were migrating from blood to lymph and were being detected as ACC after traversing the lamina propria. To investigate this point, a second experiment was done, in which all the large lymphocytes in the donor lymph were prelabelled with tritiated thymidine. The results of the experiment were unexpected in that none of the ACC recovered in the recipient's thoracic duct lymph were labelled, indicating that they had arisen from recirculating small lymphocytes which had differentiated into ACC in the absence of further antigenic stimulation. A strong candidate for this precursor is the population of small lymphocytes in thoracic duct lymph which carry surface IgA (Williams & Gowans 1975).

These experiments suggest that the generation of a local intestinal immune response is due to the migration into the gut of large lymphocytes which pre-existed in the lymph as ACC, to the antigen-dependent recruitment of recirculating small lymphocytes which differentiate into plasma cells after entry into the lamina propria, and to the antigen-dependent proliferation of either or both locally. The varying contributions of these three processes have yet to be assessed, but it is first essential to show that B lymphocytes carrying surface IgA do circulate from blood to lymph through the lamina propria of the small intestine and are thus, in principle, available for recruitment into the local immune response. Evidence on this point has now been obtained by cannulating the lacteals which carry lymph from lengths of small intestine that either bear Peyer's patches or are devoid of them.

Anaesthesia was induced in normal non-immunized Lewis rats by a subcutaneous injection of 1 ml of 10% Nembutal and supplemented during the course of the experiment with ether as necessary. The abdominal thoracic duct was ligated, and the small intestine was mobilized through a mid-line incision, and placed in warm saline in a transilluminated Petri dish. Siliconized glass needles were micromanipulated into the distended lacteals, and lymph was withdrawn under gentle suction into a small quantity of heparinized DAB (Dulbecco A + B buffered salt solution). A measured quantity of the diluted lymph was expelled into a measured volume of DAB/10% fetal calf serum containing a known concentration of sheep red blood cells and the mixture was centrifuged onto a coverslip placed at the bottom of a flat-ended tube. The number of lymphocytes in the sample of lymph was then estimated from a differential count of cells on the coverslip. The mean output of cells in single lymphatics from gut with Peyer's patches was about 10 times that of lymph from other regions of the small intestine (5.2 $\times$ 10^6/h in 14 experiments and 5.3 $\times$ 10^5/h in 13 experiments respectively) but owing to the stasis in the lacteals which followed ligature of the thoracic duct, the reliability of the estimates is open to serious question. It was clear, however, that substantial numbers of cells were emerging in lymphatics which drained the non-lymphoid regions of the small intestine.

Since it is known that Peyer's patches are a source of the precursors of IgA-secreting cells in the gut (Craig & Cebra 1971), it was important to determine the characteristics of the lymphocytes in lymph from Peyer's patches and from other regions of the gut. Table 1 records the proportion of lymphocytes in lymph from intestinal lacteals which carry surface Ig, surface IgA or internal IgA. Contrary to what might be expected, the relative proportions of

TABLE 1

Percentages of cells (mean and range) in lacteal lymph from anaesthetized Lewis rats (number of samples examined in parentheses).

| | % Cells carrying Ig or IgA in lacteal lymph from: | |
	Peyer's patches	Non-Peyer's patch
Surface Ig	25.2 (3)	21.3 (5)
	(18.3–29.2)	(16.7–27.3)
Surface IgA	16.4 (4)	16.3 (5)
	(14.2–19.2)	(15.0–17.5)
Internal IgA	0.95 (9)	1.3 (7)
	(0.5– 1.6)	(0.7– 1.9)

Rabbit F (ab')$_2$ anti-rat antibodies employed for detection: surface Ig, ^{125}I-anti-rat Fab; surface IgA, ^{125}I-anti-rat IgA; internal IgA, FITC-conjugated anti-rat IgA (preparation in Williams & Gowans 1975). Autoradiographs prepared with Ilford K2 emulsion and exposed for five days; quantitation by photometric grain-count method using a Vickers M74 photometer.

these three classes of cells are similar in lymph from the two sources. Compared with thoracic duct lymph (see Williams 1975), each contains a smaller proportion of surface Ig cells; that is, a lower proportion of B cells. Presumably, efferent intestinal lymph is augmented with surface IgM cells from the mesenteric lymph nodes. On the other hand, lymphocytes carrying surface IgA make up a much higher proportion of all B cells in lacteal lymph than in thoracic duct lymph; namely, 65% of all B cells in lymph from Peyer's patches and 77% of all B cells in lymph draining other areas from the small intestine. It can be concluded that the lymphocytes which normally recirculate through the lamina propria of the small intestine are predominantly T cells, but that the B cell component is enriched with respect to cells carrying surface IgA.

DISCUSSION

Studies in the sheep have established that peripheral lymph draining the intercellular spaces of a number of organs and tissues contains few cells. For example, up to about 80 times as many lymphocytes leave the prescapular and prefemoral nodes as enter it from the afferent lymphatics (Smith et al 1970a) and a similar disparity has been noted for the popliteal node (Miller & Adams 1977). Peripheral lymph in sheep also contains a lower proportion of lymphocytes with surface Ig than do central lymph or blood and is distinctive in that 10–20% of all the cells are macrophages (Miller & Adams 1977, Scollay et al 1976b). Thus the flux of lymphocytes through the normal tissues drained by these nodes is low and B lymphocytes are either preferentially excluded during the migration from the blood or are retained in the tissues, presumably the former.

The 'round cell' infiltrates which are seen in sections of pathological tissues are a static view of a traffic of cells through the tissues which can reach dramatic proportions. As many lymphocytes may be discharged into the afferent lymph from a chronic granuloma as normally leave in the efferent lymph from a lymph node (Smith et al 1970b). Similarly, high outputs of cells have been recorded in afferent lymph from renal allografts (Pedersen & Morris 1970) and hydronephrotic kidneys (Smith et al 1970b). The nature of the attraction which these pathological tissues hold for circulating lymphocytes is not known; little is known about factors which are chemotactic for lymphocytes nor have any changes yet been identified in the venules which may give them properties similar to those of the high-endothelial venules (HEV) of the lymph nodes. Their properties would not be identical because the HEV allow both B and T lymphocytes to enter the nodes from the blood while renal homografts carry a predominantly T cell traffic. Afferent lymph from renal homografts is, like normal peripheral lymph, deficient in cells with surface Ig (Miller & Adams 1977) and there is no evidence that B cells are retained within the kidney.

It is well known that large numbers of lymphocytes pass along the intestinal lymphatics (Mann & Higgins 1950, Beh & Lascelles 1974). In rats in which lymphatic regeneration has repaired the damage which follows complete removal of the mesenteric lymph nodes, the output of lymphocytes in the intestinal lymph is reduced by up to half, showing that substantial numbers of cells are derived from the intestine itself (G. Mayrhofer & J.L. Gowans, unpublished). The present study, in which individual lacteals draining areas of gut with and without Peyer's patches have been cannulated, has shown that there is normally a traffic of cells through the lamina propria of the small intestine (additional to that which occurs through the Peyer's patches) and that the B cells in this circulating population are enriched with lymphocytes with surface IgA. The magnitude of this traffic is difficult to measure in the rat but it is likely to be both considerably larger than that through other normal tissues and distinctive with respect to the distribution of Ig classes on the lymphocytes.

The immunological studies suggest that the cells with surface IgA which recirculate through the lamina propria can be recruited into a local intestinal response where they give rise to IgA-secreting plasma cells. This suggestion raises additional questions. How is the locally deposited antigen presented to lymphocytes in the lamina propria when the main trap for intraluminal antigen is thought to be Peyer's patches? Does antigen increase the flux of cells through the lamina propria or is recruitment limited to the normal migrants? How is the bias in favour of cells with surface IgA achieved? What is the function of the dominant T cell traffic through the lamina propria, apart from providing local help for B cells? Finally, since the secretory immune system is not limited to the small intestine, does a similar circulation of lymphocytes occur through other tissues in which IgA antibody is synthesized locally—for example, in the mammary gland?

References

Baker RD 1932 The cellular content of chyle in relation to lymphoid tissue and fat transportation. Anat Rec 55:207-219
Beh KJ, Lascelles AK 1974 Class specificity of intracellular and surface immunoglobulin of cells in popliteal and intestinal lymph from sheep. Aust J Exp Biol Med Sci 52:505-514

Cahill RNP, Poskitt DC, Frost H, Trnka A 1977 Two distinct pools of recirculating T lymphocytes: migratory characteristics of nodal and intestinal T lymphocytes. J Exp Med 145:420–428

Craig SW, Cebra JJ 1971 Peyer's patches: an enriched source of precursors for IgA-producing immunocytes in the rabbit. J Exp Med 134:188-200

Ford WL, Atkins RC 1971 Specific unresponsiveness of recirculating lymphocytes after exposure to histocompatibility antigen in F₁ hybrid rats. Nature New Biol 234:178-180

Gowans JL 1957 The effect of the continuous re-infusion of lymph and lymphocytes on the output of lymphocytes from the thoracic duct of unanaesthetized rats. Br J Exp Pathol 38:67-78

Gowans JL 1959 The recirculation of lymphocytes from blood to lymph in the rat. J Physiol (Lond) 146:54-69

Gowans JL, Knight EJ 1964 The route of recirculation of lymphocytes in the rat. Proc R Soc Lond B Biol Sci 159:257-282

Gowans JL, McGregor DD, Cowen DM, Ford CE 1962 Initiation of immune responses by small lymphocytes. Nature (Lond) 196:651-655

Guy-Grand D, Griscelli C, Vassalli P 1974 The gut-associated lymphoid system: nature and properties of the large dividing cells. Eur J Immunol 4:435-443

Hall JG, Morris B 1963 The lymph-borne cells of the immune response. J Exp Physiol 48:235-247

Halstead TE, Hall JG 1972 The homing of lymph-borne immunoblasts to the small gut of neonatal rats. Transplantation 14:339-346

Husband AJ, Gowans JL 1978 The origin and antigen-dependent distribution of IgA-containing cells in the intestine. J Exp Med 148:1146-1160

Mann JD, Higgins GM 1950 Lymphocytes in thoracic duct, intestinal and hepatic lymph. Blood 5:177-190

Miller HRP, Adams EP 1977 Reassortment of lymphocytes in lymph from normal and allografted sheep. Am J Pathol 87:59-80

Morris B 1972 The cells of lymph and their role in immunological reactions. In: Meeson H (ed) Handbuch der Allgemeinen Pathologie, Vol 3, part 6. Springer-Verlag, Berlin, Heidelberg, New York, p 405-456

Ogra PL, Karzon DT 1969 Distribution of poliovirus antibody in serum, nasopharynx and alimentary tract following segmental immunization of lower alimentary tract with poliovaccine. J Immunol 102:1423-1430

Parrott DMV, Ferguson A 1974 Selective migration of lymphocytes within the mouse small intestine. Immunology 26:571-588

Pedersen NC, Morris B 1970 The role of the lymphatic system in the rejection of homografts: a study of lymph from renal transplants. J Exp Med 131:936-969

Pierce NF, Gowans JL 1975 Cellular kinetics of the intestinal immune response to cholera toxoid in rats. J Exp Med 142:1550-1563

Scollay R, Hopkins J, Hall J 1976a Possible role of surface Ig in non-random recirculation of small lymphocytes. Nature (Lond) 260:528-529

Scollay R, Hall J, Orlans E 1976b Studies on the lymphocytes of sheep. II. Some properties of cells in various compartments of the recirculating lymphocyte pool. Eur. J Immunol 6:121–125

Smith JB, McIntosh GH, Morris B 1970a The traffic of cells through tissues: a study of peripheral lymph in sheep. J Anat 107:87-100

Smith JB, McIntosh GH, Morris B 1970b The migration of cells through chronically inflamed tissues. J Pathol 100:21-29

Williams AF 1975 IgG₂ and other immunoglobulin classes on the cell surface of rat lymphoid cells. Eur J Immunol 5:883-885

Williams AF, Gowans JL 1975 The presence of IgA on the surface of the rat thoracic duct lymphocytes which contain internal IgA. J Exp Med 141:335-345

Discussion

Hall: Blast cells don't recirculate to a great extent but we usually find about 10% coming back into the intestinal lymph and I was surprised that none of yours did. Could the fairly heavy labelling have interfered with their recirculation and even possibly their subsequent development?

Gowans: That is a possibility. In the original experiments with Julie Knight (Gowans & Knight 1964) we found a few that recirculated but it was very few.

Williams: Does 6 h seem short for those cells to live? How long does the gut keep on secreting IgA after irradiation?

Gowans: For a long time but our animals were not being provided continuously with precursors of intestinal plasma cells as would be the case normally. The precursors are being drained away continuously through a thoracic duct fistula.

Williams: Serum IgA drops off quite quickly in the irradiated mouse but secretory IgA comes into the gut for a long time after irradiation, implying that long-lived plasma cells are present (Bazin et al 1971). The cells you put into the animal are very immature plasma cells so why do they survive for only 6 h after settling in the lamina propria?

Gowans: Has the half-life of a plasma cell been calculated?

Hall: In mice it is about three to six days (Mattioli & Tomasi 1973).

Gowans: I agree that it is difficult to reconcile the disappearance of our plasma cells from the gut in 6 h with the evidence from the radiation experiments and the existing estimates of plasma cell half-life. On the earlier point about the recirculation of large lymphocytes it should be added that even if the transfused cells were not detected in thoracic duct lymph (which was the observation) they may have recirculated into the lacteals but have been trapped in the mesenteric nodes. Is there any evidence for such trapping in other species?

Hall: I don't know. They can certainly recirculate through the wall of the gut.

Morris: They recirculate in the sheep. Large cells can be detected much sooner than the labelled small cells, in the intestinal lymph. I am not saying that they all recirculate but certainly some labelled large cells come back in the lymph very quickly.

Howard: I agree. When I was doing double labelling experiments on the recirculation of B lymphocytes in rats I used uridine as a marker (Howard 1972). There was an early peak of recirculating uridine at about 9 h which was entirely associated with large lymphocytes and which had all gone by about 15 h, long before the B lymphocytes had started to come out. There were not very many lymphocytes but the finding was unambiguous. We recovered both counts and cells.

Does keeping a continuous fistula in the thoracic duct abbreviate or abort the gut IgA response?

Gowans: Pierce and I (1975) found that the IgA intestinal response to cholera toxoid was much reduced but not abolished if a thoracic duct fistula was established at the same time as intestinal challenge with antigen.

A residual response of this kind is not difficult to explain if it is accepted that part of the pool of precursors which gives rise to intestinal plasma cells consists of recirculating small B lymphocytes. I imagine that these 'memory' cells would be generated by the priming dose of antigen and considerable numbers might be in transit through the gut and available for recruitment into the local response.

Morris: I take the point that there is extensive recirculation of cells in the follicular areas but one of the characteristics of the Peyer's patch follicles is that all the cells in there will incorporate thymidine and they are all dividing. But what is the evidence that these cells incorporating thymidine are liberated into the intestinal lymph and are responsible for the very large output of cells from the intestine?

Gowans: There is no direct evidence and I have already drawn attention to the traffic of cells through the lamina propria which contributes to this output.

Morris: I have some experimental evidence that is against this interpretation. The cannulated intestinal duct of the lamb carries a very large output of cells. The output of cells from the gut falls after the first three to four days to low levels. If most of the cells in the intestinal lymph were coming from the Peyer's patches as newly formed cells, one would expect that the output might be sustained. And when we remove the last 3 m of the gut of lambs and make the intestine functional again by reanastomosing the jejunum to the caecum, the output of cells in the intestinal lymph isn't reduced statistically, yet 95% of the Peyer's patches have now gone. I fully agree that there is a vast cell traffic through the lamina propria all along the gut. That is one of the things that has not been understood until recently.

Weissman: But you have removed 95% not only of the Peyer's patches but also of the small intestine.

Morris: No, we leave a bit of small intestine! The small intestines of the lamb are somewhat longer than 3 m.

McConnell: Is the surface IgA on the small cells endogenous and do the cells have an immunoglobulin phenotype characteristic for memory cells? In many cases memory cells only express the committed isotype, which in this case would be α. Virgin cells often simultaneously express several isotypes, including δ. Is there any co-capping information about these cells?

Williams: It isn't actually known that the IgA is endogenous for the small cells. It is for the large cells. If you want to be extreme you could say it is all acquired. To eliminate this possibility someone would have to make anti-IgA allotype antibody and look for allelic exclusion on the small cells in an F1 animal. Nothing further is known on the small IgA cells except that they exist.

Gowans: Do we know anything about δ in the rat?

Williams: Most of the Ig-positive thoracic duct lymphocytes have δ but that doesn't say that the small number of IgA cells have δ.

Morris: I cannulated intestinal lymphatics at different levels to collect peripheral lymph, allowing the animal to recover and collecting lymph chronically. Peripheral lymph from the ileum has only about twice the concentration of cells that peripheral lymph from the duodenum or the jejunum has. I can't make any comment about the cell output because the concentration of cells is a function of the amount of lymph being produced. But a concentration of 1.3×10^8 cells/ml is unheard of, so I suspect there is something quite abnormal about your samples of Peyer's patch lymph. Our experiments suggest strongly that the concentration of cells in lymph collected from all parts of the small intestine is quite high.

Gowans: What is the distribution of lymphoid tissue along the gut of the sheep? In the rat there are about 20 discrete Peyer's patches with no organized lymphoid tissue between them.

Morris: In the sheep there is one Peyer's patch about 2 m in length; there are also disassociated Peyer's patches that extend cranially from the continuous Peyer's patch in the ileum for about another metre. One can take those out too, but a few Peyer's patches are always left in the jejunum. Certainly one can take out about 30 g of Peyer's patches.

Cahill: The average length of the sheep's small intestine is about 15 m, so there is still plenty of small intestine left.

Morris: These animals don't just curl up and die but do quite well without their terminal 3 m of ileum.

Davies: It is always a tricky matter to extrapolate from one genus to another. Nevertheless it is quite clear that the Peyer's patch of the mouse consists mainly of large germinal centres within which divisions can be found, but it has various other areas in which divisions are not to be found. In the conventional mouse, those cells that are dividing in the Peyer's patches are not T lymphocytes. It is extremely rare, in an animal where the T lymphocyte is chromosomally marked, to find a dividing cell in a Peyer's patch which is unambiguously a T cell, according to its genetic marker. This changes a little in intestinal infections with, for example, *Trichinella spiralis*. In certain circumstances one can then increase the proportion of T cells dividing in the Peyer's patch. You said, on the basis of indirect evidence, that the intestinal lymph had a high proportion of T lymphocytes. I say that the cell populations dividing in the Peyer's patches are not T lymphocytes. Therefore, on logistic grounds, one might get away from this conundrum that you have been going over. It is not surprising that these cells don't seem to be newly produced: the dividing cell population there is a B cell population, not a T cell population. The principal cell going out is a T cell, not a B cell.

Williams: We need a chart comparing Peyer's patches in the sheep and rat. The Peyer's patches may have a completely different function in the two species. In the adult sheep there are no Peyer's patches, are there?

Morris: Peyer's patches in the sheep regress at almost the same time as the thymus regresses.

Humphrey: Is there a deficit of lymphocytes in the mesenteric veins or the portal veins compared with the arteries? In other words, do so many cells leave from the arterial circulation and return via the lymphatics that there would be a deficit in the venous blood? There is a lot of argument about whether intact proteins, usually bacterial toxins, in so far as they are absorbed from the gut, travel via the veins in the lamina propria or via the lymph, and it appears that it is different for different sorts of protein.

Gowans: There is almost certainly no deficit. There are quantitative studies on the fraction of blood lymphocytes extracted by lymph nodes and spleen.

Ford: An alternative approach would be to study lymphocyte migration in an isolated perfused segment of intestine. The proportion of DNA-synthesizing lymphocytes from thoracic duct lymph entering an isolated, perfused mesenteric lymph node is almost twice that of small lymphocytes (Sedgley & Ford 1976). Presumably they fail to enter these lymph nodes in large numbers after i.v. injection because the large blood flow through the gut gives it a great competitive advantage for the capture of these cells.

Weissman: When you injected the labelled small lymphocytes from the thoracic duct lymph and looked at small lymphocytes entering the gut—if you saw any—did the vessels where they lodged three or four minutes later show any particular characteristics?

Gowans: We have not studied this important point. In conventional sections the vasculature in the lamina propria looks unremarkable and in studies with labelled lymphocytes we have not noticed the lodging of cells within particular vessels. But this is something that ought to be looked at systematically.

Woodruff: In some preliminary experiments with frozen sections of the small intestine that were overlaid with the thoracic duct lymphocytes the binding to HEV of Peyer's patches was unimpressive. However, lymphocytes were found adhering to the lamina propria and mucosa.

Gowans: So you don't find binding of lymphocytes to the high endothelial cells in the Peyer's patches?

Woodruff: That is correct—or rather we see trivial levels in comparison to that observed with high endothelial cells of lymph nodes. The clustering of lymphocytes in the lamina propria along the mucosa was very unexpected. It suggested that there might be something in the interstitial spaces which is involved in the accumulation of lymphocytes in this area.

References

Bazin H, Maldague P, Schonne E, Crabbe PAC, Bauldon H, Heremans JF 1971 The metabolism of different Ig classes in irradiated mice. Immunology 20:571

Gowans JL, Knight EJ 1964 The route of recirculation of lymphocytes in the rat. Proc R Soc Lond B Biol Sci 159:257-262

Howard JC 1972 The life span and recirculation of marrow-derived lymphocytes in the rat thoracic duct. J Exp Med 135:185-199

Mattioli CA, Tomasi TB 1973 Life span of IgA plasma cells from the mouse intestine. J Exp Med 138:452-460

Pierce NF, Gowans JL 1975 Cellular kinetics of the intestinal immune response to cholera toxoid in rats. J Exp Med 142:1550

Sedgley M, Ford WL 1976 The migration of lymphocytes across vascular endothelium. I. The entry of lymphocytes into the isolated mesenteric lymph node of the rat. Cell Tissue Kinet 9:231-243

Reassortment of cell populations within the lymphoid apparatus of the sheep

WENDY TREVELLA and BEDE MORRIS

Department of Immunology, John Curtin School of Medical Research, Australian National University, P.O. Box 334, Canberra City A.C.T.

Abstract   As lymphocytes recirculate through the blood, tissues and lymph they are sorted into populations which have varying morphological and functional characteristics. Lymphocytes are added, deleted and transformed within the lymphoid apparatus as a consequence of non-random migration and antigenic stimulation.

There is evidence that the physiological characteristics of peripheral and central lymph nodes vary as a result of differences in the origins of the cells entering the nodes. Lymphocytes enter the lymph nodes from the blood and lymph in varying numbers; consequently the cell population in the efferent lymph of central and peripheral lymph nodes contains different proportions of blood-borne and lymph-borne cells. Cells arriving in lymph nodes by way of the blood or the lymph migrate differently within the node. Those entering from the blood go principally to the paracortex and the follicular areas. Lymphocytes entering in the lymph are distributed through both the cortex and the medulla.

In humoral antibody responses and in the response that occurs during the rejection of a renal allograft, lymph-borne cells populate the medullary cords, cortex and germinal centres of the nodes they enter. Within these nodes, new populations of cells are generated which have different functional attributes from the cells which provoked their formation.

There is a continual migration of lymphocytes in the sheep between the blood, tissues and lymph (Smith et al 1970, Morris 1972). This traffic of lymphocytes occurs in the fetus *in utero,* in the absence of extraneous antigen and in the absence of immunoglobulins (Ig). Its establishment has nothing to do with the immune response, but is an inherent physiological property of lymphocytes and vascular endothelium (Pearson et al 1976). The extent of the cell traffic through a tissue is influenced by a variety of physiological factors: the amount of blood and the number of lymphocytes flowing to the tissue; the area and prevalence of the blood capillaries in the tissue through which the

© *Excerpta Medica 1980*
Blood cells and vessel walls: functional interactions
(Ciba Foundation symposium 71) p 127-144

cells may escape; the nature of the tissue itself; the affinity between the lymphocyte and the vascular endothelium at the sites where migration occurs; the 'permeability characteristics' of the blood vessels in the tissue to lymphocytes and the ease with which the lymphocytes can cross the interstitium and re-enter the regional lymphatics. The normal traffic of lymphocytes through a tissue can also be influenced by external stimuli which act independently or through various physiological determinants which influence the number of cells entering and leaving a tissue.

The migrant lymphocytes have several possible fates once they leave the blood: they may pass through the tissues and enter the lymph, they may pass directly from the tissue fluid back into the bloodstream, or they may remain for a variable time in the tissues, and become part of the fixed lymphoid elements. Additionally the lymphocytes may become involved in humoral or cell-mediated immune responses, if they encounter an appropriate stimulus and begin to divide and differentiate into various types of effector cells (Morris & Courtice 1977).

DIFFERENCES IN CELL TRAFFIC THROUGH VARIOUS TISSUES

The migration of lymphocytes between the blood and lymph is not random. Because of this a continual physiological reassortment of cells takes place in the body, due to the different patterns of lymphocyte migration into different tissues and to differences in the proportion of different classes of lymphocytes entering and leaving the various tissues (Hay & Morris 1976). The non-random escape of lymphocytes from the bloodstream suggests that some recognition mechanism exists between the vascular endothelium and the membrane of the lymphocyte which identifies the cell's position in the vascular system. This recognition mechanism may relate to some specific anatomical element in the blood vasculature or be due to some particular physiological property of a tissue. Some such recognition mechanism evidently causes the bulk of lymphocyte traffic to take place through lymphoid tissue, but further arrangements must exist to enable specific classes of lymphocytes to discriminate not only between lymphoid and non-lymphoid tissue but between lymphoid tissues in different sites of the body. The fact that most migration of lymphocytes from the blood takes place through lymph nodes is reflected in the high concentration of lymphocytes in efferent lymph (Hall & Morris 1962, 1965). There are around 5 to 8 $\times$ 10^6 cells/ml in efferent lymph from lymph nodes in the sheep; these are predominantly small lymphocytes and normally about 25–30% of them bear IgM on their surface membranes (Beh & Lascelles 1974, H. Gerber, W. Trevella & B. Morris, unpublished work, 1977).

The traffic of cells from the blood through skin and muscle and through the kidney, lungs, heart, ovaries and uterus is small compared to that through lymph nodes and quite different in character. Lymph coming from these tissues (peripheral lymph) carries between 2 and 8×10^5 cells/ml and of these, up to 20% may be macrophages. Of the lymphocytes in lymph from the skin, muscles and kidney (and almost certainly from the ovaries, lungs, heart and uterus), between 5 and 10% bear surface Ig (Scollay et al 1976a, Miller & Adams 1977).

The number of lymphocytes in peripheral lymph from the liver and intestines is significantly higher than from other visceral organs. There are about 4×10^6 cells/ml in peripheral hepatic lymph and these cells are typical of peripheral lymph from other parts of the body: a small proportion (less than 10%) of the lymphocytes have surface Ig and there are about 10 to 20% of macrophages present.

Peripheral intestinal lymph collected from different regions of the gut contains a large number of cells, irrespective of the presence or absence of fixed lymphoid tissue. For example, lymph from the duodenum of lambs has $2-3 \times 10^7$ cells/ml and lymph from the jejunum has about the same number, while lymph from the ileum has about 5.0×10^7 cells/ml. About 30% of small lymphocytes in peripheral and central intestinal lymph have surface Ig. While most of these have surface IgM, a few have surface IgA. Large lymphocytes bearing IgA and IgM are present in peripheral and central intestinal lymph in about equal numbers (Beh & Lascelles 1974; H. Gerber, W. Trevella & B. Morris, unpublished work, 1977).

KINETICS OF LYMPHOCYTE RECIRCULATION IN THE SHEEP

Labelled sheep lymphocytes, injected intravenously, appear in both peripheral and central lymph from most tissues within the first hour. Their numbers in the lymph increase to a peak around 20–30 hours after injection and then decrease. If the specific activities of the cell populations in lymph collected from several sites in the body are monitored simultaneously the curves describing the appearance of labelled lymphocytes from the different tissues are generally similar, although there is variation in the specific activity–time relationships in different tissues, because of differences in the rate of lymph formation and in the size and turnover-time of the pool of lymphocytes into which the migrant cells pass (Fig. 1). Although there is a vast disparity in the size of the lymphocyte pools in the popliteal lymph node and in the skin and muscles of the lower hindlimb of the sheep, the equilibration of labelled lymphocytes in these tissues, as measured by the

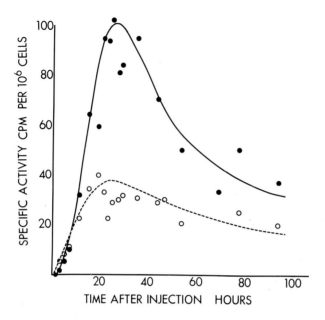

Fɪɢ. 1. The specific activity of cell populations in efferent lymph from the prefemoral node (●————●) and afferent lymph from the hind limb of a sheep (○- - - -○) after intravenous injection of ^{51}Cr-labelled isologous lymphocytes obtained from the efferent prefemoral duct.

specific activity curves of peripheral and central leg lymph, is not all that different. This suggests that the lymphocyte pool in the node in which cells arriving from the blood mix is much smaller than the total pool of cells within the node. In peripheral nodes such as the popliteal, prescapular and prefemoral nodes of the sheep this pool is largely in the cortico-medullary region and in the lymph sinuses.

Migrating lymphocytes do not appear to pass through any vascular structures in sheep similar to the high-endothelial venules through which lymphocytes migrate in species such as the rat and mouse. Cells which escape from the blood in the lymph nodes of sheep appear to move away rapidly from the blood vessels into the surrounding lymphoid tissue and enter the lymph sinuses. The impression is that the transit of lymphocytes across the blood vascular endothelium in lymph nodes of the sheep is a rapid event and not a rate-limiting step in the migration process.

TRAFFIC OF LYMPHOCYTES THROUGH CENTRAL LYMPH NODES

Any consideration of the recirculation of lymphocytes from blood to lymph must take into account the fact that lymphocytes which escape from the blood in the tissues and in the peripheral lymph nodes will be conveyed in the lymph stream to other lymph nodes more centrally placed along the lymphatic chain. The fate of recirculating cells entering lymph nodes by the lymphatic route does not seem to have been studied experimentally to any extent, although there is no reason to assume that cells which arrive at a lymph node in this way will have the same fate in the node as cells entering from the blood. By a similar argument, central lymph nodes experiencing this very different cell traffic may function differently from peripheral lymph nodes which receive cells mostly from the blood. There is evidence that the rate of cell proliferation and the number of germinal centres in the lumbar, deep cervical, mediastinal and mesenteric nodes of sheep and other species is higher than in the popliteal, prefemoral and prescapular nodes (B. Morris, unpublished work, 1960, Yoffey & Olson 1967).

The inflow of lymphocytes into a node in the afferent lymph may be important in regulating the extent of lymphocyte recirculation into and out of the node by way of the blood and lymph. When the mesenteric lymph nodes are removed from lambs and the intestinal lymphatics allowed to rejoin there is no statistically significant reduction in the cell output from the intestinal lymph (Reynolds 1976; H. Gerber, W. Trevella & B. Morris, unpublished work, 1977). The mesenteric nodes of lambs weigh between 20 and 30 g and it would be expected that if as large a proportion of the recirculating cells in intestinal lymph were entering from the blood as in the popliteal node (Hall & Morris 1965), the removal of the mesenteric nodes would create a substantial deficit in the cell output in intestinal lymph. This does not happen. This suggests either that the large number of cells entering the mesenteric nodes in the lymph suppressed the migration of cells into these nodes from the blood, or that many of the lymph-borne cells entering the nodes failed to leave them in the efferent lymph. The only feasible fate for these cells would be their return to the blood directly from within the nodes. To establish whether the traffic of cells through a lymph node is regulated by the number of cells entering it we investigated the effect of the arrival of cells in a lymph node in the afferent lymph on the output of cells in the efferent lymph. This question is of particular significance in deciding the role played by the central lymph nodes in reordering the traffic of cells within the lymphatic system.

When lymphocytes were collected from the efferent prescapular duct of the sheep and infused into the lumbar nodes by way of the lymph there was no increase in the cell output from the node over a period of several days, even though some of the infused cells passed through the node and entered the efferent lymph (Fig. 2). Similarly, in experiments in which an external loop

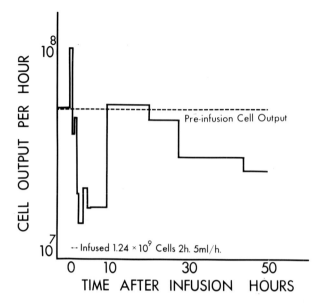

Fig. 2. The effect of infusing 1.24×10^9 cells into the lumbar lymph nodes on cell output in the lumbar lymph. The infused cells were obtained from the efferent prescapular lymph duct.

joining the afferent and efferent lymphatic of the lumbar nodes was established and lymph recirculated through the lumbar nodes for periods of up to 10 days no sustained increase occurred in the cell output (Fig. 3).

FATE OF LYMPHOCYTES ENTERING NODES IN THE LYMPH

Lymphocytes which normally enter central lymph nodes in the lymph stream will be either newly formed cells produced in more peripherally placed lymphoid tissues or recirculating cells derived from the blood in non-lymphoid tissues or in the peripheral lymph nodes.

The fate of these cells has been studied in the sheep by collecting the cellular effluent from a peripheral node quantitatively and returning it continuously to the central lumbar lymph nodes through a cannula in the efferent lymph duct of the popliteal node. The lymph leaving the central nodes is then diverted from the animal, collected quantitatively, and analysed. This experimental approach can be adapted to study the modifications occurring in a population of cells on their passage sequentially through more than one node or to study the accumulation or disappearance of labelled cells in a closed recirculating lymphatic loop into which a series of lymph nodes are connected.

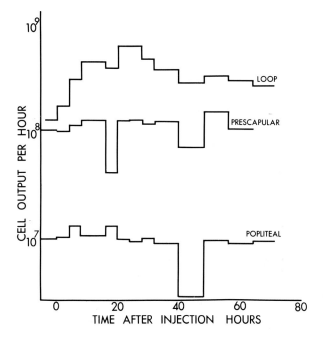

FIG. 3. The cell output per hour from the popliteal, prescapular and lumbar lymph nodes. The lumbar nodes were connected by a closed loop so that the lumbar lymph could be reinfused continuously into the lumbar nodes.

If lymphocytes are infused into a node through a lymph duct, many of the cells are held up in the node for a long time—in this respect lymphocytes behave quite differently from isologous red cells or plasma proteins, which pass rapidly through the nodes and can be recovered quantitatively in the lymph (Fig. 4).

It seems certain from our recent experiments that some of the lymphocytes that enter lymph nodes by way of the lymph pass directly into the blood. Such a pathway has been proposed by Sainte-Marie for lymphocytes in the rat (Sainte-Marie et al 1967, Sainte-Marie 1975). Our evidence for this is not conclusive in that we have not identified labelled lymphocytes in the blood or in lymph from other parts of the body when these cells were infused into a node. We have, however, shown that labelled cells injected into the bloodstream accumulate in a closed recirculating lymphatic loop at the same rate as in lymph from other sites in the body and that the number of lymphocytes in the loop does not continue to increase with time (Fig. 5). We have also shown that when very large numbers of lymphocytes are infused into central nodes over periods of several days, many of them fail to appear in

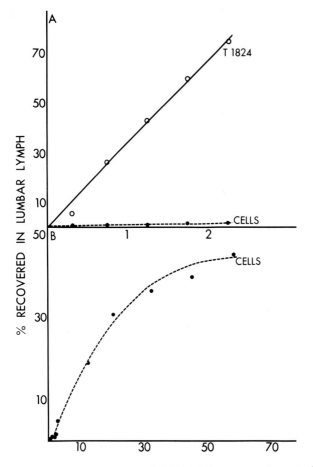

FIG. 4. A. The recovery of T-1824 isologous proteins and [³H]cytidine-labelled isologous lymphocytes during a 2-hour period after they were infused together into the lumbar lymph node of a sheep. B. The recovery in the efferent lymph of the same cells as in A, monitored over a period of 70 hours after the infusion.

the efferent lymph. At the end of the perfusion these nodes are not enlarged, suggesting that the ingress and egress of lymphocytes had reached a stead state.

The recirculating cells which enter lymph nodes by way of the lymph are distributed much more widely in the nodes than are blood-borne cells. Lymph-borne cells pass into the medullary cords, the cortico-medullary region and the deep and superficial cortex. This extensive distribution may be an important factor in determining differences in the physiology of peripheral

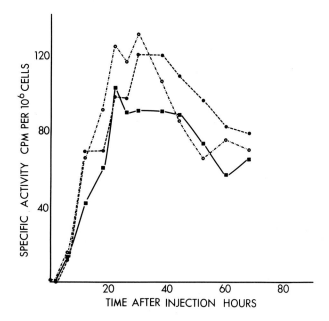

FIG. 5. The specific activity of the cells in the popliteal (●----●) and prescapular (○—·—○) lymph and in a closed recirculating loop (■————■) connecting the afferent and efferent ducts to the lumbar nodes after injection of 5.2 × 10⁹ [³H]cytidine-labelled cells intravenously.

and central lymph nodes. The fate of recirculating lymphocytes is thus bound up not only with events in the primary node in which they leave the bloodstream but, probably more importantly, with events in other nodes in the lymphoid apparatus to which they are conveyed.

MIGRATION PATTERNS OF LYMPHOCYTES

Certain categories of cells in the migrant lymphocyte population recirculate from the blood to the lymph at different rates from others. For example, the cells present in lambs thymectomized *in utero* recirculate much more slowly than do the lymphocytes in intact lambs (Pearson et al 1976). This is also the case in rats (Ford & Simmonds 1972) and in mice (Sprent 1973). There is a high proportion of Ig-positive lymphocytes in the blood and lymph of these animals and it is suggested that these cells recirculate slowly.

Some types of lymphocyte appear to have a propensity to accumulate in certain tissues more readily than in others. For example intestinal lymph of lambs contains large numbers of lymphocytes, of which many are blast cells which label readily with [³H] thymidine. When injected intravenously most of these cells accumulate in the mesenteric nodes, the lamina propria and the interfollicular areas of the Peyer's patches; smaller numbers enter the peripheral lymph nodes and the spleen.

The small lymphocytes in intestinal lymph have a less well-defined pattern of localization after being injected intravenously. These cells are found 48 hours after injection in the spleen and mesenteric lymph nodes and in the interfollicular areas of the Peyer's patches, particularly in the jejunum and to a lesser extent in the peripheral lymph nodes (Reynolds 1976).

Some of these blast cells and small lymphocytes reappear in the lymph after they leave the blood. Both classes of cells are found in greater numbers in the intestinal lymph than in lymph coming from the peripheral tissues; in this respect they show a predilection for recirculating through the tissues of the gut. Whether this preference is a property of the cell population in general or of a specific subclass of lymphocytes is not known. The large cells in intestinal lymph pass out of the blood and enter the lymph much more rapidly than do the small lymphocytes.

Studies by Scollay et al (1976) and Cahill et al (1977) have shown further physiological distinctions between populations of cells in intestinal lymph and in lymph coming from peripheral nodes. Small lymphocytes from intestinal lymph recirculate in greater numbers through the intestines than through peripheral nodes when injected back into the bloodstream, whereas the opposite holds for cells collected from the popliteal lymph. Whatever the physiological mechanisms are which determine these differences in migration patterns, the outcome is that the composition of the migrant lymphocyte populations is modified during passage through the tissues, the lymphoid apparatus and the blood by the addition and removal of specific classes of cells: the final population of cells re-entering the blood is very different from the original migration.

EFFECT OF IMMUNE RESPONSES ON LYMPHOCYTE RECIRCULATION

Humoral and cell-mediated immune responses alter patterns of lymphocyte migration and lead to the reassortment of the recirculating cell populations of lymph (Hay & Morris 1976). The anatomy of the lymph is modified further by the proliferative response that invariably occurs in lymph nodes and tissues where antigen or activated cells become localized. The selective addition to and deletion of certain types of cells from the free-floating populations in lymph superimposes a further degree of physical and functional heterogeneity on the recirculating lymphocyte populations, such that cells collected from one particular site in the lymphatic system will have a distinctive range of re-activities not present in the cell population in lymph at another site.

Humoral immune responses

Primary and secondary immune responses to antigens are accompanied by a succession of changes in the cell population of the lymph coming from the site where the antigen first localizes (Hall & Morris 1963). The immediate effect of antigen is a dramatic reduction in the cell output in the lymph from the node. This is followed by a protracted rise in the number of cells in the efferent lymph and the appearance of blast cells and antibody-forming cells after about 72 hours in a primary response and about 60 hours in a secondary response.

In nodes more centrally placed in the lymphatic chain a similar sequence of events occurs as the immune response is propagated along the lymphatic chain. The timing and intensity of the response in the central nodes is related to the physical nature of the antigen and the strength of the immune response it evokes. Many of the stimulated cells that pass from the peripheral to the central nodes are retained there or fail to appear in the efferent lymph. It is likely that some enter the bloodstream directly from within the central nodes.

We have shown that the immune response can be confined to the popliteal node if antigen is introduced into this node via an afferent lymphatic, provided that all the cells leaving the node in the efferent lymph are diverted from the body (Hall & Morris 1965, Morris 1966). This suggests that if cells involved in the immune response enter the bloodstream directly within the primary lymph node, they have no capacity for propagating the immune response systemically, as they are able to do when they pass to central lymph nodes along the lymphatic chain.

Many cells retained within the central nodes enter the medullary cords and differentiate into plasma cells; a further population becomes incorporated into germinal centres in the outer cortex and cortico-medullary areas of the node. There does not seem to be any component of the blood-borne lymphocytes which has this particular distribution within lymph nodes; the finding that lymph-borne cells are involved in the formation of germinal centres provides an explanation for the greater frequency of germinal centres in central lymph nodes than in peripheral nodes and can be seen as an important part of the processes whereby the immune response is amplified within the lymphoid apparatus. This redistribution of lymph-borne cells is a further example of the physiological reordering of the migrating lymphocyte populations through the incorporation of some of these cells into structural components of fixed lymphoid tissue.

Cell-mediated responses

The immune response to a renal allograft involves the large-scale redirection of lymphocytes from the bloodstream into the graft and out of it by way of the lymph. The vast numbers of lymphocytes which are drawn from the blood and subsequently leave the graft are predominantly cells which carry no surface Ig and produce no antibodies (Pedersen & Morris 1970, Pedersen et al 1975, Miller & Adams 1977).

The cell output in lymph from a renal allograft increases from around $1-2 \times 10^6$/h during the first 24 hours after the graft is installed to $4-5 \times 10^8$ cells/h 5 to 7 days after grafting; the percentage of cells bearing surface Ig coming from the graft, however, does not change and is always less than 10% of the total lymphocytes (Miller & Adams 1977).

Macrophages are present in lymph from a renal allograft at all times, although their numbers increase as rejection of the graft becomes imminent. These cells together with lymphocytes are conveyed from the lymph to the regional node and become incorporated into the phagocytic continuum lining the subcapsular, intermediate and medullary sinuses (V.A. Fahy, C.Zukowski & B. Morris, unpublished work, 1977).

The cells which leave the graft have no demonstrable reactivity against the graft donor *in vivo*. However, when they enter the regional node they stimulate a vigorous and protracted immune response which gives rise to a population of cells in the efferent lymph that are specifically reactive against the graft donor when tested *in vivo* (Pedersen 1971, Hay & Morris 1976).

The appearance of a population of specifically reactive cells in lymph from the regional node distinguishes this free-floating population of cells from the anergic population which leaves the graft in the peripheral lymph. The reactive cells may either be cells newly formed in the node in response to interactions between the cells in peripheral lymph from the graft and a class of cell already present in the node; or they may be the same cells as those that left the graft, or their progeny, modified in terms of their reactivity against the graft by exposure to the environment of the regional lymph node.

The proliferative response in the regional node is associated with the redistribution of many of the lymph-borne cells coming from the graft throughout the medulla and the germinal centres. In contrast to the structural rearrangement of cells that occurs in central nodes during humoral antibody responses, fewer cells localize in the cortico-medullary areas of the regional node during the transplantation reaction (V.A. Fahy, C. Zukowski & B. Morris, unpublished work, 1977).

References

Beh KJ, Lascelles AK 1974 Class specificity of intracellular and surface immunoglobulin of cells in popliteal and intestinal lymph from sheep. Aust J Exp Biol Med Sci 52:505-514

Cahill RNP, Poskitt DC, Frost H, Trnka Z 1977 Two distinct pools of recirculating T lymphocytes: migratory characteristics of nodal and intestinal T lymphocytes. J Exp Med 145:420-428

Ford WL, Simmonds SJ 1972 The tempo of lymphocyte circulation from blood to lymph in the rat. Cell Tissue Kinet 5:175-189

Hall JG, Morris B 1962 The output of cells in lymph from the popliteal node of the sheep. J Exp Physiol Cogn Med Sci 47:360-369

Hall JG, Morris B 1963 The lymph-borne cells of the immune response. Q J Exp Physiol Cogn Med Sci 48:235-247

Hall JG, Morris B 1965 The origin of the cells in the efferent lymph from a single lymph node. J Exp Med 121:901-910

Hay JB, Morris B 1976 Generation and selection of specific reactive cells. Br Med Bull 32:135-140

Miller HRP, Adams EP 1977 Reassortment of lymphocytes in lymph from normal and allografted sheep. Am J Pathol 87:59-80

Morris B 1966 Lymphoid cells and their role in the establishment of systemic immunity. In: Walsh RJ, Pitney WR (eds) Proc XIth Congr Int Soc Haematol. Victor CN Bright, Govt printer, Sidney, p 25-36

Morris B 1972 The cells in lymph and their role in immunological reactions. In: Handbuch der Allgemeinen Pathologie. Springer, Berlin, p 405-486

Morris B, Courtice FC 1977 Cells and immunoglobulins in lymph. Lymphology 10:62-70

Pearson LD, Simpson-Morgan MW, Morris B 1976 Lymphopoiesis and lymphocyte recirculation in the sheep fetus. J Exp Med 143:167-186

Pedersen NC 1971 Studies in transplantation immunity. PhD thesis, Australian National University, Canberra

Pedersen NC, Morris B 1970 The role of the lymphatic system in the rejection of homografts: a study of lymph from renal transplants. J Exp Med 131:936-969

Pedersen NC, Adams EP, Morris B 1975 The response of the lymphoid system to renal allografts in sheep. Transplantation (Baltimore) 19:400-409

Reynolds JD 1976 The development and physiology of the gut-associated lymphoid system in lambs. PhD thesis, Australian National University, Canberra

Sainte-Marie G 1975 A critical analysis of the validity of the experimental basis of current concept of the mode of lymphocyte recirculation. Bull Inst Pasteur 73:255-279

Sainte-Marie G, Sin YM, Chan C 1967 The diapedesis of lymphocytes through post-capillary venules of rat lymph nodes. Rev Can Biol 26:141-151

Scollay R, Hall J, Orlans E 1976a Studies on the lymphocytes of sheep II. Some properties of cells in various compartments of the recirculating lymphocyte pool. Eur J Immunol 6:121-125

Scollay R, Hopkins J, Hall J 1976b Possible role of surface Ig in non-random recirculation of small lymphocytes. Nature (Lond) 260:528-529

Smith JB, McIntosh GH, Morris B 1970 The traffic of cells through tissues: a study of peripheral lymph in sheep. J Anat 107:87-100

Sprent J 1973 Circulating T & B lymphocytes of the mouse. I. Migratory properties. Cell Immunol 7:10-39

Yoffey JM, Olson IA 1967 The formation of germinal centres in the medulla of lymph nodes. In: Cottier H et al (eds) Germinal centres in immune responses. Springer, Berlin, p 40-48

Discussion

Weissman: Did you say that the cells coming out of an antigen-activated lymph node, whether it be for a humoral or a cellular immune response, may be largely Ig⁻ cells?

Morris: In a humoral response all the blast cells in the lymph are Ig positive. From the renal homograft they are almost all Ig negative.

Weissman: Whether they are mainly Ig⁺ or Ig⁻, when they are labelled extracorporeally and infused into the next lymph node up the chain both will be major contributors to the germinal centres. Have you double-labelled the cells within the germinal centre in any way to see whether they are indeed Ig⁻ cells within the germinal centre?

Morris: Both Ig⁺ and Ig⁻ blast cells will enter germinal centres. We haven't double-labelled cells in germinal centres but we have shown that in the allograft experiments the blast cells, as far as we can identify them, are Ig⁻. None of them make antibodies. They are still not doing so when they enter the germinal centre. It might be possible for them to begin making antibody there but I have no comment to make on that. My point is that, from those experiments, germinal centres may be derived from either Ig⁺ or Ig⁻ cells reaching the node in the lymph.

Weissman: Do you have a marker for T cells or do you believe there are T cells in the sheep?

Morris: No, I don't believe there are T cells or B cells in sheep in the sense that other people believe in these cells. There are cells that come from the thymus and there are cells that come from elsewhere. There are cells with surface Ig and there are cells without surface Ig.

Weissman: If you infuse those cells, not into the next lymph node up the chain but into a lymph node far away, in the afferent lymph, will germinal centres also develop? Is there a requirement for something else draining from the antigen-bearing area?

Morris: There certainly is a requirement in relation to the propagation of the immune response. Unless the next node along the chain has received the first 48 h of lymph, an asynchronous immune response develops in the lumbar lymph. There are no cells in this early lymph that are capable of propagating the immune response unless they subsequently interact with 48–72-h effluent material; so that first 48 h is not irrelevant. Some people might say that reactive cells had been selected out. Probably there are cells in there but they are unreactive at the time they enter the lymph. The material in lymph appearing early after an antigenic challenge is important in propagating the response but this material is not necessarily antigen. You cannot generate an immune response with this material by transferring it to another node.

Cahill: On the question of a marker for T lymphocytes in the sheep, Ivor Heron in our laboratory has produced two monoclonal antibodies. One of these stains a subpopulation of peripheral T lymphocytes and the other stains 95% of thymocytes and most of the peripheral T cells.

Weissman: Have you stained germinal centres with this monoclonal antibody?

Cahill: No, so far we have only looked at cells in lymph or cells in suspension.

Weissman: We did that in the mouse, which has cells with these markers that are related to function. In germinal centres raised by antigens that were not as clearly delineated as yours, we were quite surprised to see that T cell blasts, or blasts bearing T cell markers, were a significant population.

Howard: Some evidence about T cells from the rat may be essentially the same thing as you described, Professor Morris—namely the requirement for a second contact with antigen for the full development of the immune response action. Ann Marshak, Darcy Wilson and I used what was then a very fashionable system, developed more or less simultaneously by Bill Ford and John Sprent. In this one induces a graft-versus-host reaction by injecting cells of one major histocompatibility type A into an irradiated A × B F1 hybrid. The injected A cells react against the unshared B alloantigens, are arrested initially in the lymphoid tissues, then begin to transform and appear in the thoracic duct lymph as blasts.

With this system Ford & Atkins (1971) got a population of rapidly recirculating cells which are depleted of reactivity to B. The anti-B cells have been in the nodes initiating an immune response. After about 40 h the anti-B reactive cells begin to come out again as a population of blast cells. These blast cells are antigen-dependent—that is, their existence depends on this confrontation. If this were a syngeneic transfer the recovery of cells from the recipient thoracic duct would just carry on down.

We were working on the functions of the blast cells in these studies because we thought that they would be antigen-specific, antigen-selected, and of considerable immunological interest. In fact we could never demonstrate any immunologically specific function for these cells at all (Wilson et al 1976). We tried a number of things—graft-versus-host reactions, the ability to kill a relevant target, the ability to transfer memory for recognition of B strain alloantigens. These cells were inert, as far as we could discover.

However, Sprent, doing rather similar experiments in mice, was able to recover a population in an F1 hybrid that was very active indeed in allogeneic immune responses (Sprent & Miller 1972). Ann Marshak had cannulated the thoracic duct at time 0 after injection and collected all the cells that came out. The cellular response dwindled away by about 72 h. John Sprent injected at time 0 but he cannulated at 4−5 days. That seemed to be the only difference in the two protocols so we did the experiment of reinfusing the cells that were collected on early days back into the animal. The blast cell response then went up and up, reaching a peak at exactly the same time as John Sprent said it should. These cells were immensely active immunologically. In other words, the early peak was a necessary condition for the existence of the second peak. They were both recovered from the same efferent lymphatic in our experiment, and presumably the second contact was either in the mesenteric bed where most of the cells come from, or perhaps in the spleen. But a second contact with the lymphoid system was necessary for the full development of function. These results have never been written up, sadly.

McConnell: Wouldn't you reinterpret these experiments in view of the experiments of Dorsch & Roser (1977) which show that T suppressor cells are present in the thoracic duct lymph which suppress *in vivo* allogeneic reactions? These cells might be making the early populations unresponsive.

Howard: All I meant about that population was that somewhere in that early peak there is a population which is necessary for the full manifestation of that particular immune response. We did not use them as suppressor cells in an attempt to demonstrate that they would inhibit the initiation of the immune response. They were without function as far as we could tell, yet they must in some way determine the existence of the main response.

Bede Morris made the point that something travels from the primary to the secondary lymph node whose nature is unknown but which is not sufficient to induce a primary response. I would guess it is antigen but in a different form.

Morris: The reality is that in the response to a renal transplant we can propagate the allogeneic response to the regional node with the material that comes out of the graft, right from the outset. This early lymph is a very powerful stimulant. That is different from the humoral antibody situation where the material that comes out in the first 48 h is incapable of propagating the response.

Howard: When you say material, do you mean everything that is solid?

Morris: I mean everything—the total lymph.

Weiss: Wasn't the lymph node very dark in your last slide?

Morris: Yes, it is an enormous reaction.

Weiss: These are not haemolymph nodes, are they?

Morris: No, these nodes are about 4−5 cm long, while haemolymph nodes are tiny in the sheep—usually about 2−4 mm in diameter.

Weiss: Are they haemorrhagic?

Morris: There is usually no blood in the lymph from these nodes. The lymph nodes responding to allogeneic cells in the sheep enlarge and have a meaty appearance—inflamed, if you like.

Weissman: For people who aren't working directly on the genesis of germinal centres, I think this is the first clear demonstration of the kinds of cells that are involved in their genesis. All the other lymphocyte traffic patterns described avoid the germinal centres.

Morris: These experiments provide some new insights into the interpretation of the domains occupied by cells in lymph nodes. A lot of these cells had previously come from the blood in the primary node, so they exhibited one kind of distribution there. In the secondary node, cells stimulated by antigen and delivered to it via the lymphatic route exhibit a different distribution and, if you like, enter another domain.

Weissman: The question I was trying to get to was that I think there is a slight asymmetry. In the cellular immune response you were looking at lymph coming directly from the kidney. In the humoral immune response you were picking it up out of the popliteal node. What about the afferent lymph to a popliteal node where the antigen is present in the skin or some place like that? Would you again have cells that can contribute to a germinal centre in the afferent lymph?

Morris: Yes. This was described by Kelly (1970). The lymphatic system has developed to cope with two distinctly different situations. There is a system of absorbing vessels which is coupled with a system of phagocytic reticular cells in the lymph nodes, and the immune response can be handled in one of two ways. The antigen can be absorbed—and transported to the reactive cells via the lymph; or, if the antigen is fixed, such as with a transplanted kidney, the reactive cells can be transported to the antigen. The immune response will also eventually develop in the regional node by the subsequent transfer of reactive cells from the site of antigen deposition to the node.

Weissman: That gets us back to whether there is peripheral sensitization in these circumstances.

Morris: There is no doubt that there is peripheral sensitization. The only thing that is different to Medawar's original proposition is that the regional lymph node is not essential in the process. The original peripheral sensitization hypothesis meant that cells that were sensitized in the periphery then had to go to the regional lymph nodes, where they developed into effector cells which were transported back to the graft to bring about its destruction. But everything that is needed to destroy that graft goes on in the graft itself. The regional lymph node is not essential to the process.

Weissman: It is extremely important to know whether an antigen like myelin that is put in the skin really produces peripheral sensitization.

Cahill: Dr Jack Hay and I did some work (unpublished) on peripheral sensitization, injecting BCG subcutaneously into sheep. We found PPD-reactive cells emerging from the afferent lymph draining the site of injection first. Some 24 h later they were present in the efferent lymph of the draining lymph nodes, and then, a day or so later, in the efferent lymph draining distant lymph nodes.

Davies: Are the cells that get into the germinal centres being recruited to existing germinal centres or actually starting new ones?

Morris: Many of these germinal centres seem to be made up almost entirely of labelled cells, so my guess is that in many instances they represent the aggregation of a particular class of labelled cells in the lymph and the formation *de novo* of a new centre. There are also established germinal centres into which some of the infused cells penetrate. Those centres had formed 24–48 hours after the transfusion of the labelled cells, so it is a rapid event. Infused cells migrate rapidly out of the lymph channels, the pericapsular sinus and the intermediary and medullary sinuses, and they do go into these areas quite quickly.

References

Dorsch S, Roser B 1977 Recirculating, suppressor T cells in transplantation tolerance. J Exp Med 145:1144-1157

Ford WL, Atkins RC 1971 Specific unresponsiveness of recirculating lymphocytes after exposure to histocompatibility antigen in F_1 hybrid rats. Nat New Biol 234:178

Kelly RH 1970 Localization of afferent lymph cells within the draining node during a primary immune response. Nature (Lond) 227:510-513

Sprent J, Miller JFAP 1972 Interaction of thymus lymphocytes with histocompatible cells. III. Immunological characteristics of recirculating lymphocytes derived from activated thymus cells. Cell Immunol 3:213

Wilson DB, Marshak A, Howard JC 1976 Specific positive and negative selection of rat lymphocytes reactive to strong histocompatibility antigens: activation with alloantigens *in vitro* and *in vivo*. J Immunol 116:1030-1040

Lymphocyte recirculation in the sheep fetus

R.N.P. CAHILL, I. HERON, D.C. POSKITT and Z. TRNKA

Basel Institute for Immunology, 487, Grenzacherstrasse, Postfach 4005 Basel 5, Switzerland

Abstract The numbers of circulating thymus-derived and surface Ig-bearing lymphocytes in the fetal lamb increase exponentially over the last third of gestation. Experiments in which [³H]thymidine was continuously infused into fetal lambs have established that these cells are long-lived in the fetus.

The migration of ⁵¹Cr-labelled autologous lymphocytes from intestinal or prescapular lymph was compared in fetal lambs and adult sheep. A subpopulation of thymus-derived lymphocytes present in intestinal lymph of adults which migrated preferentially to the small intestine was not found in fetal intestinal lymph. There were marked differences in the migration of fetal and adult lymphocytes to the lungs and liver. In spite of the absence of circulating antibodies or immunoglobulins and of extrinsic antigen in the immunologically virgin sheep fetus, the circulation of lymphocytes through the spleen and lymph nodes of fetal lambs was more intense than in the adult, indicating that the pathways of recirculation and the capacity of cells to recirculate arise as a physiological process independently of antigenic stimulation.

Recirculating lymphocytes are long-lived cells which migrate continuously between the bloodstream, tissue fluids and lymph via the lymph nodes (Gowans & McGregor 1965); they are of central importance in the development of a normal immune response in the intact animal (reviewed by Morris 1972, Ford 1975, Sprent 1977). It has become increasingly obvious that the cells which make up this pool of recirculating lymphocytes possess considerable functional heterogeneity and that they undergo a continuous process of reassortment as they move between the various lymphoid compartments which collectively form the lymphoid apparatus (Morris & Courtice 1977). The interaction between circulating lymphocytes and the blood vascular endothelium is crucial to the recirculation and reassortment of lymphocytes, since the vascular endothelium is the first barrier which migrating

© *Excerpta Medica 1980*
Blood cells and vessel walls: functional interactions
(Ciba Foundation symposium 71) p 145-166

lymphocytes must negotiate as they move from the blood into the various lymphoid compartments of the body. There are at least three major questions which relate to this interaction between migrating lymphocytes and vascular endothelium. (1) What is the nature of the receptors on lymphocytes and specialized endothelial cells which presumably exist to permit lymphocytes to recirculate through lymph nodes? When one considers that in sheep about 14 000 recirculating lymphocytes per second are passing through a single resting popliteal lymph node (weighing about 1 g) and that this can increase to about 140 000 lymphocytes per second in response to antigen—to the exclusion of non-recirculating lymphocytes, granulocytes and other free-floating cells in the bloodstream and without any alteration in transit time for the majority of lymphocytes (Cahill et al 1976)—then obviously some special arrangement must exist for their transfer from the blood across the endo-thelium into the lymph node. (2) Do there exist separate or additional recognition mechanisms between subpopulations of recirculating lymphocytes or other free-floating cells such as blast cells and the vascular endothelium (Scollay et al 1976, Cahill et al 1977, Hall et al 1977, Husband et al 1977) which result in selective migration streams of lymphocytes between particular tissues where lymphocytes are concentrated? And (3) what is the influence of antigen on the interaction between lymphocytes and vascular endothelium? Can immunological specificity, for example, be imposed on the vascular endothelium? The influence of antigen on the recirculation of lymphocytes is considerable but also complicated (Morris 1972, Ford 1975, Sprent 1977), and it would be illuminating to study lymphocyte circulation in an environment unconditioned by antigenic stimulation and circulating immunoglobulins. If lymphocytes do recirculate in such an environment, this would pose questions about the basic properties and function of recirculating lymphocytes. Knowledge of the tempo and pathways of such recirculation could be useful in approaching the questions asked earlier about the nature of the interactions between recirculating lymphocytes and the vascular endothelium. The introduction of one defined antigen into such an environment and a com-parison of any subsequent perturbations provoked in the recirculation of lymphocytes with those produced in the mature animal should similarly be useful.

The fetal lamb provides such an opportunity. The epitheliochorial nature of the ovine placenta ensures that the developing fetus has no contact with maternal or any other extrinsic antigen; the fetus is indeed agammaglobu-linaemic and so provides a novel opportunity for studying the growth and development of the lymphoid system in a milieu where the fetus remains immunologically virgin (Fahey & Morris 1974). Morris and his colleagues have developed surgical techniques for thymectomizing fetal lambs (Cole & Morris 1971) and for establishing chronic lymphatic fistulas in sheep fetuses *in utero* (Smeaton et al 1969). They used these techniques (Pearson et al 1976) to show for the first time that lymphocyte recirculation from blood to thoracic duct lymph is established at least as early as 75 days of gestation (the normal gestation period in the sheep being 150 days), indicating that pathways of recirculation and the capacity of cells to recirculate arise as a physiological process independent of antigenic stimulation.

The experiments described here have been done on normal healthy outbred sheep and are concerned with (1) the development and lifespan of the cells of the fetal pool of recirculating lymphocytes, (2) with differing patterns of migration of lymphocytes which are observed between the fetus and the adult, and (3) briefly with the relationship between fetal and adult pools of recirculating lymphocytes.

DEVELOPMENT OF THE RECIRCULATING POOL OF LYMPHOCYTES IN THE FETUS

Surgical techniques were developed (Cahill et al 1979a) which allowed lymph to be collected continuously from a single prescapular lymph node and the intestines of fetuses *in utero* during the last third of gestation (Days 100–150). Table 1 shows the hourly output of lymphocytes from the prescapular lymph node and the intestinal duct from fetuses of varying gestational ages. There is an exponential increase in the output of cells from both ducts between 100 and 150 days. This is comparable with the exponential increase in thoracic duct cell output in fetal lambs observed by Pearson et al (1976). These workers also demonstrated that thymectomy between 61 and 80 days of gestation caused a decrease in thoracic duct cell output of between 77 and 86%, when measured between 124 and 126 days after conception. After drainage for 14 days the thoracic duct output from thymectomized fetuses was about 10% of the output of normal fetuses. On the basis of the results in Table 1 we assume that the majority of the cells, which are surface Ig-negative, are thymus-derived (T) lymphocytes and that the surface Ig-bearing cells are B lymphocytes, and therefore that the output of recirculating T and B lymphocytes from intestinal and prescapular lymph increases exponentially between 100 and 150 days' gestation.

It is interesting to note (see Table 3, p. 150) that in the fetus the output of lymphocytes from a single prescapular lymph node and the intestinal duct is roughly the same. In young sheep, however, the output from the intestinal duct is about 10 times greater than that from one prescapular lymph node. It is not known to what extent the great increase in output from the intestinal

TABLE 1

Hourly output of cells from the prescapular efferent lymph duct and the intestinal duct of fetal lambs

Prescapular duct			Intestinal duct		
Days of gestation	Cell output	% surface Ig-bearing cells	Days of gestation	Cell output	% surface Ig-bearing cells
100	8.7×10^6	3	110	4.5×10^7	2
116	2.3×10^7	3			
125[a]	2.1×10^7	4	125[a]	2.5×10^7	3
			130	4.7×10^7	1.5
135[b]	2.8×10^7	3	135[b]	3.8×10^7	3.0
140[c]	3.0×10^7	2.5	140[c]	3.3×10^7	5.0
			140	1.3×10^8	1.5
144	5.3×10^7	9			
150	1.2×10^8	4			

a, b, c, Cell outputs were measured simultaneously in the same fetus. Surface immunoglobulin was detected using a rhodamine-conjugated polyvalent rabbit anti-sheep immunoglobulin. Except for fetus c, whose conception date was known, fetal ages were estimated from forehead–rump length. Some of these results have been published previously (Cahill et al 1979a).

duct is influenced directly by the massive antigenic load in the intestines. The number of recirculating lymphocytes in the sheep fetus is nevertheless considerable. The pool of recirculating T lymphocytes in a fetus of 130 days' gestation was estimated by the fall in cell output which occurs during chronic lymphatic drainage to be 6×10^9, which is comparable to the total pool size of $5.7 \pm 1.2 \times 10^9$ in fetuses of 130–135 days of age given by Pearson et al (1976). The sheep fetus in fact contains a huge pool of recirculating immunologically virgin lymphocytes, and in a fetus near to term the pool size is $1.2 \pm 0.3 \times 10^{10}$ (Pearson et al 1976). In terms of body weight this is at least equal to its maximum size in the postnatal animal.

LIFESPAN OF CELLS IN THE FETAL POOL OF RECIRCULATING LYMPHOCYTES

Since the majority of recirculating lymphocytes in adult rats and mice are long-lived (evidence reviewed by Sprent 1977) the question arises of whether recirculating lymphocytes in the fetus constitute a rapidly turning over pool of short-lived cells or whether they are long-lived, at least during fetal life. In an attempt to answer this question a series of sheep fetuses ranging in age from

120 to 150 days were infused intravenously (i.v.), via a cannula inserted in the jugular vein of the fetus, with [³H]thymidine for periods of between 3 and 17 days. [³H]Thymidine of high specific activity (20 Ci/mmol) was infused at a rate of 125 μCi/h, a dose comparable to that used by Sprent & Basten (1973) in experiments which established that thoracic duct lymphocytes in young adult mice are long-lived.

In order to avoid depleting the fetus of recirculating lymphocytes during the period of infusion we did not cannulate lymphatic ducts but instead examined autoradiographs of Leishman-stained peripheral blood smears for the presence of [³H]thymidine-labelled cells. Autoradiographs were exposed for 21 days; no increase in the percentage of labelling was seen after 14 days' exposure.

Table 2 shows the results of one such experiment in which a fetus was

TABLE 2

The appearance of [³H]thymidine labelled cells in the peripheral blood of a fetal lamb

Days after [³H]thymidine infusion started	% labelled small lymphocytes	% labelled granulocytes
1	1.3	11
7	5.9	75
9	9.8	100
17	19.0	100

The fetus was continuously infused i.v. with [³H]thymidine (specific activity 20 Ci/mmol) at a rate of 125 μCi/h from Day 125 to Day 142 of gestation. Autoradiographs were exposed for 21 days.

infused intravenously with [³H]thymidine for 17 days between Days 125 and 142 of gestation. Although 100% of granulocytes were labelled with [³H]thymidine by nine days, which is comparable with the time required to label 100% of granulocytes in the peripheral blood of young adult rats after continuous i.v. infusion of [³H]thymidine (Robinson et al 1965), only some 10% of small lymphocytes were labelled after nine days and after 17 days some 80% of peripheral blood small lymphocytes remained unlabelled.

In order to discover what percentage of peripheral blood lymphocytes were recirculating we next examined the [³H]thymidine-labelling patterns of recirculating lymphocytes directly by cannulating a prescapular lymph node and collecting the efferent lymph cells during a continuous infusion of [³H]thymidine. The number of [³H]thymidine-labelled cells collected was determined from autoradiographs of Leishman-stained smears. Autoradiographs of anti-Ig-stained smears were examined in the phase-contrast microscope to detect [³H]thymidine-labelled B cells. The results of one such experiment, in which a fetus was infused for seven days, between Days 130 and 137 of gestation, are shown in Table 3.

As a result of the relatively high output of cells from the prescapular node in the fetus the cell output fell by over 80% during the period of drainage, indicating that the majority of the fetal pool of recirculating lymphocytes had, by seven days, been drained off. During this time 3.7×10^9 lymphocytes were collected, of which 4.6×10^8 (12.3%) were labelled with [³H]thymidine. The output of [³H]thymidine-labelled cells fluctuated between 1.7×10^6 and 4.1×10^6/h, but overall the output rose somewhat during the seven days drainage and the average output of labelled cells was 2.7×10^6/h for the whole period. At the beginning of the experiment 3.5% of the cells were B lymphocytes and after seven days' drainage 9.2% of the cells emerging were B lymphocytes, of which 16% were [³H]thymidine-labelled. This observation suggests that either B lymphocytes enter the recirculating pool more slowly than T lymphocytes or that they are more slowly mobilized during chronic lymphatic drainage than T lymphocytes in the fetus. At this time 30% of T lymphocytes were [³H]thymidine labelled. The percentage of blast cells remained low (1%) during the whole period of drainage, as reported previously by Pearson et al (1976), which is in marked contrast to the relative increase in large cells caused by chronic thoracic duct drainage in rats and mice (Sprent 1977). Taken together, the experiments illustrated in Tables 2 and 3 show that recirculating T and B lymphocytes in the fetal lamb are long-lived cells which perhaps continue to recirculate from blood to lymph from the time they enter the lymph stream at least until birth.

TABLE 3

The appearance of [³H]thymidine-labelled cells in the prescapular efferent lymph of a fetal lamb

Days of gestation	Total cell output/h	[³H]-Thymidine labelled cell output/h	% [³H]thymidine labelled cells	% blast cells
130	5×10^7	1.7×10^6	3.5	1
137	9.3×10^6	2.7×10^6	28.7	1

The fetus was continuously infused i.v. with [³H]thymidine (specific activity 20 Ci/mmol) at a rate of 125 μCi/h from Day 130 to Day 137 of gestation. Autoradiographs were exposed for 21 days.

DIFFERING PATTERNS OF MIGRATION OF RECIRCULATING LYMPHOCYTES IN
FETAL AND POSTNATAL LAMBS

Non-random patterns of migration of lymphocytes between intestinal
lymph and the efferent lymph which drains peripheral lymph nodes have been
observed in adult sheep (Scollay et al 1976, Cahill et al 1977). When [51]Cr-
labelled small T lymphocytes from intestinal lymph are infused i.v. their
recovery in intestinal lymph is about twice that in lymph draining a prescapular lymph node; when ^{51}Cr-labelled T cells from prescapular lymph are
infused i.v. the opposite pattern of recovery is found (Cahill et al 1977). From
these observations we proposed that two subpopulations of small T lymphocytes recirculate in adult sheep, one population circulating preferentially
through the small intestine and the other preferentially through lymph nodes
(Cahill et al 1977). We suggested that these two subpopulations might exist in
the fetus. This suggestion proved to be incorrect (Cahill et al 1979b). When
[51]Cr-labelled T lymphocytes (containing 3% B lymphocytes) from the
intestinal duct of a 135-day fetus (Fig. 1) were infused i.v. and the appearance
of labelled cells in intestinal and prescapular lymph was followed
simultaneously over the next three days, the concentration of labelled cells in
the efferent lymph of both ducts was virtually the same at all times, in contrast
to the much higher relative recovery of intestinal lymphocytes in the intestinal
lymph of adult sheep (Cahill et al 1977). Peak concentrations of labelled cells
occurred around 18 hours in prescapular and intestinal lymph, indicating a
modal transit time of around 18 hours; this is about the same modal transit
time as in adult sheep (Cahill et al 1976, 1977) and as observed from blood to
thoracic duct lymph by Pearson et al (1976) in sheep fetuses.

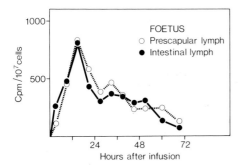

FIG. 1. The recirculation of ^{51}Cr-labelled intestinal lymphocytes in a 135-day-old sheep fetus.
Lymphocytes collected from the intestinal duct were labelled with ^{51}Cr and infused i.v. Their
reappearance in intestinal lymph (●) and in lymph draining a single prescapular node (○) was
followed simultaneously.

When ⁵¹Cr-labelled T lymphocytes (containing 2.5% B lymphocytes) obtained from prescapular lymph of a 140-day-old fetus (Fig. 2a) were infused i.v. and their reappearance was followed simultaneously in prescapular and intestinal lymph, the concentration of labelled cells was, if anything, a little higher in intestinal than prescapular lymph. The modal transit time was again about 18 hours. After birth the lamb gained weight normally and 11 weeks after birth the prescapular and intestinal ducts were recannulated. The output of lymphocytes from the prescapular node was found to have increased from 3.5×10^7 before birth to 8×10^7/h at 11 weeks of age, while the output of lymphocytes from the intestinal duct increased from 3.5×10^7 to 10^9/h, showing that the large increase in thoracic duct output that occurs in the first few months after birth is largely due to an increased output from the intestinal duct. The experiment of infusing ⁵¹Cr-labelled T lymphocytes from prescapular lymph i.v. and following their reappearance simultaneously in intestinal and prescapular lymph was then repeated (Fig. 2b). We used nylon wool columns to deplete prescapular lymphocytes of B lymphocytes (Cahill et al 1978). This time, in contrast to what we had found in the fetus, the recovery of ⁵¹Cr-labelled lymphocytes in prescapular lymph was much higher than in intestinal lymph, the peak concentration of ⁵¹Cr-labelled cells in prescapular lymph being about twice that found in intestinal lymph. The adult pattern had therefore been established in the 11-week-old lamb.

Further differences between the migration of recirculating lymphocytes in the fetus and that in the adult are revealed when the organ distribution of ⁵¹Cr-labelled autologous lymphocytes from fetal or adult sheep is examined

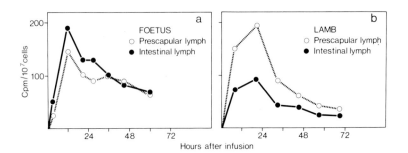

FIG. 2. (a) The recirculation of ⁵¹Cr-labelled prescapular lymphocytes in a 140-day-old sheep fetus. Lymphocytes collected from the prescapular duct were labelled with ⁵¹Cr and infused i.v. Their reappearance in intestinal lymph (●) and in lymph draining a single prescapular lymph node (○) was followed simultaneously. (b) The recirculation of nylon-wool-separated, ⁵¹Cr-labelled T lymphocytes from prescapular lymph in the same animal as shown in (a) 11 weeks after birth. After i.v. infusion the reappearance of the lymphocytes in intestinal lymph (●) and in prescapular lymph (○) was again followed simultaneously.

3 h after their reinfusion i.v. (Table 4). Fetuses ranged in age between about 100 and 150 days postconception. Fetuses of this age are able to produce both IgM and IgG_1 when challenged *in utero* with antigen, although the development of specific immunological competence does not appear simultaneously for all antigens (Fahey & Morris 1974). When ^{51}Cr-labelled, nylon-wool-separated intestinal T lymphocytes from adult sheep (containing less than 3% surface Ig-bearing cells and less than 3% blast cells) were infused i.v. the mean percentage radioactivity recovered in the small intestine was nearly 10 times that found when prescapular T cells were infused. This was not the case with fetal lymphocytes, where the absolute recoveries in the small intestine were the same with both prescapular and intestinal lymphocytes. Labelled cells do, however, localize in the small intestine of the fetus to a significant extent and the recovery of labelled lymphocytes per gram of fetal small intestine was high compared with that found in the adult sheep.

TABLE 4

Tissue distribution of ^{51}Cr-labelled autologous lymphocytes from intestinal or prescapular lymph three hours after their i.v. infusion in fetal lambs (Days 100–150 of gestation) and adult sheep

	Fetus		Adult	
	Intestinal lymphocytes	Prescapular lymphocytes	Intestinal lymphocytes	Prescapular lymphocytes
Spleen	12.2[a]	11.7	20.0	25.7
	250[b]	241	21.5	21.4
Mesenteric lymph node	1.2	1.6	3.4	2.2
	66	125	6.5	4.6
Peripheral lymph nodes	1.2	3.2	2.8	3,6
	78	82	8.0	7.0
Small intestine	3.0	2.5	12.9	1.4
	3.8	4.5	2.2	0.3
Liver	32.6	23.7	4.4	2.8
	34.1	27.6	0.7	0.5
Lung	8.4	9.6	24.0	28.3
	7.3	8.0	6.0	4.9

[a]Mean percentage radioactivity recovered (three animals in each group). [b]Specific activity (mean percentage radioactivity recovered per gram of tissue × 100). Procedures for obtaining nylon-wool-separated T cells from adult sheep and for labelling lymphocytes with ^{51}Cr, together with methods of killing the animals and gamma counting, have been published (Cahill et al 1977, 1978). Some of these results have been published previously (Cahill et al 1979b).

The failure of intravenously infused fetal prescapular and intestinal lymphocytes to migrate differently to the small intestine in the way they do in adult sheep, and the absence of any asymmetry in the recirculation of fetal lymphocytes between prescapular and intestinal lymph, suggest that there is no special circulation in the fetus of a subpopulation of small T lymphocytes from the blood through the small intestine and back into the blood via the intestinal and thoracic ducts. Further experiments are needed to determine whether (a) there is a selective deficiency in the fetus of a subpopulation of T lymphocytes which only appears after birth, perhaps under the influence of antigen or perhaps due to functional differences in the lymphoid tissues of the small intestine at various stages of development, or whether (b) such cells exist in the fetus but the mechanism allowing them to migrate to the small intestine is not functional before birth. A population of T lymphocytes migrating preferentially through the adult small intestine might be involved in the expression of immunoglobulins of the IgA class and it is interesting that IgA has not been detected in fetal lambs undergoing an immune response *in utero* (Fahey 1974).

The mean percentage radioactivity recovered after infusing ^{51}Cr-labelled autologous lymphocytes (Table 4) was much lower in the lungs of fetal lambs than in adult sheep although the recovery of labelled lymphocytes per gram of lung was much the same in the fetus. In adult sheep the lungs appear to sequester dead and damaged cells (Frost et al 1975) and it is possible that the higher absolute recovery of cells from the lungs of adult sheep is due to changes in the pulmonary capillary vascular bed that coincide with the onset of respiration after birth. An interesting difference in the migration of lymphocytes in fetuses and adult sheep was that there were many more labelled cells in the fetal liver (Table 4). This could be explained by the presence of large numbers of damaged cells but this would require fetal lymphocytes to be more readily damaged by *in vitro* manipulation and labelling with ^{51}Cr and for the roles of the lung and liver in removing damaged cells from the circulation to be reversed in the fetus, a possibility for which we could find no evidence either way. Another explanation might be that since the fetal liver is a haemopoietic organ, a large-scale circulation of T lymphocytes exists in fetal liver which is somehow connected with the maturation of the lymphoid apparatus and the immune system.

The localization of labelled cells in the fetal spleen was lower in absolute terms than in the adult, although localization in lymph nodes was much the same in fetuses and adults (Table 4). The recovery of labelled cells per gram of tissue was however around 10 times higher in fetal lymph nodes and spleen, suggesting that the recirculation of lymphocytes is more intense in the fetal lymph nodes and spleen than it is in adult sheep.

RELATIONSHIP BETWEEN FETAL AND ADULT POOLS OF RECIRCULATING LYMPHOCYTES

The experiments reported here demonstrate that lymphocyte recirculation is established during fetal development in the sheep.   The fetal pool of recirculating lymphocytes increases exponentially over the last third of gestation and consists predominantly of T lymphocytes, although it contains considerable numbers of B lymphocytes as well. The cells of the fetal pool are not a rapidly turning over cell population but a high proportion are long-lived cells which possibly continue to recirculate from the time they enter lymph throughout fetal life. The existence of the fetal pool of recirculating lymphocytes indicates that pathways of recirculation and the capacity of cells to recirculate arise independently of antigenic stimulation.   The factors regulating the development of this fetal pool and the function *in utero,* if any, of such cells is obscure.

The hypothesis has been advanced by Sprent (1977) that the cells of the recirculating pool are predominantly memory cells which arise as a result of direct contact with antigen. Sprent has proposed that cells leaving the thymus do not enter the recirculating pool directly but migrate primarily to the spleen where they soon die unless they meet specific antigen.   According to his scheme, virgin B lymphocytes migrate predominantly to the spleen as well, where they also soon die unless they meet specific antigen. If either virgin T or virgin B lymphocytes encountered specific antigen they would proliferate and differentiate into effector cells and memory cells. Memory lymphocytes, either T or B, would then enter the recirculating pool of lymphocytes, where they would recirculate through lymph nodes as long-lived cells carrying specificities for previously encountered antigens. Sprent's hypothesis has the merit that it is simple and sensible. It is sensible because it would obviously be economical to have a small number of mobile memory lymphocytes recirculating through lymph nodes, which on encountering antigen of the appropriate specificity could assist in promoting a more efficient and prompt response to antigen, rather than to have vast numbers of fixed memory cells. But Sprent's hypothesis is clearly wrong when applied to the fetal lamb. The fetal lamb meets no extrinsic antigen, is immunologically competent very early in gestation (Fahey & Morris 1974), is agammaglobulinaemic and yet contains a huge pool of long-lived recirculating immunologically virgin T and B lymphocytes. However, since the fate of the cells of the fetal pool after birth is unknown and since the life history of the cells which constitute the pool of recirculating lymphocytes which is expanding so rapidly after birth is unknown, Sprent's notion that the recirculating lymphocytes of the adult pool are predominantly memory cells is still possible. The relationship between the fetal and adult pools of recirculating lymphocytes, the role played by antigen in the rapid expansion of the pool of recirculating lymphocytes which occurs soon after birth, and the extent to which the changing patterns of lymphocyte recirculation observed soon after birth are influenced by antigen, are matters which require further experiments before they are resolved.

References

Cahill RNP, Frost H, Trnka Z 1976 The effects of antigen on the migration of recirculating lymphocytes through single lymph nodes. J Exp Med 143:870-888

Cahill RNP, Poskitt DC, Frost H, Trnka Z 1977 Two distinct pools of recirculating T lymphocytes: migratory characteristics of nodal and intestinal T lymphocytes. J Exp Med 145:420-428

Cahill RNP, Poskitt DC, Frost H, Julius MH, Trnka Z 1978 Behaviour of sheep-immunoglobulin-bearing and non-immunoglobulin-bearing lymphocytes isolated by nylon wool columns. Int Arch Allergy Appl Immunol 57:90-96

Cahill RNP, Poskitt DC, Heron I, Trnka Z 1979a The collection of lymph from single lymph nodes and the intestines of fetal lambs in utero. Int Arch Allergy Appl Immunol 59:117-120

Cahill RNP, Poskitt DC, Hay JB, Heron I, Trnka Z 1979b The migration of lymphocytes in the fetal lamb. Eur J Immunol 9:251-253

Cole GJ, Morris B 1971 The growth and development of lambs thymectomized in utero. Aust J Exp Biol Med Sci 49:33-53

Fahey KJ 1974 Immunological reactivity of the foetus and the structure of fetal lymphoid tissues. In: Brent L, Holborow J (eds) Progress in immunology II. North-Holland, Amsterdam, vol 3:49-60

Fahey KJ, Morris B 1974 Lymphopoiesis and immune reactivity in the fetal lamb. Ser Haematol 7:548-567

Ford WL 1975 Lymphocyte migration and immune responses. Prog Allergy 19:1-59

Frost H, Cahill RNP, Trnka Z 1975 The migration of recirculating autologous and allogeneic lymphocytes through single lymph nodes. Eur J Immunol 5:839-843

Gowans JL, McGregor DD 1965 The immunological activities of lymphocytes. Prog Allergy 9:1-78

Hall JG, Hopkins J, Orlans E 1977 Studies on the lymphocytes of sheep. III. Destination of lymph-borne immunoblasts in relation to their tissue of origin. Eur J Immunol 7:30-37

Husband AJ, Monié HJ, Gowans JL 1977 The natural history of the cells producing IgA in the gut. In: Immunology of the gut. Excerpta Medica, Amsterdam (Ciba Found Symp 46), p 29-42

Morris B 1972 The cells of lymph and their role in immunological reactions. In: Handbuch der Allgemeinen Pathologie. Springer-Verlag, Berlin p 405-456

Morris B, Courtice FC 1977 Cells and immunoglobulins in lymph. Lymphology 10:62-70

Pearson LD, Simpson-Morgan MW, Morris B 1976 Lymphopoiesis and lymphocyte recirculation in the sheep fetus. J Exp Med 143:167-186

Robinson SH, Brecher G, Lourie IS, Haley JE 1965 Leukocyte labelling in rats during and after continuous infusion of tritiated thymidine: implications for lymphocyte longevity and DNA reutilization. Blood 26:281-295

Scollay RG, Hopkins J, Hall JG 1976 Possible role of surface Ig in non-random recirculation of small lymphocytes. Nature (Lond) 260:528-529

Smeaton TC, Cole GJ, Simpson-Morgan MW, Morris B 1969 Techniques for the long-term collection of lymph from the unanaesthetized foetal lamb *in utero*. Aust J Exp Biol Med Sci 47:565-572

Sprent J 1977 Recirculating lymphocytes. In: Marchalonis JJ (ed) The lymphocyte structure and function. Dekker, New York, p 43-112

Sprent J, Basten A 1973 Circulating T and B lymphocytes of the mouse. II. Lifespan. Cell Immunol 7:40-59

Discussion

Weissman: The development of different patterns from fetus to newborn lamb to adult animal may be either an inherent property of the cells that you are sampling or something to do with the maturation of the organs through which they pass. Have you put fetal cells into adult sheep to see whether you can find the adult pattern? I know it is an allogeneic transfer but with short-term homing it may not be critical if one crosses histocompatibility barriers.

Cahill: I would be worried about using allogeneic lymphocytes because so few of them recirculate. In one fetus we tried freezing cells down and then, after thawing, reinfusing them into the same animal after birth. Unfortunately the cells did not recirculate but we are repeating these experiments.

Davies: You spoke of the fetus as immunologically virgin, with no reaction to antigen, but are you totally satisfied that nothing happens to stimulate the fetal lymphoid system?

Cahill: By antigen of course I mean extrinsic antigen. You aren't worried about all those sheep red blood cells, are you?

Davies: I might be. If the fetus is genetically different from the mother, then the red blood cells going across the placenta might constitute an antigenic stimulation.

Cahill: The nature of the sheep placenta is such that the sheep fetus is not exposed to maternal red blood cells. If the fetus was exposed to foreign antigen one would expect to find Ig in the circulation, but one cannot. The fetus is agammaglobulinaemic in the sense that immunoradiodiffusion techniques with a sensitivity of about 1 μg/ml don't detect immunoglobulin, except that from about day 120 trace amounts can be found in some fetuses. The proportion of fetuses in which trace amounts of immunoglobulin can be found increases towards term to 30–40%.

Davies: The exponential increase in the output of recirculating lymphocytes occurs over quite a long period of time, doesn't it?

Cahill: The experiments I showed were from days 100–150. Bede Morris's original work was from day 75 to day 150.

Davies: This presumably doesn't correlate with something simple like the increase in body size or size of the lymph nodes at this particular time?

Cahill: The nodes get bigger. For example a prescapular lymph node weighs about 250 mg at 100 days' gestation and about 2.5 g at term.

Morris: The size of the recirculating pool of lymphocytes is getting bigger in proportion to the size of the lymphoid tissue and the size of the body too. It is only after birth that we can clearly demonstrate a disproportionate increase in the size of certain lymph nodes, particularly the gut-associated ones, relative to the increase in body size.

Davies: So it is at least possible that a series of other isometric growth curves in the animal relate to lymphoid organ size. Your experiments lasted 17 days, which is about a third of the time that you are looking at the exponential increase in the cells which showed relatively little labelling. Where do all these lymphocytes come from? Are they produced by mitosis or is there some other way of getting them? Or is the time over which you transfused with thymidine too short to be significant in relation to the exponential increase of recirculating cells?

Cahill: It is possibly a question of how long it takes after a labelled cell emerges from the thymus or the bone marrow before it enters the compartments that we looked at. However one should also consider changes in blood volume. The blood volume of the fetus increased from about 300 ml to 600 ml during the 17 days of infusion, so that in absolute terms 8×10^6 lymphocytes in the blood were labelled after one day of infusion compared with 2.3×10^8 labelled cells after 17 days, which is about a 28-fold increase.

Morris: When fetuses are drained of their thoracic duct lymph for a few days the recirculating pool of lymphocytes is depleted but as you continue draining the lymph the concentration of cells in the residual pool begins to increase. The mother during this time loses a lot of weight, while the fetus, which may have lost 10–15 litres of lymph and 200 g or more of plasma protein, is born thoroughly well nourished.

Davies: But the mother doesn't give cells to her fetus.

Morris: No, but she gives it the wherewithal to keep making cells. With continuing drainage the cell output from the thoracic duct can be shown to increase as the lymphoid apparatus grows. In a free-living animal, of course, if you drain the thoracic duct for more than a few days the animal perishes.

Gowans: So it is not an occasion for surprise that the number of cells in the recirculating pool and the size of the lymph nodes is increasing *pari passu* with the growth of the animal. This is simply part of the physiology of development and requires no driving by antigen.

Morris: One would want to be quite clear that these fetuses are quite competent immunologically. When you challenge them with an antigen they produce perfectly good immune responses.

Gowans: Do sheep fetuses swallow *in utero*?

Morris: Yes. There are no germinal centres in the lymphoid tissue of normal fetal sheep, except when you put antigenic material into the amniotic fluid or the gut or challenge them by parenteral injection of antigen.

Gowans: Is the amniotic fluid absorbed in the gut as part of the fetal nutrition?

Morris: It is swallowed and some of it is absorbed. I don't know how much it contributes to the nutrition of the fetus.

McConnell: On the question of the source of antigen which might drive the repertoire of the immune system, Jerne put forward the view that all antigen is already 'internal'. This internal antigen is either the animal's own MHC present, say, on the thymic epithelial cells, or alternatively it might be the idiotype of developing B cells. If Jerne is right, no external antigen is needed. The proliferation caused by internal antigen occurs at a very early stage, perhaps even before the 100-day stage.

Cahill: What you say is certainly true but at the moment I wouldn't like to speculate on any relationship between cells recirculating in fetal lymph and theories for the generation of diversity.

Morris: If there is any endogenous antigen affecting the lymphoid system of the fetus, it doesn't produce the structural changes that exogenous antigen produces, namely germinal centres, blast cells and Ig-positive cells. If you challenge these fetuses deliberately you get a significant expansion in the number of Ig-positive cells and the germinal centres appear in the lymph nodes.

McConnell: If internal antigen is to develop the repertoire from a few germ-line specificities then clearly the process takes place within the primary lymphoid organs, thymus and bone marrow. It is distinct from the conventional response seen in secondary lymphoid tissue.

Born: Presumably the positive control experiment has been done of introducing antigen into the fetus?

Morris: We have monitored antibody-forming cells in the fetus and looked at the Ig classes and so on after intentional antigenic challenges. Fetuses give good immune responses to a wide range of antigens after 60–65 days' gestation.

Owen: You mentioned that the sheep immune system was immature at the time of your experiments but it has a long gestation period. Max Cooper reported that B lymphocytes in the thymus were present before 60 days' gestation (Owen et al 1976). So the first formation of cells which are recognizable as B cells must be earlier than your experiments. Whether the generation of the antibody repertoire is antigen-dependent or antigen-independent is another matter but from studies in other species on the primary differentiation of lymphocytes, after a proliferative phase and the appearance of the cells with specificity for antigen the cells go out of cell cycle. The 'mature' lymphocytes, for example, in thymus and in bone marrow, are not in cell cycle. Hence, the fact that you see long-lived lymphocytes in the fetus doesn't mean that they are equivalent to long-lived lymphocytes in adult animals. They may be non-cycling 'virgin' cells which have not been triggered by antigens.

Cahill: Yes, exactly. Further experiments are needed to determine the life history of the cells of the fetal pool after birth.

Owen: I think that very early stages of gestation must be examined to see the inception of the immune system. What you are looking at in your fetal lamb work is a late developmental stage. The controversy about whether immune responses occur before the thymus is lymphoid in the fetal lamb and the incomplete effect of thymectomy hinges on looking carefully to see when the thymus becomes lymphoid. Jordan's studies of the development of the thymus in the fetal lamb show that the thymus is lymphoid by 36 days, i.e. at least 30 days before thymectomies have been performed, which may well explain why thymectomy doesn't have a complete effect (Jordan 1976).

Morris: The thymus in the fetal lamb was no more lymphoid when we did the thymectomy than is the thymus in the newborn mouse. The point at issue is that the thymectomy is done before a particular range of immune reactivities can be demonstrated in the fetus. Subsequently all those immune responses appear in thymectomized animals.

Owen: I accept that, but the responses don't appear before the thymus is lymphoid.

Morris: Of course not.

Owen: One could argue that immune responses appear in thymectomized fetal lambs because thymus-derived lymphocytes having migrated from the thymus before thymectomy undergo further maturation in peripheral lymphoid organs.

Williams: In the rat the memory cells for the IgG anti-DNP response constitute a tiny fraction—1 or 2%—of the total recirculating population (Mason 1976). These are the B cells with surface IgG. The total B cell pool constitutes 45% of thoracic duct lymphocytes and it is not known what most of these cells do. It seems that all the memory responses of the IgG type are mediated by 1–2% of the thoracic duct lymphocytes.

Cahill: I agree. One thinks of recirculating lymphocytes as memory cells which arose after stimulation by antigen. But they exist in the immunologically virgin fetus and, as you say, B cell memory has only been shown to be the property of a small minority of recirculating B cells. It is puzzling.

Williams: If you estimate that a rat has 2×10^8 circulating B cells, 1% is 2×10^6 cells. You can't think in percentage terms. The big puzzle in the rat is, what are the IgM/IgD cells doing? They are presumably primary cells but that has been difficult to demonstrate.

Cahill: Perhaps, although my feeling is that a very large number of cells are recirculating, the function of which has yet to be determined.

Weissman: Several years ago Graham Mitchell and I infused either tritiated thymidine or tritiated adenosine directly into the thymus of a 4 or 5-day-old rat and cannulated the thoracic duct 1, 2, or 3 days later. Surprisingly, the labelled cells constituted up to 4 or 5% of the lymph node cells but less than 1 in 1000 cells in the thoracic duct lymph were labelled. If recent thymic immigrants are as close to a virgin T cell as might exist, and if you don't feel troubled that they have a slow recirculation pattern, that seems to say that those cells, at the time when the lymphoid tissues are expanding massively, do not contribute significantly to the pool of cells that you find in thoracic duct and therefore to the recirculating pool. There are some flaws in that experiment in that thymus cell migrants could get stuck in tissue just for a day or two days longer and you could postulate that you had lost the marker by dilution or nucleic acid turnover. But those cells are not contributing rapidly to the rat recirculating pool.

Ford: How many of these cells were in the spleen?

Weissman: There were lots of labelled cells in the periarteriolar white pulp. We couldn't do the percentages because of the red pulp cells. We did the same experiment with adult rats in 1964 (unpublished) and got the same result.

Owen: Again the point is that if you infuse after birth the thymus is already lymphoid or has been lymphoid for some days of fetal life, which is quite a long time in the rodent. So there is plenty of opportunity to generate a thoracic pool of cells from lymphocytes which came from the thymus some days before the labelling was introduced.

Howard: If one is prepared to allocate the whole of primary responsiveness to a non-recirculating lymphocyte population, then one has to think about a whole new mechanism for immunization. The conventional attitude to the function of recirculation is that it provides the mechanism for immunization *in vivo*. Everything becomes much more problematic if you say there is a non-recirculating primary population which is equipped to deal with the problems of primary immunization.

Weiss: How does the sheep lymphatic system differ from that of the mouse? Is it simply a matter of size?

Morris: I have an idea that the lymphatic system of the sheep is quantitatively different from that of the mouse. But there are many other differences—such as the intrinsic contractility of the lymphatic vessels. In sheep the vessels show rhythmic contractions, which is one reason why the flow is good, and more easily sustained than in smaller animals. In the fetus lymph can be collected for a month or more without problems. Cattle on the other hand have very big lymphatics but the flow isn't as good as in sheep. You have to use a different range of techniques to get the same sort of results over the same periods of time. However, sheep aren't a unique species whose lymphoid apparatus is unrelated to that of any other animal.

Davies: In that case the mouse is a small sheep as far as you are concerned.

Weiss: Are the haemal nodes significant?

Morris: There is a difference between haemal and haemolymph nodes. The haemolymph nodes are not attached to the lymphatic system in the sheep whereas haemal nodes are. In most species there is a traffic of red cells through the kidneys and as a consequence the renal node in the rat, mouse and sheep contains red cells and is a haemal node. But this is a very different proposition to the haemolymph node which is in fact a small spleen attached to the lymphatic system. You can't drain lymph from a haemal lymph node as far as I know.

Gowans: Is there any evidence for differences in principle in the generation of lymphocytes, the anatomy of the lymphatic system and the generation of immune responses at the cellular level between rats, mice and sheep? Would you like to list any differences for us?

Cahill: Small animals such as mice and rats have relatively more lymphoid tissue with respect to total body weight than the sheep. There are also differences in placentation and in the stages at which they become immunologically mature but I don't know of any fundamental differences.

Howard: Bede Morris mentioned the question of the direct return of lymphocytes to the blood. The pig probably does all its lymphocyte recirculation that way (Binns 1979), although that has never been formally demonstrated. There is no evidence that this happens in rats or mice but whether that is significant is difficult to say.

Owen: From what you have said, the great inflow of lymphocytes in the lymph to the node might mean that there is a shut-off in the entry of lymphocytes from the blood to the lymph.

Morris: That is a real possibility. The theory behind the recirculating loop experiments is that you can collect all the lymphocytes that are in the recirculating pool by sequestering them in the loop. Provided that the cells can't get back into the bloodstream directly from the lymph, they must keep going round and round. In fact, after a day or so the number of cells in the loop doesn't change much. That seems to me to suggest that the cells coming from the blood into that system are not confined there. There may be an autoregulatory phenomenon in which the number of cells entering the lymph from the blood in a lymph node is related to the number entering the node in lymph. When you infuse labelled cells you are lucky to get 30% of them back out of the node over a period of 48 hours. What happens to the rest I don't know but I don't believe all of them stay in the node forever.

Gowans: If you put labelled cells into the bloodstream do they enter the closed loop?

Morris: They get into the loop but the kinetics of arrival is almost exactly the same as the kinetics of arrival in other situations in the body. That is difficult to interpret because clearly other cells are coming into the loop as well. The labelled cells start to disappear from the loop but they could be being held in the node or they could actually be leaving the loop and migrating to other parts of the body. The experiments that demonstrate differing patterns of recirculation between cells derived from peripheral nodes and from the intestine also demonstrate that a proportion of cells from these two different sites recirculate through specific sites rather more readily than through other sites. It is in no way a specificity phenomenon. There is quite a substantial overlap between the cell populations. A lot of the cells from the popliteal node don't recirculate back through the popliteal node; some go back through the intestine. We need to establish whether they maintain this differential recirculation pattern in a steady state. In other words is it an immutable property of the cell that remains throughout its life?

Hall: From short-term experiments we have no grounds for saying that this is an immutable property of the cell. At best it is a biased recirculation, an asymmetry, but it is not an all or nothing effect. Gut-derived immunoblasts show a dramatic predilection for the gut and one naturally asks whether their small lymphocyte precursors show the same predilection. I didn't think they would and I was surprised that we could show the degree of bias that we did (Scollay et al 1976).

Gowans: What subpopulations of lymphocytes are involved in the bias?

Hall: We used cells from whole lymph which were a mixture of T and B cells. We didn't analyse them during the course of the experiments but Ross Cahill had virtually identical results with cells depleted of B lymphocytes. I am led to assume that both B and T lymphocytes would show this bias; but it is only a bias.

Cahill: The bias varies. The absolute recoveries of ^{51}Cr-labelled cells from one prescapular lymph node are often not much less than those from the whole thoracic duct. Since the thoracic duct puts out about 2×10^9 cells/h and the prescapular lymph node is putting out an order of magnitude less, that is a very large bias.

Morris: The bias could be explained in another way. From your experiments you show that some 90% of the cells that you recover from the intestinal lymph don't go back through the prescapular node. This is a bias in another sense.

McConnell: Most of the cells in the intestinal lymph are specific for gut-type antigens. Cells in the peripheral nodes perhaps respond to quite different environmental antigens. If you immunize a popliteal node with an antigenic cocktail representative of what might occur in the gut, do the gut-seeking cells now recirculate to the peripheral node? If so this would indicate that the differential pattern of migration was antigen-directed.

Hall: I have only done one such experiment. I immunized one popliteal node with a selection of killed *E.coli* organisms and then compared the recirculation of the efferent lymphocytes with those from a node immunized with oxazolone. I found absolutely no difference but I don't think that answers the question. If Ross Cahill is right and this differential, biased recirculation doesn't occur with either prescapular or intestinal cells before birth, whereas it does occur after birth, when we know that there is an explosive expansion of the gut-associated lymphoid tissue, then it seems reasonable to postulate that the bias may have something to do with the postnatal antigenic experience of the recirculating, gut-associated lymphocytes.

Ford: In rats and mice there is so far no evidence that any subset of small lymphocytes migrates preferentially into particular lymph nodes, although de Freitas et al (1977) looked carefully for any such discrimination. In rats the localization of thoracic duct lymphocytes in different groups of lymph nodes (per unit weight) is extraordinarily uniform within an individual recipient. The small coeliac lymph nodes are exceptional in this and other respects (Smith et al 1979). It might be argued that this uniformity is a coincidence arising from the conglomerate nature of the thoracic duct population. However in secondary transfer experiments involving the injection of ^{51}Cr-labelled thoracic duct lymphocytes into primary recipients and their re-transfer one hour later from spleen, lymph nodes or blood there were no major differences in distribution within the secondary recipient related to the recent environment of the cells. These observations contrast with the migratory behaviour of activated lymphocytes in small rodents which show remarkable preferences for different lymph nodes (Griscelli et al 1969). Apparently the sheep acquire after birth a subset of small lymphocytes which migrate according to similar rules as do activated lymphocytes in both sheep and rodents.

Hall: There is no reason to assume that results from rats, mice and sheep should be identical. The secretory immune system in sheep is not confined, as it is in the rodent, to IgA. IgGs are involved to a much greater extent, and this of course might be reflected in different patterns of lymphocyte migration.

Vane: I am surprised to hear the discussion about whether a sheep is a giant mouse, or vice versa. Which of these species better represents an unintelligent human being?

Morris: Sheep are genetically constituted to survive in nature, which is more than most of the fancy strains of mice used for experiments in immunology are able to do.

Cahill: Sheep tend to have large single lymph nodes with only one efferent duct which makes them technically accessible. Human beings have a meshwork of lymphatic vessels with considerably higher numbers of lymph nodes.

McConnell: Human lymph nodes occur in chains and often have multiple efferents. In patients about to receive renal allografts Tilney et al (1970) performed thoracic duct drainage as a method of immunosuppression. The recirculating lymphocyte pool was substantially reduced after about 16 days of thoracic duct drainage. In this respect therefore humans are more like sheep than pigs, where very little of the recirculating lymphocyte pool is present in the thoracic duct.

Howard: There are very few cells in the pig efferent lymphatics.

Born: I didn't know that the pig was as different as Jonathan Howard says. It seems very relevant.

Hall: The anatomy of the pig lymph nodes is rather different from that of most mammals. They have the 'cortex' in the middle where the medulla should be. The circulatory function of the lymphatic system seems to operate in a perfectly normal manner in the pig but when the lymph flows through the lymph nodes it doesn't pick up a significant number of lymphocytes (Binns & Hall 1966). One has to conclude that if lymphocyte recirculation is necessary for immune responsiveness in the pig, the cells must come into the node from the blood, do what they do, and then go out again by the same route. Alternatively, lymphocyte recirculation doesn't occur at all in the pig.

Howard: Richard Binns (1979) has looked at the fate of labelled pig peripheral blood lymphocytes that were infused through the afferent lymphatics. They did not reappear in the efferent lymph but at least some activity was recovered in the venous effluent. There is obviously a great deal more to be done in this system but the general impression that recirculation must be blood–node–blood seems to be right.

Hall: The pig has something like 12 000 lymphocytes/mm³ blood, which is one of the highest levels in mammals.

Ford: The spleen is the predominant site of lymphocyte recirculation in the pig by an even greater margin than in the rat. In splenectomized pigs the rate of migration of lymphocytes from the blood is reduced to about 25% of normal values and in these recipients many cells probably migrate into the bone marrow (Pabst & Trepel 1976).

Morris: The afferent lymphatics of the pig lymph nodes come out of the hilum and the efferents go into the pericapsular area. The only other animal where that happens, as far as I know, is the elephant!

References

Binns RM 1979 Pig lymphocytes - behaviour, distribution and classification. In: Trnka Z, Cahill RNP (eds.) Essays on the anatomy and physiology of lymphoid tissues. Karger, Basel (Monogr Allergy) vol 16:19-37

Binns RM, Hall JG 1966 The paucity of lymphocytes in the lymph of unanaesthetised pigs. Br J Exp Pathol 47:275-280

de Freitas AA, Rose ML, Parrott DMV 1977 Mesenteric and peripheral lymph nodes in the mouse: a common pool of small recirculating T lymphocytes. Nature (Lond) 270:731

Griscelli C, Vassalli P, McCluskey RT 1969 The distribution of large dividing lymph node cells in syngeneic recipient after intravenous injection. J Exp Med 130:1427-1451

Jordan RK 1976 Development of sheep thymus in relation to in utero thymectomy experiments. Eur J Immunol 6:693-698

Mason DW 1976 The class of surface immunoglobulin of cells carrying IgG memory in rat thoracic duct lymph: the size of the subpopulation mediating IgG memory. J Exp Med 143:1122-1130

Owen JJT, Raff MC, Cooper MD 1976 Studies on the generation of B lymphocytes in the mouse embryo. Eur J Immunol 5:468-473

Pabst R, Trepel F 1976 The predominant role of the spleen in lymphocyte recirculation. II. Pre- and post-splenectomy retransfusion studies in young pigs. Cell Tissue Kinet 9:191

Scollay RG, Hopkins J, Hall JG 1976 Possible role of surface Ig in the non-random recirculation of small lymphocytes. Nature (Lond) 260:528-529

Smith ME, Martin AF, Ford WL 1979 The migration of lymphoblasts in the rat. The preferential localization of DNA-synthesizing lymphocytes in particular lymph nodes and other sites. In: Trnka Z, Cahill RNP (eds.) Essays on the anatomy and physiology of lymphoid tissues. Karger, Basel (Monogr Allergy), vol 16:203-232

Tilney NL, Atkinson JC, Murray JE 1970 The immunosuppressive effect of thoracic duct drainage in human kidney transplantation. Ann Intern Med 72:59-64

Lymphocyte traffic through lymph nodes during cell shutdown

IAN McCONNELL, JOHN HOPKINS and PETER LACHMANN

MRC Group on Mechanisms in Tumour Immunity, The Medical School, Hills Road, Cambridge CB2 2QH

Abstract Antigenic challenge of lymph nodes in sheep has marked effects on lymphocyte traffic through lymph nodes. The non-specific effects include a marked reduction in lymphocyte output in efferent lymph without a corresponding decrease in lymph flow—a phenomenon known as cell shutdown. With certain antigens there is a total disappearance of B lymphocytes during cell shutdown. The phenomenon can be reproduced in unprimed lymph nodes whenever localized complement activation occurs within the node. This also induces the release of prostaglandins, particularly PGE_2. These results suggest that cell shutdown might be a two-step process involving both complement and prostaglandins.

Repeated stimulation of nodes with antigen also has considerable effects on the traffic of antigen-specific lymphocytes. Antigen localized within the node can promote the selective entry into the node of T lymphocytes specific for the challenge antigen. Consequently there is a net loss from the whole animal of T cells reactive to the challenge antigen. These results are discussed in relation to lymphocyte recirculation through antigen-stimulated lymph nodes.

Many physiological and immunological changes occur in antigen-stimulated lymph nodes. Physiological changes are reflected in alterations to the nodal microvasculature which mediates gross changes in the overall flux of lymphocytes through the node. Immunological responses, however, are dictated by the strategic localization of antigen within the node. The initial function of antigen is to select cells of the appropriate specificity from the recirculating lymphocyte pool and subsequently to induce their differentiation into memory and effector cells. This paper is concerned with possible initiating mechanisms for physiological and immunological events within lymph nodes.

Blood cells and vessel walls: functional interactions
(Ciba Foundation symposium 71) p 167-195

KINETICS OF LYMPHOCYTE TRAFFIC THROUGH ANTIGEN-STIMULATED LYMPH NODES

The kinetic changes in lymphocyte traffic through lymph nodes are best observed in sheep with the lymphatic cannulation model of Morris and collaborators (Lascelles & Morris 1961, Hall & Morris 1962, 1963). The characteristic sequence of events after challenge of nodes with particulate antigens (Hall & Morris 1965), viruses (Smith & Morris 1970), soluble antigens (Hay et al 1973) and homografts (Pedersen & Morris 1970) is well documented. The changes in lymphocyte output in efferent lymph begin with a reduced cellular output from the node (cell shutdown) followed by a characteristically biphasic lymphocyte output over the next 100 h. The first wave of cells to appear consists entirely of small lymphocytes, whilst blast cells and early antibody secreting cells appear in the second peak (Hall & Morris 1965, Cahill et al 1974).

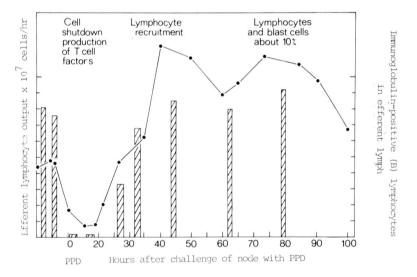

FIG. 1. Kinetics of total and B lymphocyte output from cannulated popliteal node in efferent lymph after local challenge of the node with 100 μg PPD in primed sheep. Lymphocyte output per hour. Vertical bars are % Ig-positive cells detected by direct immunofluorescence using rabbit anti-sheep F(ab')₂.

FIG. 2. Kinetics of total and B lymphocyte output from cannulated popliteal lymph node to PPD and T4 phage in primed and unprimed sheep. (A) 100 μg PPD to drainage area of node; (B) 100 μg PPD infused via afferent lymphatic to node; (C) 100 μg T4 phage infused via afferent lymphatic in T4 phage primed sheep; (D) as (C) but in unprimed sheep.

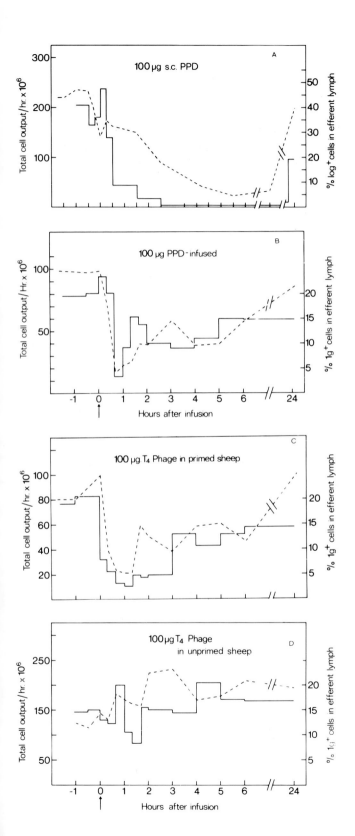

Fig. 2.

In our studies we have been investigating the response to tuberculin-purified protein derivative (PPD) (Central Veterinary Laboratories, Weybridge, England), and T4 coli phage in primed and unprimed animals. PPD challenge of cannulated nodes in Bacille Calmette Guérin (BCG)-primed sheep produces cell shutdown, lymphocyte recruitment and blast cell formation (Figs. 1 and 2) in efferent lymph. In unprimed animals PPD has little effect, largely because PPD is a 'T cell hapten' and, unlike other antigens, does not induce an immune response. Subcutaneous injection of PPD into the drainage area of the node produces cell shutdown within 6−20 h. Shutdown occurs within 1 h if antigen is directly infused into the node via an afferent lymphatic. Lymphocyte output in efferent lymph drops from a resting level of $40-150 \times 10^6$ lymphocytes/h to about $4-8 \times 10^6$/h without any reduction in lymph flow. Together with this marked drop there is almost total disappearance of immunoglobulin-bearing B lymphocytes. During cell shutdown there is increased synthesis of the soluble mediators of delayed hypersensitivity, notably antigen-specific and non-specific migration inhibition factor (MIF). In the response to PPD these factors are easily detected in shutdown lymph (Hay et al 1973, Scott et al 1978). More recently we have also detected potent suppressor factor(s) in shutdown lymph during the response to PPD which suppress the *in vitro* antigen-induced transformation of lymphocytes to PPD (J. Hopkins & I. McConnell, unpublished work). Cell shutdown and B lymphocytes similarly disappear after infusion of 100 μg T4 phage via an afferent lymphatic (afferent infusion) to a cannulated node in sheep primed to T4. With unprimed sheep there is no shutdown. Thus T4 phage is quite unlike influenza virus which will produce shutdown and acellular lymph in unprimed animals (Smith & Morris 1970, Cahill et al 1974).

The disappearance of immunoglobulin-bearing cells from efferent lymph is presumably due to retention of these cells within the node and not to the release in shutdown lymph of proteases which might degrade surface immunoglobulin on B lymphocytes. Incubation of normal efferent lymphocytes from a resting node in shutdown lymph has no effect on the percentage of surface immunoglobulin-bearing cells (Table 1).

Contrary to expectations there is a net inflow of cells into lymph nodes during cell shutdown (Hall 1974). Cahill et al (1976) have shown very clearly that if radiolabelled autologous lymphocytes were returned intravenously the labelled cells increased by three- to fourfold in cannulated lymph nodes undergoing cell shutdown compared to the number seen in non-stimulated nodes of the same sheep. These studies also convincingly demonstrated that the first wave of cells appearing after shutdown were derived from cells which had entered the node during shutdown. The increased input during shutdown

TABLE 1

Effects of shutdown lymph on surface immunoglobulin on B lymphocytes

Lymphocytes[a] incubated in	% Immunoglobulin-positive cells
Medium alone (RPMI)	15
Normal sheep serum	20
Normal efferent lymph	15
PPD shutdown lymph (6−24 h)	16
CVF[b] treated normal sheep serum	17.5
CVF-treated normal lymph	15

[a] 2.10^6 efferent lymphocytes from unstimulated node incubated in 50 μl of serum or lymph at 37°C for 1 h. Cells then washed 3 times in medium and treated with fluoresceinated rabbit anti-sheep F(ab′)$_2$.
[b] Lymphocytes incubated in 50 μl of serum or lymph previously treated at 37 °C for 15 min with 10 μg CVF (see text).

is mainly due to the considerable increase in blood flow known to occur in antigen-stimulated lymph nodes (Herman et al 1972). By injecting ^{85}Sr-labelled microspheres into the arterial circulation and measuring their localization in capillary networks Hay & Hobbs (1977) have shown that there is a fourfold increase in blood flow to the node during cell shutdown, and they calculate that 60% of the recirculating lymphocyte pool will pass through the node in the blood in a five-day period. Immunologically, cell shutdown may represent an important mechanism whereby a large number of cells become available for selection by antigen. It also affords an opportunity for the interaction between rare antigen-specific cells that must occur in T−B cell collaboration.

EFFECTS OF LOCALIZED COMPLEMENT ACTIVATION WITHIN LYMPH NODES ON LYMPHOCYTE TRAFFIC THROUGH THE NODE

Since cell shutdown was more frequently observed in primed rather than unprimed animals we next investigated the possibility that localized complement activation within nodes might be one of the initiating events in cell shutdown.

Cell shutdown after afferent infusion of antigen−antibody complexes

In these experiments immune complexes made with sheep IgG anti-T4 phage and antigen as well as antibody or phage alone were infused via an

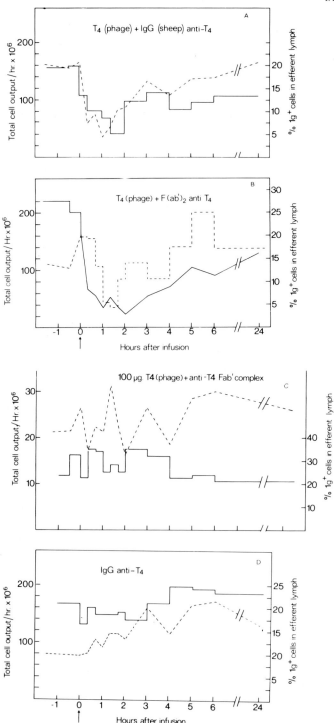

FIG. 3.

afferent lymphatic to a cannulated popliteal lymph node. Fig. 3 shows that infusion of either IgG anti-phage or T4 phage alone does not produce shutdown. Phage neutralizing activity as well as phage particles were recovered in the efferent lymph, showing that both had passed through the node. Immune complexes made with T4 phage and whole IgG anti-phage or the F(ab')$_2$ fragment produced marked shutdown. Complexes made with Fab' anti-T4 gave no shutdown. Similarly, soluble immune complexes of sheep IgG anti-human C3 and human C3 prepared in antibody excess were also found to induce cell shutdown and B cell disappearance (Fig. 4d).

These results suggest that the presence of immune complexes and presumably complement activation within nodes can initiate cell shutdown and B cell disappearance. It is not surprising that immune complexes made with F(ab')$_2$ antibody fragments also produce shutdown, as sheep F(ab')$_2$ is known to activate the alternative pathway of complement (Hobart 1970).

Effect of afferent infusion of complement activators on lymphocyte traffic through lymph nodes

Cobra venom contains a potent activator of the alternative complement pathway known as cobra venom factor (CVF), which is now known to be cobra C3b (Alper & Balavitch 1976). Its unique ability to activate the alternative pathway is due to the fact that it forms an alternative pathway convertase with serum Factor B $\overline{\text{(CVF-Bb)}}$ which is resistant to the mammalian C3b inactivators β1H and KAF, thus producing massive breakdown of C3 (see review by Lachmann 1979). Inulin is also a potent activator of the alternative pathway and acts by a different mechanism, involving protection of bound C3b (Fearon & Austen 1977). Both substances can be used to produce alternative pathway activation *in vivo* and *in vitro*.

When either CVF or inulin was infused via an afferent lymphatic to a cannulated popliteal node, cell shutdown occurred within 15 min and lasted for several hours (Fig. 4). Heat-inactivated CVF was without effect. This is of interest since purified CVF often contains phospholipase, which is heat-resistant. Thus it is unlikely that the shutdown action of CVF is related to its phospholipase content. CVF convertase activity $\overline{\text{(CVF-Bb)}}$ as assayed by a haemolytic technique (Lachmann & Hobart 1978) can be detected in the efferent lymph 10 min after afferent infusion of CVF (Fig. 5). Shutdown

---

←

FIG. 3. Kinetics of total and B lymphocyte output from cannulated popliteal lymph nodes challenged with immune complexes infused via an afferent lymphatic to cannulated node. Complexes preformed at 4 °C by preincubation of 100 μg of T4 phage with 100 μg antibody (as IgG, F(ab')$_2$ or monovalent Fab') for 12 h. 50% phage neutralization titre of IgG anti-phage 1 in 10^6.

174

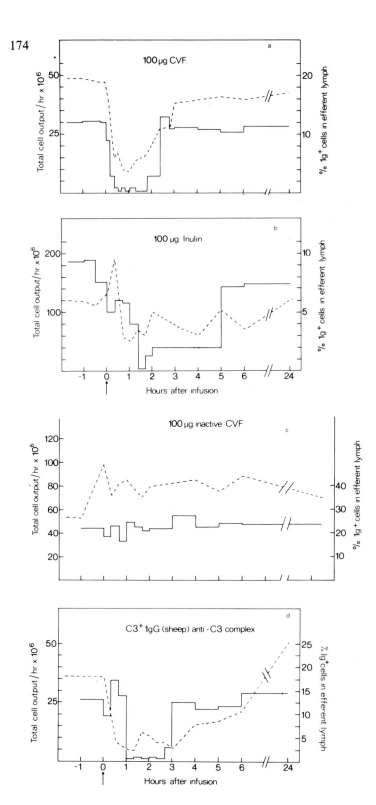

Fig. 4.

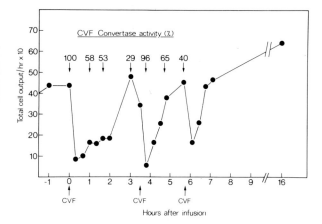

FIG. 5. <u>Effect</u> of repeated afferent injections of 100 μg CVF on lymphocyte output from the node. CVF-Bb convertase by haemolytic assay.

produced by CVF does not show marked tachyphylaxis since repeated injections of the CVF elicit second and third shutdowns (Fig. 5). After each injection increased CVF convertase activity was detected in efferent lymph, showing that within the node sufficient Factor B is derived from the plasma passing through the node and peripheral lymph to form more convertase with CVF.

Many of the breakdown products of complement activation are anaphylatoxic and induce the release of histamine and other vasoactive mediators from mast cells. It seems unlikely, however, that shutdown is related to this since afferent infusion of histamine or bradykinin does not produce shutdown (Fig. 6a).

Complement activity in efferent lymph

The efferent lymph contains all the circulating plasma proteins at a lower concentration than is seen in plasma (Smith et al 1970). Most of the plasma complement components are present in efferent lymph and can be detected functionally and antigenically. Table 2 summarizes some of the findings on C3 and total alternative pathway activity in efferent lymph during shutdown to various complement activators. In efferent lymph draining a non-stimulated (resting) node there is a pre-existing level of 12−23% conversion of

---

FIG. 4. Kinetics of total and B lymphocyte output from cannulated popliteal lymph nodes challenged with cobra venom factor (CVF), inulin or immune complexes infused via afferent lymphatic. Inactive CVF heated at 72 °C for 1 h.

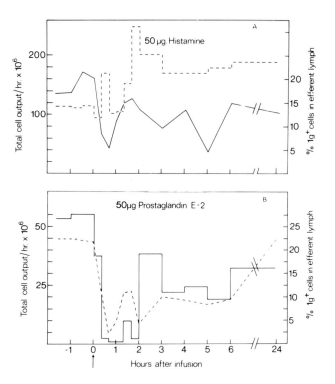

F<small>IG</small>. 6. Kinetics of total and B lymphocyte output in efferent lymph from cannulated popliteal node after afferent infusion of 50 μg histamine and 50 μg prostaglandin E₂ (PGE₂).

C3 as measured by two-dimensional crossed immunoelectrophoresis. After infusion of CVF via an afferent lymphatic the level of converted C3 rises to 80% and then slowly declines. After 24 h about 40% of the C3 remains converted. With inulin, immune complexes and PPD there is no increase above the pre-existing level.

The presence of converted C3 in efferent lymph may be a consequence of the cannulation due either to C3-cleaving tissue proteases at the site of the indwelling cannula or to C3 conversion within the cannula during lymph collection. Alternatively the presence of C3b in efferent lymph may indicate that a certain amount of C3 is always converted naturally in lymph nodes and that this may be relevant to lymph node function *in vivo*.

PROSTAGLANDINS AND PHYSIOLOGICAL CHANGES IN ANTIGEN-STIMULATED LYMPH NODES

Prostaglandins are intimately involved in a wide variety of tissue changes,

TABLE 2

C3 conversion and alternative pathway (AP) activity in 'shutdown lymph'

Time after challenge (min)		% C3 conversion			Total AP activity (% of normal lymph)			
	CVF	Inulin	Immune complex	PPD	CVF	Inulin	Immune complex	T4
Pre	15	12	18	23	100	100	100	100
0 – 20	87	28	20	17	0	66	62	60
20 – 40	87	23	22	18	0	112	88	97
40 – 60	98	24	22	19	0	122	112	109
120 – 140	77	24	28	17	0	101	100	116

particularly those associated with inflammatory responses. In collaboration with Drs J. Pearson and J. Gordon at the ARC Institute of Animal Physiology, Babraham, we have begun to investigate the role of prostaglandins in the physiological changes associated with antigen stimulation of lymph nodes. Using a sensitive radioimmunoassay, Dr J. Pearson has assayed shutdown lymph for PGE_2, 6-oxo-$PGF_{1\alpha}$ (the stable derivative of prostacyclin) and thromboxane B_2. So far PGE_2 is the only prostaglandin to show a marked increase in level in efferent lymph which

TABLE 3

Prostaglandin (PGE_2) levels in efferent lymph during shutdown (data provided by Dr J. Pearson)

Agent inducing shutdown	Shutdown	Peak[b] PGE_2 levels during shutdown (pg/100 μl lymph)
PPD, 100 μg	+	1800
T4 phage, 100 μg (primed sheep)	+	1160
T4 phage, 100 μg (unprimed sheep)	–	< 20
Histamine 50 μg	–	400
CVF[c], 100 μg	+	2900
Inactive CVF, 100 μg	–	125

[a]All infused via afferent lymphatic to cannulated popliteal lymph node.
[b]Measured at time of maximum shutdown.
[c]Cobra venom factor. Inactive CVF heated to 72 °C for 1 h.

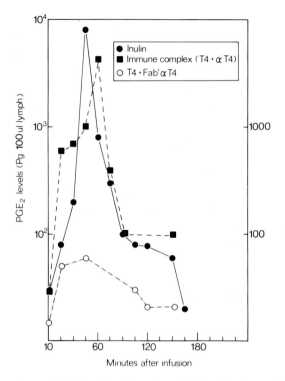

FIG. 7. PGE_2 levels in efferent lymph during shutdown. PGE_2 levels assayed by radioimmuno-assay (data provided by Dr J. Pearson).

correlates with the events of cell shutdown induced either by antigen, CVF, inulin or immune complexes (Table 3, Fig. 7). No shutdown can be induced with T4 phage plus Fab'-anti-T4; similarly there is little increase in PGE_2 levels in efferent lymph. 6-Oxo-$PGF_{1\alpha}$ is low throughout shutdown whereas thromboxane B_2 (derived from thromboxane A_2) shows an initial rise in every lymph tested, including those where no shutdown occurs. The involvement of PGE_2 in shutdown is shown by the observation that shutdown occurs and B cells disappear when 50 μg PGE_2 is infused via an afferent lymphatic to a cannulated popliteal lymph node (Fig. 6b). Recently Johnston et al (1978) have also established a role for prostaglandins in cell shutdown.

At present we have little idea of the precise effect of the complement and the prostaglandin systems in physiological events within lymph nodes. These experiments raise the possibility that one or both systems are involved in the physiological changes in antigen-stimulated lymph nodes, although we have no precise idea yet of how they are involved. PGE_2 is believed to be released by polymorphonuclear leucocytes, monocytes and possibly the endothelial cells of blood vessel walls. Lymphocytes are not thought to be capable of PGE_2 release. Polymorphonuclear leucocytes and monocytes are not seen in resting lymph but can appear transiently during the early stages of the first wave of cells released by the node after shutdown. Thus, initiation of cell shutdown might be a two-step phenomenon including both complement activation, which induces permeability changes in the microvasculature within the node, and subsequent entry of polymorponuclear leucocytes and monocytes, which release PGE_2 and thereby induce shutdown. Alternatively there are cell types within the node which under certain circumstances release quite substantial quantities of prostaglandins, thereby inducing shutdown.

Although these experiments suggest possible initiating events in cell shutdown the actual mechanism which promotes the temporary stasis in lymphocyte flow through an antigen-stimulated lymph is unknown. This stasis might be due to an induced change in the surface properties of the lymphocytes within the node which inhibits their traffic through the node; alternatively, some of the fixed cells may form a physical barrier to lymphocyte outflow from the node.

EFFECTS OF ANTIGEN ON THE TRAFFIC OF SPECIFIC ANTIGEN-REACTIVE CELLS THROUGH LYMPH NODES

The frequency of lymphocytes specific for a given antigen is of the order of 1 in 10^5 cells. The probability of this low frequency of antigen-specific cells ever interacting with antigen would be very low if it were not for lymphocyte recirculation (Gowans & Knight 1964, see also review by Ford 1975). Antigen is known to be concentrated at selective sites within organized lymphoid tissue, particularly in cortical regions of lymph nodes and marginal zones in the spleen (Ford 1975).

Several studies have shown that antigen localized within lymphoid tissue selects out antigen-specific cells from the passing stream of recirculating lymphocytes (Rowley et al 1972, Sprent & Miller 1974). In rats, cells specific for certain antigens, including alloantigens, become specifically trapped at sites of antigen localization in the spleen. The recirculating pool becomes transiently depleted of the reactive cells, as shown by the fact that the thoracic duct is devoid of cells specific for the spleen-localized antigen. Similarly in sheep, shortly after challenge of a cannulated node with PPD, cells reactive to this antigen disappear from the efferent lymph of the stimulated node (Cahill et al 1974) for about three days after challenge. The absence of reactivity from either thoracic duct or efferent lymph is unlikely to be due to tolerance or suppression since if alloreactive or antigen-reactive cells are radiolabelled and injected they subsequently become localized by antigen *in vivo* (Atkins & Ford 1975, Thursh & Emeson 1972). In these experiments the specific antigen-reactive cells may enter the lymphoid tissue at random and the findings may be due to 'immunoadsorbent' capacity of antigen for the appropriate specific cells in the recirculating lymphocyte pool.

There are however two sets of experiments which indicate that the entry of antigen-specific cells into lymph nodes may not be simply a question of random movement from the blood into the node. There is an antigen-specific selective mechanism in stimulated lymph nodes which enhances the entry into the node of cells specific for the challenge antigen (McConnell et al 1974). As shown in these experiments, chronic drainage of efferent lymph from a repeatedly stimulated node in sheep resulted in the whole animal becoming unresponsive to the challenge antigen. The specificity of the unresponsiveness was shown by a normal reaction to other antigens such as chicken globulin and the hapten NIP (4-hydroxy-5-iodo-3-nitrophenacetyl). Repeatedly injected control sheep did not become unresponsive. Using a similar experimental system Cahill et al (1974) have also shown in sheep that challenge of a cannulated node with allogeneic lymphocytes leads to the disappearance of alloantigen-reactive cells from the efferent lymph draining a contralateral node. These experiments imply that the presence of antigen in cannulated lymph nodes can select out the appropriate antigen-reactive cells from those present in the intravascular compartment of the node.

In both these experiments the stimulated nodes are cannulated and antigen or specifically reactive cells present in the node are therefore removed from the animal. There is no apparent systemic spread either of antigen or of cells. In our own experiments with PPD it was necessary to drain the efferent lymphocytes for at least three weeks. Chronic lymphatic drainage for three weeks from a node repeatedly challenged with antigen is not easy to maintain so the number of experiments is small. In those animals where cannulation and repeated stimulation with PPD for longer than three weeks has been achieved the abrogation of the response to PPD is a consistent experimental finding.

We have recently repeated these experiments using the protocol outlined in Table 4. The complete results for one test sheep and a repeatedly injected but non-cannulated control are shown in Figs. 8, 9 and 10 and Table 5. With sheep 019 repeated stimulation and drainage of cells from the node led to the animal becoming unresponsive to PPD by skin testing (Fig. 8). The skin test response of the injected control was normal at 24 h and slightly reduced at 48 h compared to the initial test. When sheep 019 was re-tested eight weeks later the reactivity to PPD had returned. This has been described before

TABLE 4

Experimental protocol for producing selective removal of PPD reactive cells *in vivo*

1.	1–2 year old Finnish Landrace sheep, primed to BCG and NIP-chicken globulin.
2.	After three weeks baseline reactivity to PPD established by skin testing (delayed hypersensitivity reaction) and *in vitro* transformation of peripheral blood lymphocytes to PPD and allogeneic cells (MLC).
3.	Cannulation of afferent lymphatic to and efferent lymphatic from popliteal node. Repeated challenge of node with 100 μg PPD (× 5) every four days for 20 days. Second challenge of PPD given as NIP–PPD and both efferent lymph and serum tested for anti-NIP antibody. Repeated testing of peripheral blood lymphocytes for *in vitro* reactivity to PPD and MLC.
4.	Cannulation stopped and systemic response to PPD measured both by skin reactions, and by testing for helper cell activity to PPD by systemic challenge with NIP–PPD. Twelve days later systemic challenge with NIP–chicken globulin. Serum anti-NIP measured. Repeated lymphocyte transformation assays on peripheral blood lymphocytes to PPD and MLC.

(McConnell et al 1974) and is possibly due to the fact that BCG is a 'living' antigen which persists *in vivo*. Thus immunologically virgin cells which are generated *in vivo* become primed. The reappearance of PPD reactivity is not due to cells emerging from the repeatedly stimulated node, since in this experiment the node was removed when the cannulation ceased. The specific failure of the test sheep to respond to PPD was also shown by its failure to make an anti-NIP response on challenge with NIP–PPD. Challenge with NIP-chicken ·globulin 12 days later, however, produced an anti-NIP response (Fig. 8).

The results for lymphocyte transformation on peripheral blood lymphocytes during the experiment are shown in Fig. 9 and Table 5. In the test animal the ability of peripheral blood lymphocytes to specifically transform in response to 10 μg and 5 μg of PPD disappears shortly after the last injection of PPD into the cannulated node. The transformation response to 1 μg of PPD is reduced but is never completely negative. This unresponsiveness is specific since the response to allogeneic lymphocytes (MLC) does not change. In a repeatedly injected (but not cannulated) control sheep reactivity to PPD diminishes only slightly. In a current experiment, repeated challenge (× 5) of a sheep with PPD at a site distant from a cannulated node which has been repeatedly stimulated with an irrelevant antigen has so far failed to reduce the response to PPD either *in vivo* or *in vitro*. This makes it unlikely that the failure to respond is due to tolerance induction in a depleted lymph node (Ford 1975).

One possible explanation for these results is that there is a gradual loss of PPD-reactive cells from the circulation during repeated challenge and drainage of cells from the node. With fewer PPD-reactive cells it might be

TABLE 5

Lymphocyte transformation of peripheral blood lymphocytes during cannulation and repeated stimulation of the popliteal lymph node in sheep.

Days after PPD challenge of cannulated node [a]	In vitro transformation[b] of peripheral blood lymphocytes to:			
	10 μg PPD	5 μg PPD	1 μg PPD	MLC[c]
–	4.16	nt	nt	nt
	4.10	3.71	3.76	4.10
4 (PPD × 1)	4.29	4.33	4.49	nt
10 (PPD × 2)	2.16	3.48	4.12	4.59
20 (PPD × 4)	4.01	2.40	2.45	3.58
26 (PPD × 4)	3.96	3.88	3.41	4.34
31 (PPD × 5)	0.00[d]	3.62	3.74	3.80
40 (PPD × 5)	0.00	0.00	2.87	3.49
51 (PPD × 5)	0.00	0.00	3.12	3.52
79	2.91	3.26	3.26	nt
90	nt	3.91	3.81	3.35
Control sheep[e]	4.9	4.79	4.78	nt
	4.92	4.88	4.88	3.67
	4.81	4.79	4.59	4.16
	4.49	4.59	4.66	3.13
Control sheep (after PPD × 5)	4.26	4.27	4.14	3.12

[a]Cannulated popliteal node challenged with 100 μg PPD at intervals over 26 days. Cannulation ceased after fifth injection of PPD.
[b]Response of 1.25 × 10⁵ peripheral blood lymphocytes to 10.5 and 1 μg PPD in 200 μl cultures. Cultures labelled 12 h before harvesting with 1μCi [³H]thymidine. Result expressed as the log of mean c.p.m. in control cultures without antigen.
[c]MLC. 1.25 × 10⁵ peripheral blood lymphocytes (Finnish Landrace) cultured with 0.62 × 10⁵ mitomycin C treated stimulator lymphocytes from Clun sheep. Cultures as for PPD transformation.
[d]Transformation in cultures with these values: < transformation seen in antigen-free control.
[e]Control sheep not cannulated but repeatedly injected (× 5) with 100 μg PPD into drainage area of lymph node.

expected that the optimal concentration of PPD required to produce stimulation would shift in favour of lower antigen concentrations—higher concentrations being tolerogenic for the low number of cells. This is observed in the response of normal sheep lymphocytes to concentrations of PPD used in the 50–100 μg range. The alternative approach to quantitating the disappearance of PPD-reactive cells would be to use limiting dilutions of cells for the transformation assays.

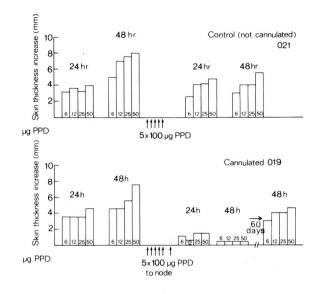

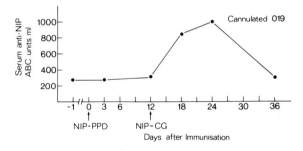

FIG. 8. Delayed hypersensitivity skin reactions to PPD in control and cannulated sheep. Sheep tested with 6, 12, 25 and 50 μg PPD injected intradermally and response measured at 24 and 48 h. In cannulated sheep repeatedly injected with PPD reactivity to PPD was tested 60 days after the end of cannulation and the last injection of PPD. Bottom part of the figure shows the serum hapten binding capacity (ABC, or antigen binding capacity) measured by the Farr assays using 10^{-8} M ^{125}I-labelled NIP caproic acid as hapten. The ABC was calculated from the linear portion of the binding curve between 3% binding (1 unit of antibody/ml) and 30% binding (10 units of antibody/ml). The figure shows the serum anti-NIP response of sheep 019 which was unresponsive to PPD after the fifth injection of PPD to the cannulated node. NIP–PPD was given systemically to sheep 019 and 12 days later the animal was rechallenged with NIP–CG. Repeatedly injected but non-cannulated control sheep had serum ABC of 60 × 10^3 units/ml six days after challenge with NIP–PPD and 140 × 10^3 units/ml 12 days after NIP–PPD challenge.

These preliminary studies support our previous results (McConnell et al 1974) and those of Cahill et al (1974) on the systemic depletion of antigen-reactive cells by repeated antigenic challenge of cannulated nodes. These

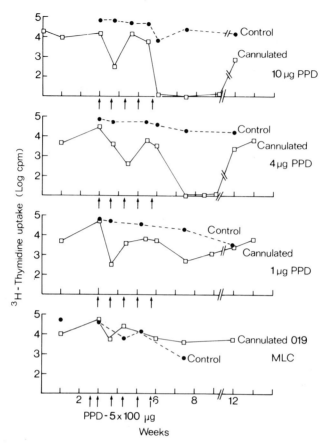

FIG. 9. Lymphocyte transformation to PPD and mixed lymphocyte cultures (MLC) in cannulated and repeatedly injected test sheep (019). For experimental details see Table 5. Lymphocyte transformation plotted as log c.p.m at various intervals throughout experiment. Control = lymphocyte transformation from a BCG-primed sheep not injected with PPD but tested simultaneously with sheep 019.

studies suggest that an antigen-selective mechanism may be operating at the level of stimulated lymph nodes.

Other explanations that the unresponsiveness to PPD is due to tolerance induction or the presence of blocking factors seem unlikely. The observations on lymphocyte transformation rule out a role for serum blocking factors, and for tolerance induction to occur antigen would have to reach the systemic circulation via the efferent lymphatic. Challenge of the cannulated node always produces an anti-NIP response in the lymph, with little response in the serum, indicating that no substantial concentrations of antigen reach the systemic circulation from the stimulated node (Fig. 11).

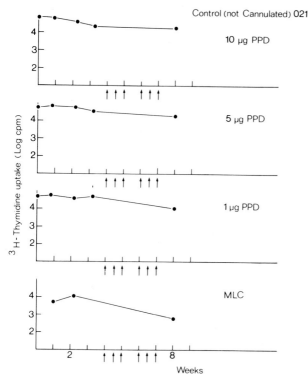

Fɪɢ. 10. Lymphocyte transformation to PPD and MLC in sheep 021 repeatedly injected with PPD into the drainage area of an uncannulated node.

We have suggested at least two possible mechanisms to explain abrogation of the response to PPD in this system (Fig. 12). If lymphocytes entered nodes at random the ratio of PPD to chicken globulin (or other antigen) would remain constant and depletion, when it did occur, would be non-specific. Specific depletion could be explained on the basis of antigen being attached to the luminal surface of the high-endothelial (HE) cells of the postcapillary venule (PCV), which might preferentially select PPD-reactive cells into the node. At each passage through the node only a fraction of the specific cells would be selected at any one time.

Hay & Hobbs (1977) have shown that 60% of the recirculating lymphocyte pool passes through the vascular compartment of a node in the five days after antigen stimulation and about 25% of these cells actually enter the node at each passage. From this it follows that 15% of the total recirculating pool is removed by the node every five days.

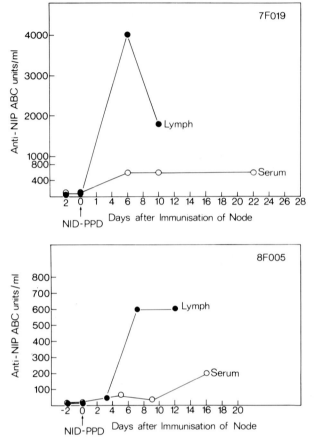

FIG. 11. Serum and lymph anti-hapten responses in cannulated sheep injected with NIP–PPD into the drainage area of cannulated popliteal lymph node. Anti-NIP response measured as in Fig. 7 in both lymph and serum at intervals after challenge of the cannulated node with antigen.

In our experiments five cycles of antigen stimulation were given over about 20 days. If we ignore the formation of new cells over this period this would leave about 40–50% of the total recirculating pool. If antigen on the HE cells can increase the entry of PPD-specific cells by a factor of 2 (i.e. 50% of specific cells traffic from blood to lymph), then after five days only 24% of specific cells will be left in the recirculating pool. If the extraction rate were three times as efficient then 9% of cells would be left in the recirculating pool after about 20 days, the time at which PPD unresponsiveness is first observed.

Antigen-presenting cells such as macrophages clearly present antigen to T

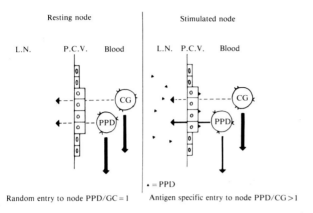

(a) Specific entry to lymph node

Random entry to node PPD/GC = 1   Antigen specific entry to node PPD/CG > 1

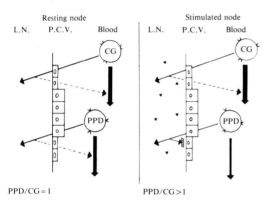

(b)   Random entry with specific trapping

PPD/CG = 1   PPD/CG > 1

FIG. 12. Hypothetical models to explain specific depletion of antigen-reactive lymphocytes from the whole animal. (a): Antigen specific entry to lymph nodes. Antigen present on the luminal surface of the high endothelial cells of the post-capillary venule preferentially selects PPD-reactive cells. Random entry of all cells still occurs. (b): Random entry with specific trapping. Most traffic of cells is from blood to lymph but a smaller number of cells re-enter the intravascular compartment from the node. If PPD-reactive cells are stimulated by PPD within the node their traffic is altered and they fail to re-enter the intravascular compartment.

cells in close association with their own major histocompatibility antigens (Paul et al 1976, Benacerraf 1978). It is conceivable that HE cells might also be antigen-presenting cells and in this respect it would be of interest to know whether they express Ia-like antigens on their surface, whether they have antigen bound to their luminal surfaces, or whether they have any *in vitro* antigen-handling capacity.

Although the bulk movement of lymphocytes is clearly from blood to lymph (Gowans & Knight 1964, Marchesi & Gowans 1964) there is little experimental evidence to refute the possibility that lymphocytes re-enter the blood directly from the node. Using isolated, perfused nodes Sedgley & Ford (1976) have clearly shown that lymphocytes do not return to the blood from the node. However *in vivo* there are two situations which indicate that retrograde traffic of lymphocytes from node to blood may occur: in the pig there is a clear contrast to our accepted views of lymphocyte recirculation in lymph nodes, in that few lymphocytes seem to leave the node via the efferent lymphatic and the thoracic duct contains a low level of lymphocytes. In sheep, Morris and his colleagues (Trevella & Morris, this symposium) have shown that if efferent lymphocytes from a popliteal node are labelled and infused via an efferent to the next node in the chain (lumbar), not all the injected labelled cells can be recovered from either the lumbar node or its efferent output. On balance there is little evidence to suggest that retrograde traffic of lymphocytes cannot occur and, given the very marked physiological changes which take place in lymph nodes responding to antigen (e.g. cell shutdown), the possibility of retrograde traffic has to be considered.

ACKNOWLEDGEMENTS

We gratefully acknowledge the support of the Cancer Research Campaign of Great Britain. We also thank Dr J. Pearson for permission to publish the information in Table 3 and Fig. 7.

References

Alper CA, Balavitch D 1976 Cobra venom factor. Evidence for its being altered C3b (third component of complement). Science (Wash DC) 191:1275-1276
Atkins RC, Ford WL 1975 Early cellular events in a systemic graft versus host reaction. I. The migration of responding and non-responding donor lymphocytes. J Exp Med 141:664-680
Benacerraf B 1978 A hypothesis to relate the specificity of T lymphocytes and the activity of I region specific Ir genes in macrophages and B lymphocytes. J Immunol 120:1809-1812
Cahill RNP, Hay JB, Frost H, Trnka Z 1974 Changes in lymphocyte circulation after administration of antigen. Haematologia 8:321-334
Cahill RNP, Frost H, Trnka Z 1976 The effects of antigen on the migration of recirculating lymphocytes through single lymph nodes. J Exp Med 143:870-888
Fearon DT, Austen KF 1977 Activation of the alternative pathway with rabbit erythrocytes by circumvention of the regulatory action of endogenous control proteins. J Exp Med 146:22-33
Ford WL 1975 Lymphocyte migration and immune responses. Prog Allergy 19:1-59
Gowans JL, Knight EJ 1964 The route of recirculation of lymphocytes in the rat. Proc R Soc Lond B Biol Sci 159:257-262

Hall JG 1974 Observations on the migration and localisation of lymphoid cells. In: Brent L, Holborow J (eds) Progress in immunology II. North-Holland, Amsterdam, vol 3:15-24

Hall JG, Morris B 1962 The output of cells from the popliteal node of sheep. Q J Exp Physiol Cogn Med Sci 47:360-369

Hall JG, Morris B 1963 The lymph-borne cells of the immune response. Q J Exp Physiol Cogn Med Sci 48:235-247

Hall JG, Morris B 1965 The immediate effect of antigens on the cell output of a lymph node. Br J Exp Pathol 46:450-454

Hay JB, Hobbs BB 1977 The flow of blood to lymph nodes and its relation to lymphocyte traffic and the immune response. J Exp Med 145:31-44

Hay JB, Lachmann PJ, Trnka Z 1973 The appearance of migration inhibition factor and a mitogen in lymph draining tuberculin reactions. Eur J Immunol 3:127-131

Herman PG, Yamamoto I, Mellins HZ 1972 Blood microcirculation in the lymph node during the primary immune response. J Exp Med 136:697-714

Hobart MJ 1970 The structure and activity of ovine immunoglobulins. PhD thesis, University of Cambridge

Johnston MG, Hay JB, Movat HZ 1978 Kinetics of prostaglandin production in lymph draining inflammatory lesions. Fed Proc 37:711 (Abstr 2615)

Lachmann PJ 1979 Complement. In: Sela M (ed) The antigens. Academic Press, New York, vol 5: 283-353

Lachmann PJ, Hobart MJ 1978 Complement technology. In: Weir DM (ed) Handbook of experimental immunology. Blackwell Scientific, Oxford

Lascelles AK, Morris B 1961 Lymphocyte traffic through lymph nodes. Q J Exp Physiol Cogn Med Sci 46:199-206

Marchesi VT, Gowans JL 1964 The migration of lymphocytes through the endothelium of venules in lymph nodes: an electron microscopic study. Proc R Soc Lond B Biol Sci 159:283-290

McConnell I, Lachmann PJ, Hobart MJ 1974 The restoration of specific immunological virginity. Nature (Lond) 250:113-116

Paul WE, Shevach FM, Thomas DW, Pickeral SF, Rosenthal S 1976 Genetic restriction on T-lymphocyte activators by antigen-pulsed peritoneal cells. Cold Spring Harbor Symp Quant Biol 41:571-578

Pedersen NC, Morris B 1970 The role of the lymphatic system in the rejection of homografts: A study of lymph from renal transplants. J Exp Med 131:936-969

Rowley DA, Gowans JL, Atkins RC, Ford WL, Smith ME 1972 The specific selection of recirculating lymphocytes by antigens in normal and pre-immunised rats. J Exp Med 136:499-513

Scott DM, McConnell I, Agomo P, Lachmann PJ 1978 Purification of antigen-dependent macrophage inhibition factor (MIF) from lymph draining a tuberculin reaction. Immunology 34:591-604

Sedgley M, Ford WL 1976 The migration of lymphocytes across specialised vascular endothelium. 1. The entry of lymphocytes into the isolated mesenteric lymph node of the rat. Cell Tissue Kinet 9:231-243

Smith JB, Morris B 1970 The response of the popliteal lymph node of the sheep to swine influenza virus. Aust J Exp Biol Med Sci 48:33-46

Smith JB, Pedersen NC, Morris B 1970 The role of the lymphatic system in inflammatory responses. Ser Haematol 111:17-61

Sprent J, Miller JFAP 1974 Effect of recent antigen priming on adoptive immune responses. 11. Specific unresponsiveness of circulating lymphocytes from mice primed with heterologous erythrocytes. J Exp Med 139:1-12

Thursh DR, Emeson EE 1972 The immunologically specific retention of long-lived lymphocytes in lymph nodes stimulated by xenogeneic erythrocytes. J Exp Med 135:754-763

Trevella W, Morris B 1980 Reassortment of cell populations within the lymphoid apparatus of the sheep. In this volume, p 127-140

Discussion

van Ewijk: Is there any direct evidence that antigens are present on HEV?

McConnell: As far as I am aware no one has looked. It would also be interesting to know whether high endothelial cells express Ia antigens in common with other antigen-presenting cells.

Gowans: Is the endothelium of the HEV phagocytic?

Ford: I am not aware of any report of marked phagocytosis by HEV in the electron microscopic studies (see Andrews et al, this volume, for references).

Weissman: You had four models but the one that seems most likely to me wasn't included, Dr McConnell. All kinds of cells are going through that lymph node and only a subpopulation is being stimulated, yet eventually all the stimulated cells are collected in the recirculating pool. Shouldn't there be a specific deficit of antigen-stimulated cells when that particular duct is cannulated?

McConnell: Yes, that is one explanation we offer.

Howard: Mere random entry of recirculating lymphocytes into the stimulated node with total collection from the efferent lymph can't account for specific depletion. It can only cause a general depletion. As long as there is no return to the blood, avoiding the efferent lymph, there is no basis for the specificity of depletion in recirculation as generally understood.

McConnell: It is not general depletion. It is specific depletion from the recirculating pool of T cells involved in delayed hypersensitivity responses or in the T helper response where PPD is the carrier.

Humphrey: Bob Kelly and his colleagues (Kelly et al 1972) infused lymphocyte activation products (lymphokines) produced *in vitro* into the afferent lymphatics of guinea-pig auricular lymph nodes. Within 24 h lymphocytes piled up in the paracortex and plugged the paracortical sinuses at the paracortico-medullary border, resembling what pathologists used to call 'sinus catarrh'. By 48 h this was still present but much diminished. Similar changes, developing more slowly, were observed in the popliteal nodes of rabbits after diphtheria toxoid was injected into the footpad; these changes were correlated with the appearance of lymphokine activity in the afferent lymph. Kelly (1970) attributes the accumulation of lymphocytes at the paracortico-medullary border to functional constriction of the diameter of the exits of the paracortical sinuses. When this ceases there is an outpouring into the medullary sinuses and thence to efferent lymphatics. I wonder whether something like this is occurring in your sheep nodes, and whether you have ever infused lymphokines and seen the same thing.

McConnell: If shutdown lymph is infused via an afferent, shutdown occurs all over again. Since shutdown lymph has high levels of PGE_2 perhaps this is not surprising.

Humprey: Kelly didn't look at prostaglandins but from the way the lymphokines were prepared prostaglandins would not be expected to be present unless they were very stable.

Vane: Prostaglandin E_2 is chemically stable but if an enzyme such as prostaglandin 15-hydroxydehydrogenase is present, it will be rapidly metabolized. I don't know whether there is any in lymph. The levels of prostaglandin that you obtained, Dr McConnell, worked out at something like 30 ng/ml. This is the kind of level found in inflammatory exudates and the kind that contributes to the signs and symptoms of inflammation. But the 50 μg that you give to cause shutdown is an enormous overdose. If you want to know whether prostaglandins modulate or mediate the shutdown response you could give indomethacin, which would prevent their formation.

McConnell: In one experiment we infused aspirin before giving cobra venom factor locally via the afferent and this substantially reduced shutdown.

Vane: Was the aspirin given in a high concentration?

McConnell: No, it was used at concentrations which would be required to inhibit PGE_2 levels of about 18 ng/ml—the concentrations we detect.

Humphrey: How does anything that enters through the marginal sinus seep into the deeper layers of the lymph node in sheep?

Morris: There are multiple afferent lymphatics to the popliteal node of the sheep and other animals. If you cannulate one afferent lymphatic and infuse an antigen there, the immune response in the node is compartmentalized. Nobody has really looked at the question of the restriction of the immune response in a lymph node. The lymph node will probably prove to respond in terms of functional units. The problem, of course, in looking at the events that follow the infusion of an antigen, is that you may be influencing only a segment of the node. I am sure that when different antigens are infused into different parts of the node there will be all sorts of competition effects and so on. Of the material that arrives in the node some will probably come out via the efferent duct without ever permeating the node at all, but most of it will go through the cortical and medullary sinuses and out through the hilum.

Gowans: What is the source of the prostaglandins in lymph nodes? Polymorphs have been suggested, although they are not normally a component of the lymph or of the nodes.

Vane: Brune et al (1978) have shown that macrophages produce thromboxane A_2 when activated.

McConnell: Bede Morris said yesterday, and some of our own data show, that polymorphs often appear in the first wave of cells that come out after shutdown. Polymorphs entering the node during shutdown could be the source of prostaglandins found in the lymph.

Zigmond: If complement was present one would expect polymorphs to be there.

Vane: Several reports (Higgs et al 1976, Goldstein et al 1977, Davison et al 1978) suggest that phagocytosing polymorphs produce rather more thromboxane A_2 than prostaglandin E_2 so again we would expect to detect thromboxane B_2.

Morris: Dr McConnell, have you seen high capillary venules in the lymph nodes of sheep? Their existence seems to be an implicit part of your hypothesis.

McConnell: I don't think high endothelial cells are an implicit part of it at all. It is antigens attached to high or not so high endothelial cells (as in sheep) that are important. As discussed earlier, sheep do not appear to have high endothelial cells, at least not in resting nodes. They can however be induced in granuloma formation, as you yourself have shown. Perhaps high endothelial cells appear in repeatedly stimulated nodes. They are certainly something we should look for.

Zigmond: Can C3b itself cause lymphocyte aggregation?

McConnell: Nascent C3b can bind to cells in two ways. It can bind to any surface via its short-lived binding site (covalent binding) or alternatively if this binding site decays it can bind as C3b to cells with C3b receptors. A possible explanation for the disappearance of B cells during shutdown is that they adhere to C3b bound to other cells in the node or that their traffic is altered when C3b is bound to their receptors. Perhaps C3b can cause B cells to become localized in and around germinal centres.

Gowans: In what pathologists used to call sinus catarrh, the lymphatic sinuses are plugged up with cells. Where you infuse by the afferent lymphatic and get shutdown do you see something similar in sections of the nodes?

McConnell: We have never done that but I agree that we should look at the histology.

Morris: The histological basis of cell shutdown was examined by Dr Roger Moe in 1962 (personal communication). He looked at both the light and electron microscopic features of the phenomenon. He saw sinus catarrh in these nodes, as you said, or at least I think that was his interpretation of it. The magnitude, intensity and duration of the shutdown bears no close relation to the subsequent intensity of the immune responses, or the antigenicity of the substance. So this phenomenon isn't a good measure of how effective a substance is as an antigen.

Gowans: Are the cells sticking together or sticking to the margins of the sinuses?

Morris: I am afraid I can't give you a good account of it.

McConnell: There are two possibilities. Either there is physical obstruction to the cells or 'factors' are released within the node which have a transient effect on their mobility.

Morris: There is no change in the lymph flow, which refutes the possibility of obstruction.

McConnell: It might be obstruction for cells but not for lymph—it all depends on what sort of obstruction existed.

Gowans: But is this process reversible? Is it possible that prostacyclin is responsible for reversing the process of shutdown by 'unsticking' the lymphocytes? Does prostacyclin disaggregate cells other than platelets?

Vane: Prostacyclin certainly prevents leucocytes from rolling along the vessel wall. I don't know whether it does anything to lymphocytes.

van Ewijk: Macrophages stick to reticular cells in the sinuses. Are they activated in some way by your prostaglandins?

McConnell: I don't know. In the presence of active complement components macrophages show altered behaviour.

Born: If the microvilli are a sign of activity, prostacyclin might round them off and you might be able to see this as inhibitory activity.

Gryglewski: Recently Bragt & Bonta (1979) reported that [^{14}C-1]arachidonic acid when continuously infused into a granuloma cavity in rats is mainly transformed to prostaglandin E_2 but not to prostacyclin or to thromboxane B_2. The generation of prostaglandins at the site of inflammation is a fairly unspecific phenomenon and it also occurs in immunologically triggered inflammatory responses.

Davies: I don't quite know what the galaxy of acute phase proteins is in the sheep. Aside from C3, do any others play some role in shutdown?

McConnell: That is an interesting question—we haven't yet looked to see whether other active components play a role in shutdown.

Vane: It is important to realize the progress that has been made in our knowledge of prostaglandin products in the last five or six years. We now know that the arachidonic acid cascade includes many more substances than just PGE_2 and PGF_2—more even than prostacyclin and thromboxane A_2. Hydroperoxides of fatty acids are also formed and the 12-hydroxy fatty acid called HETE has been shown to be chemotactic for leucocytes. So it is no longer possible to look simply at PGE_2 levels. Many other compounds with pharmacological activity might be there.

McConnell: Quite clearly these experiments are going to take us into much deeper waters or deeper lymph!

Humphrey: Is there any change in the blood flow through the draining lymph nodes with time after antigen administration?

McConnell: Yes. Dr Cahill, as well as Hay & Hobbs (1977), has shown that there is a marked increase in blood flow to nodes during shutdown.

Cahill: We measured the changes in blood flow three hours after giving influenza virus to single lymph nodes in sheep. This produces a shutdown effect similar to yours. We consistently found about a fivefold increase in the volume of blood flow compared with the contralateral unstimulated node. At the same time we measured the entry of ^{51}Cr-labelled lymphocytes into the lymph node. The entry of lymphocytes didn't correlate with the increase in blood flow at that time. The increase in blood flow may be partly due to acute inflammation caused by administration of a very large amount of virus.

McConnell: Hay & Hobbs (1977) injected labelled microspheres intra-arterially and looked at their location in nodes undergoing cell shutdown. They found a fourfold increase in blood flow during shutdown to a node. By microangiography Herman et al (1972) have observed quite dramatic changes in the capillary network after antigenic stimulation. The lymph node microvasculature begins to look like a kidney.

Humphrey: But Herman thinks that shunting occurs through vessels whose smallest effective diameter is greater than 9 μm instead of the usual 5 μm or so (Herman et al 1979). Herman regards these as arteriovenous shunts which might or might not be associated with increased flow of cells through the vessels.

Cahill: Peter Herman's work is very important because it suggests that antigen-induced increases in blood flow through a lymph node are not distributed uniformly through the vascular bed. If there are regional changes in blood flow then simple correlations between lymphocyte entry and blood flow measurements using microspheres are difficult to make.

References

Bragt PC, Bonta IL 1979 In vivo metabolism of [1-^{14}C]arachidonic acid during different phases of granuloma development in the rat. Biochem Pharmacol, in press

Brune K, Glatt M, Kälin H 1978 Pharmacological control of prostaglandin and thromboxane release from macrophages. Nature (Lond) 274:261-263

Davison EM, Ford-Hutchinson AW, Smith MJH, Walker JR 1978 The release of thromboxane B$_2$ by rabbit peritoneal polymorphonuclear leukocytes. Br J Pharmacol 63:407P

Goldstein IM, Malmsten CL, Kaplan HB, Jindahl H, Samuelsson B, Weissman G 1977 Thromboxane generation by stimulated human granulocytes: inhibition by glucocorticoids and superoxide dismutase. Clin Res 25:518A

Hay JB, Hobbs BB 1977 The flow of blood to lymph nodes and its relation to lymphocyte traffic and the immune response. J Exp Med 145:31-44

Herman PG, Yamamoto K, Mellins HZ 1972 Blood microcirculation with the lymph node during the primary immune response. J Exp Med 136:697-714

Herman PG, Utsonomiya R, Hessel SJ 1979 Arteriovenous shunting in the lymph node before and after antigenic stimulus. Immunology 38, in press

Higgs GA, Bunting S, Moncada S, Vane JR 1976 Polymorphonuclear leukocytes produce thromboxane$_?$A$_2$ activity during phagocytosis. Prostaglandins 12:749-757

Kelly RH 1970 Functional anatomy of lymph nodes 1. The paracortical cords. Int Arch Allergy Appl Immunol 48:836-849

Kelly RH, Wolstencroft RA, Dumonde DC, Balfour BM 1972 The effect of lymphocyte activation products on lymph node architecture and evidence for peripheral release of LAP following antigenic stimulation. Clin Exp Immunol 10:49-65

Effect of skin painting with oxazolone on the local extravasation of mononuclear cells in sheep

J.G. HALL

Chester Beatty Research Institute, Institute of Cancer Research, Royal Marsden Hospital, Downs Road, Sutton, Surrey

Abstract A characteristic feature of the induction of cell-mediated delayed hypersensitivity reactions by chemicals such as oxazolone is the enlargement of lymphocyte traffic areas in the paracortices of regional lymph nodes. In sheep oxazolone is a powerful immunogen but the cellular changes in lymph efferent from nodes draining areas of oxazolone-painted skin do not differ significantly from responses to conventional antigens. Specific complement-binding antibodies appear in the plasma of sensitized sheep, which respond to secondary challenges with an immediate Arthus reaction.

In studies of peripheral lymph from areas of skin painted with oxazolone the number of mononuclear cells in the lymph increased 10–50-fold two days or so after skin painting. Most of these cells were small lymphocytes lacking surface immunoglobulin (presumptive 'T' cells). This big increase in lymphocyte traffic through the skin may be a consequence of the binding to local structural proteins of myriads of oxazolone epitopes. If so, and bearing in mind the large doses of immunogen used in experiments on mice, it is easy to envisage how the traffic areas of lymph nodes expand and become congested with lymphocytes after being flooded with a highly immunogenic and reactive chemical like oxazolone. Whether this is relevant to the induction of cell-mediated immunity is unknown.

Investigation of sensitization with chemical compounds goes back to at least 1912. In the 1930s Karl Landsteiner and his colleagues did a classical series of experiments on the sensitization of animals with simple compounds (e.g. Landsteiner & Jacobs 1935). Their work, together with studies of the cutaneous sensitivity to nitrobenzene derivatives that afflicted workers making explosives during World War II (Gell 1944), presented the biomedical world with intriguing and important phenomena. The problem of delayed-type hypersensitivity and cell-mediated immunity, hitherto enmeshed in the complexities of tuberculosis, became susceptible to study in a variety of laboratory animals, and immunologists were made more aware of both the

© *Excerpta Medica 1980*
Blood cells and vessel walls: functional interactions
(Ciba Foundation symposium 71) p 197-209

practical and the theoretical advantages of working with chemically defined antigens. It is recognized now that the cellular basis of delayed hypersensitivity to chemical compounds has much in common with that of allograft rejection. An absolute distinction between these phenomena and classical humoral immunity has, however, been necessarily elusive (Gell 1967), and the puzzling hinterland between them has yet to be mapped in detail.

I became interested in this topic because of the claim that the carcinogenicity of hydrocarbons was related inversely to their potency as skin-sensitizing agents (Old et al 1963). Studies of the responses of the regional lymphatic apparatus of sheep to such agents were initiated and responses to fat-soluble substances ranging from carcinogens to cyclosporin A have been investigated. Fluorodinitrobenzene (FDNB) was used in an early study (Hall & Smith 1971) but the results were complicated by the acute inflammatory reactions attributable to the intrinsic vesicant properties of such compounds. Oxazolone (4-ethoxymethylene-2-phenyloxazol-5-one; Gell et al 1946) promised to be a less irritant immunogen and some interesting results have been produced with it.

OXAZOLONE AS AN IMMUNOGEN

Although oxazolone is recognized as a powerful skin-sensitizing agent its ability to elicit abundant production of specific humoral antibody is mentioned less often. Consequently, the histological features of responses to oxazolone, such as the paracortical enlargement of lymph nodes (Turk & Stone 1963, Oort & Turk 1965, Davies et al 1969), are often cited as characteristic qualities of 'pure' cell-mediated (as distinct from antibody-mediated) although in my view the changes are those of degree rather than of kind.

In sheep it is possible to monitor immune responses both qualitatively and quantitatively by collecting all the lymph from the lymph node for some weeks (Hall & Morris 1963). In this way it was shown that although vigorous cellular and humoral responses can be induced by painting the regional skin with a solution of oxazolone in acetone, the responses do not in general differ significantly from those to bacterial or viral antigens. During the primary response many blast cells, some containing immunoglobulin, appeared in the lymph (Hall et al 1978), and humoral antibody titres (measured by passive haemagglutination; Askenase & Asherson 1972) reached a peak 10 days after skin painting. During secondary responses these events recurred more rapidly and antibody reached its highest titre in lymph plasma four days after challenge. Nonetheless, one feature of the responses to oxazolone deserves special mention in this symposium and may be particularly relevant to Dr McConnell's observations (this symposium, pp 167-190). It became apparent that during primary challenge, and provided that the concentration of oxazolone in the acetone was below 7%, the acute 'shut-down' of lymphocyte traffic through the regional node that accompanies the administration of conventional antigens (Hall & Morris 1963, 1965) did not occur, even though vigorous cellular and humoral responses were induced. Secondary challenges, which because of pre-existing, complement-binding antibodies elicited an Arthus reaction, or primary challenges with doses of oxazolone high enough to elicit granulocyte extravasation (seen with FDNB), always caused the usual shut-down. These results are consistent with the view that the cessation of lymphocyte traffic through lymph nodes may be mediated by complement cleavage products released by either the classical or the alternative pathway. However, the extravasation and peregrination of lymphoid cells within lymphoid tissues are complex events which are thought to depend on the presence of specialized capillary and venular endothelia (Ford & Gowans 1969). It may be easier to draw general conclusions about the extravasation of mononuclear cells by looking at what goes on in more mundane areas of the microvasculature.

EXTRAVASATION OF MONONUCLEAR CELLS IN SKIN AND ITS INCREASE IN RESPONSE TO THE LOCAL APPLICATION OF OXAZOLONE

Haynes & Field (1931) gave one of the first accounts of the composition of peripheral lymph coming (mainly) from the skin, in the dog. Although their findings were confirmed by several acute experiments a systematic study of the cellular components of such lymph was not feasible until techniques for its collection from unanaesthetized sheep were perfected (Hall & Morris 1963, Smith et al 1970a,b). It is known now that lymph draining from the skin of sheep (e.g. lymph afferent to the popliteal node) contains usually only between 200 and 1000 white cells per mm³. Most of these cells are small lymphocytes but 5−15% are macrophages which can be distinguished, particularly in living preparations, by the ebullient hyaloplastic processes enclosed by their plasma membrane (Fig. 1).

The lymphocytes in peripheral lymph can be derived only from extravasation in the peripheral tissues and, as Morris and his colleagues (Smith et al 1970a, b) have emphasized, this must mean that the ability to export a few lymphocytes, and at the same time to retain most other formed elements of the blood, is a property of most microvasculature. However, for two or three days after a cannula has been placed in a peripheral lymphatic

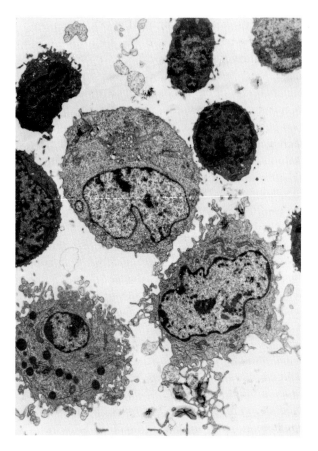

FIG. 1. Electron micrograph of macrophages and lymphocytes in peripheral lymph of the sheep, from an area of skin painted with oxazolone three days previously. × 5400.

vessel, some neutrophil polymorphonuclear granulocytes are present in the lymph, presumably denoting a response to operative trauma. Thus, to collect such lymph under physiological conditions it is necessary to wait for a few days until these neutrophils have disappeared. Then, if one wishes to study the response to a chemical such as oxazolone, one must hope that the preparation will flow for at least a further 10 days. In all, the lymph must flow without interruption for nearly three weeks for a successful experiment, and although I have studied the peripheral lymph from the legs of 20 sheep before and after painting the regional skin with oxazolone, I have succeeded on only four occasions, in two primary and two secondary responses, in following the complete cycle of events in an individual animal.

The first result to note is that the application of acetone solutions containing 7% or less (w/v) of oxazolone did not cause neutrophils to appear in the lymph of unsensitized sheep, and for this reason 5% solutions were used for the primary application and 2.5% solutions for secondary challenge. The solutions were sprayed from a syringe fitted with a 27 gauge needle onto the skin of the cannon-bone between the fetlock and the hock, which had been shorn and cleansed with alcohol−ether to remove the wool fat.   In unsensitized sheep this procedure caused no discernible swelling or increased warmth of the skin and the animals were not irritated or discomfited.   The details of the subsequent changes in the composition of the lymph are shown in Fig. 2.   Although there were significant increases in the flow rate and

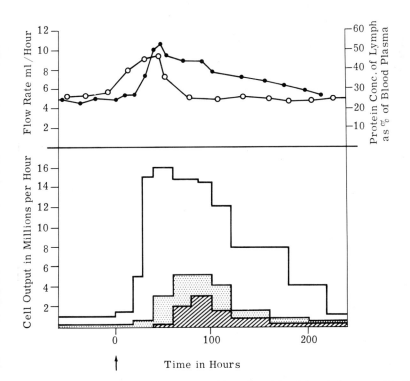

FIG. 2.   Changes in the flow and composition of peripheral lymph from one hindlimb of an unsensitized sheep after a 5% solution of oxazolone in acetone had been applied to the skin of the lateral aspect of the cannon-bone at time zero.
Upper graph: flow rate (●———●) and protein concentration (○----○) (Hartree 1972). *Lower graph:* the clear area of the histogram refers to small lymphocytes, the stippled area to macrophages, and the hatched area to immunoblasts. Differential cell counts were made using a phase-contrast microscope and confirmed by electron microscopy.

protein concentration of the lymph during the two to three days after the application of oxazolone, the most striking change was the big increase in the number of mononuclear cells, most of them small lymphocytes. In this experiment, the number of cells had increased 16-fold within two days of the stimulus, and in one experiment a 50-fold increase was observed. In spite of this, histological studies of skin sections taken at the time of maximum lymphocyte efflux did not show any striking changes under either the light or electron microscope.

The percentage of macrophages increased only slightly but at the height of the response (three days) microscopical examination indicated that some of these cells were more like the monocytes of the blood than those usually seen in normal lymph.

About three days after the stimulus, significant numbers of lymphoid blast cells, some in mitosis, appeared in the lymph and by the fourth day they accounted for 20% of all the cells present. During succeeding days these changes regressed and by the 10th day the composition of the lymph had returned to normal.

The secondary challenges were followed within a few hours by obvious oedema of the oxazolone-painted area; the protein content of the lymph more than doubled, and the flow rate quadrupled. Large numbers of neutrophils appeared in the lymph and they remained the dominant cell type for the first two days. At later times the numbers and types of cells did not differ significantly from those observed in primary responses. In sheep that had been sensitized previously the secondary challenge always elicited an Arthus-type reaction; this does not mean that delayed-type hypersensitivity reactions did not occur, but if they did they were always overshadowed by the acute inflammatory process.

The increases in the numbers of mononuclear cells in peripheral lymph after the application of oxazolone were an order of magnitude greater than those seen in response to conventional antigens (Hall & Morris 1963), allografts of skin (Hall 1967) and tumours (Hall et al 1975).

OCCURRENCE OF SURFACE IMMUNOGLOBULIN ON PERIPHERAL LYMPHOCYTES

During the periods of increased extravasation of mononuclear cells those lymphocytes in the peripheral lymph that displayed surface immunoglobulin were assayed by immunoperoxidase techniques (Scollay et al 1976). The proportion of Ig-positive lymphocytes was always significantly below 10% (Table 1). This is usually the case (Scollay et al 1976), but the relative abundance of fresh cells yielded by the response to oxazolone makes it easier

TABLE 1

The percentage of Ig-positive lymphocytes in the peripheral lymph of sheep, assayed by the immunoperoxidase technique, after the regional skin had been painted with oxazolone

Time after skin painting	% (mean ± SE) of small lymphocytes with surface Ig
2-5 days after primary challenge	8.1 ± 0.8
2-5 days after primary challenge	7.6 ± 1.2
3-4 days after secondary challenge	7.2 ± 1.3
3-4 days after secondary challenge	6.8 ± 1.4

Mean values and their standard errors (SE) were calculated from counts made on slides prepared from cells from at least three samples that had been collected for no more than six consecutive hours. Over 100 Ig-positive cells were identified in each count. Results from four sheep are shown.

to accept such results with confidence. Also, this result agrees with a thorough study by Miller & Adams (1977) of the cells in peripheral lymph coming from allografted kidneys. Thus it seems that the lymphocytes that extravasate in peripheral tissues, either normally or during an experimentally induced increase, cannot represent a random selection from the systemic recirculating pool of lymphocytes because about 25% of these bear surface Ig (Scollay et al 1976, Miller & Adams 1977).

Only a very small number of the immunoblasts in peripheral lymph showed unequivocal ultrastructural or immunohistochemical evidence of internal Ig, so it is likely that most of them were 'T' blasts.

ORIGIN OF IMMUNOBLASTS IN PERIPHERAL LYMPH

Four experiments were done in which lymph efferent from a node stimulated with oxazolone was collected at the same time as peripheral lymph coming from the area of painted skin. At the height of the response the immunoblasts from efferent lymph were labelled *in vitro* with [125I]iododeoxyuridine and returned to the sheep by i.v. injection (Hall et al 1977); the reappearance of the labelled cells in peripheral lymph and, later, in the areas of painted skin was monitored by scintillation counting. Even when generous allowances are made for experimental error, it was apparent that very few of the blast cells in peripheral lymph could be accounted for by direct extravasation. It seems fair to conclude that most of those cells must have been generated locally by the blastogenic transformation of extravasated small lymphocytes. Incidentally, it was noted also, as in similar previous studies (Hall 1967, Moore & Hall 1973), that those few immunoblasts that extravasated into the areas of painted skin did not do so because they had any immunological affinity for oxazolone.

SUMMARY AND CONCLUSIONS

The application of oxazolone to the skin of sheep has an unequivocal effect on the behaviour of the microvasculature. This effect is distinct from acute inflammation in that granulocytes are not involved and it takes 48 h to develop. It is characterized principally by a 10–50-fold increase in the extravasation of mononuclear cells; most of these cells are small lymphocytes that lack surface Ig and some of them undergo transformation into blast cells after they have extravasated. The effect lasts for about four days and might be attributed either to a particular pharmacological property of oxazolone or to the general consequences of the local fixation and recognition of antigen. The second possibility seems to be the most likely; usually, antigen is fixed in lymph nodes and whenever an immunogenic dose is fixed there is always a substantial increase in lymphocyte traffic through the node. However, because oxazolone can both penetrate the skin and react with amino groups, some of it may become fixed locally in the skin and subcuticulum. A microgram of oxazolone would express some 10^{15} epitopes, so even a small amount might be sufficient to induce the volume of lymphocyte traffic that normally occurs only in organized lymphoid tissue.

ACKNOWLEDGEMENTS

Some of the preliminary experiments on the responses to oxazolone in the efferent lymph of sheep were done in collaboration with Dr John Hopkins; electron microscope studies were kindly performed by M.S.C. Birbeck. Part of the work described was supported by an MRC/CRC project grant.

References

Askenase PW, Asherson GL 1972 Contact sensitivity to oxazolone in the mouse. VIII. Demonstration of several classes of antibody in the sera of contact sensitised and unimmunized mice by a simplified antibody assay. Immunology 23:289-296

Davies AJS, Carter RL, Leuchars E, Wallis V 1969 The morphology of immune reactions in normal, thymectomised and reconstituted mice. II. The response to oxazolone. Immunology 17:111-126

Ford WL, Gowans JL 1969 The traffic of lymphocytes. Semin Hematol 6:67-83

Gell PGH 1944 Sensitization to 'tetryl'. Br J Exp Pathol 25:174-192

Gell PGH 1967 Delayed hypersensitivity: specific cell-mediated immunity. Br Med Bull 23:1-2

Gell PGH, Harington CR, Rivers RP 1946 The antigenic function of simple chemical compounds: production of precipitins in rabbits. Br J Exp Pathol 28:267-286

Hall JG 1967 Studies of the cells in the afferent and efferent lymph of lymph nodes draining the site of skin homografts. J Exp Med 125:737-754

Hall JG, Morris B 1963 The lymph-borne cells of the immune response. Q J Exp Physiol Cogn Med Sci 48:235-247

Hall JG Morris B 1965 The immediate effect of antigens on the cell output of a lymph node. Br J Exp Pathol 46:450-455

Hall JG, Smith ME 1971 Studies on the afferent and efferent lymph of lymph nodes draining the site of application of fluorodinitrobenzene (FDNB). Immunology 21:69-79

Hall JG, Scollay RG, Birbeck MSC, Theilen GH 1975 Studies on FeSV induced sarcomata in sheep with particular reference to the regional lymphatic system. Br J Cancer 32:639-659

Hall JG, Hopkins J, Orlans E 1977 Studies on the lymphocytes of sheep. III. Destination of lymph-borne immunoblasts in relation to their tissue of origin. Eur J Immunol 7:30-37

Hall JG, Birbeck MSC, Robertson D, Peppard J, Orlans E 1978 The use of detergents and immuno-peroxidase reagents for the ultrastructural demonstration of immunoglobulins in lymph cells. J Immunol Methods 19:351-359

Hartree EF 1972 Determination of protein: a modification of the Lowry method that gives a linear photometric response. Anal Biochem 48:422-427

Haynes FW, Field ME 1931 The cell content of dog lymph. Am J Physiol 97:52-56

Landsteiner K, Jacobs J 1935 Studies on the sensitisation of animals with simple chemical compounds. J Exp Med 61:643-656

Miller HRP, Adams EP 1977 Reassortment of lymphocytes in lymph from normal and allografted sheep. Am J Pathol 87:59-80

Moore AR, Hall JG 1973 Non-specific entry of thoracic duct immunoblasts into intradermal foci of antigens. Cell Immunol 8:112-119

Old LJ, Benacerraf B, Carswell E 1963 Contact sensitivity to carcinogenic polycyclic hydrocarbons. Nature (Lond) 198:1215-1216

Oort J, Turk JL 1965 A histological and autoradiographic study of lymph nodes during the development of contact sensitivity in the guinea pig. Br J Exp Pathol 46:147-159

Scollay RG, Hall JG, Orlans E 1976 Studies on the lymphocytes of sheep. II. Some properties of cells in various compartments of the recirculating lymphocyte pool. Eur J Immunol 6:121-125

Smith JB McIntosh GH, Morris B 1970a The traffic of cells through tissues: a study of peripheral lymph in sheep. J Anat 107:87-100

Smith JB, McIntosh GH, Morris B 1970b The migration of cells through chronically inflamed tissue. J Pathol 100:21-29

Turk JL, Stone SH 1963 Implications of the cellular changes in lymph nodes during the development and inhibition of delayed type hypersensitivity. In: Amos B, Koprowski H (eds) Cell bound antibodies. Wistar Institute Press, Philadelphia, p 51-60

Discussion

Gowans: What is the percentage of Ig-bearing cells in the blood in sheep?

Hall: About 25–30%. The figures I gave in Table 1 (p 203) refer merely to small lymphocytes, not to macrophages or blasts.

Gowans: I would like to be clear about the magnitude of the regional recirculation through normal tissues. In the gut there appears normally to be a substantial traffic but your baseline level for the skin was very low. Can I take it that normally there is not much traffic through normal skin?

Hall: Yes, I think that is so. Arnfin Engeset and John Humphrey have, I believe, collected peripheral leg lymph from themselves and got results similar to those we find in the sheep, with counts well under 1000 white cells/mm³ and perhaps as low as 200. That is a rather small traffic. While the increased traffic was going on we tried to see cellular extravasation with light and electron microscopes but we really couldn't. This makes the point that a very vigorous lymphocyte traffic may leave very little in the way of microscopic stigmata.

Gowans: A 15-fold increase in a very small number still isn't a very large number.

Hall: No, but we had up to a 50-fold increase which is up to the sort of number that might come out of a lymph node and that might be called immunologically significant. Whether that has any relevance to the induction of delayed hypersensitivity is another matter, which these experiments don't show. In the FDNB experiments the amount of antigen transported to the nodes in combination with the cells in the afferent lymph was quite inconsequential. Most of it was combined with plasma proteins.

Humphrey: Could the blasts you saw have been generated peripherally by cells that were initially stimulated in the lymph node and then entered the bloodstream, recirculated and were restimulated by antigen in the skin? In other words, if you cannulated that node and took away the cells in the efferent lymph, would you have stopped the arrival of blasts?

Hall: We did a lot of experiments where we had another lymph node which had been stimulated with oxazolone. We labelled those blasts *in vitro,* injected them and tried to recover them in the afferent lymph coming from the site of skin painting. There are technical difficulties in that sort of experiment because some blasts are filtered out in the lungs and so on. Even making a very generous allowance for that sort of experimental error, we couldn't begin to account for the number of blasts in the afferent lymph in terms of recirculation from the blood. I think that they were generated locally. Most of them had the characteristics of non-immunoglobulin-forming blasts.

Gryglewski: One of the techniques used to induce exudative inflammation is injection of macromolecules such as dextran, carrageenin or glycogen into the rat peritoneal cavity. In carrageenin-induced peritonitis, for instance, polymorphs predominate. In other types of peritonitis 'monocytes' may predominate. Has this phenomenon any relation to the oxazolone-induced antigenic inflammation? Could fixed antigen induce generation of macromolecules in the tissue which will selectively attract 'monocytes' or 'polymorphs'?

Hall: I don't know how to answer that. We have certainly made the same observation. Rats and mice react very differently, according to which particular stimulus is used in the peritoneal cavity. In one situation we get polymorphs and in others, if we are lucky, we get what we are after, which is macrophages. I haven't enough data to say that it is the antigenic macromolecules that call forth the macrophages. The inherently inflammatory ones are obviously going to call forth polymorphs but they may be antigenic as well.

Gowans: The classical way of obtaining large numbers of polymorphs in a rabbit is to put saline in its peritoneal cavity. Similarly, an almost pure population of macrophages can be obtained from a rabbit by inducing a peritoneal exudate with paraffin oil. If you try these procedures in a rat only mixed populations are obtained.

Vane: Have you any evidence that tritium-labelled FDNB was sticking in the skin, and was that for a long time?

Hall: One can get a rough idea of how much oxazolone is absorbed by back-titrating it, i.e. one can use afferent lymph coming from the treated skin to inhibit an antibody—antigen reaction involving oxazolone. In this way it was found that the amount of oxazolone absorbed seemed to follow the kinetics shown by tritiated FDNB (Hall & Smith 1971). We couldn't measure it accurately.

McConnell: Have you any idea of the mitotic activity of the non-immuno-globulin-bearing cells in the afferent lymph draining to the node?

Hall: Certainly the blast cells in the afferent lymph are taking up DNA precursors. I haven't looked at their mitotic indices but the blast cells are actively synthesizing DNA.

Simionescu: Does the extravasation of these cells take place at the level of the capillaries?

Hall: As we can't see this in the microscope I can't say where it is occurring. I can only assume that the cells come out of the blood vessels.

Simionescu: But did you look specifically for the venules, for example?

Hall: The electron microscopists looked everywhere, trying to find cells that were extravasating, and were unable to do so.

Morris: I want to support your observations on the transmission of pure populations of lymphocytes by conventional endothelial vessels. The liver is a classical example. The content of cells in its peripheral lymph is about $3-4 \times 10^6$ cells/ml—not all that different from the number in efferent lymph from the portal lymph node. These cells are small lymphocytes together with a proportion of macrophages. The transmission of large numbers of lymphocytes also occurs when a kidney autotransplant is done. For some reason there is always a very large increase in the lymphocyte traffic coming out of the autotransplanted kidney and this continues for a very long time. As many as $2-3 \times 10^6$ cells/ml appear in renal afferent lymph after the autotransplantation of a kidney into the neck. If you partially occlude the ureter the same phenomenon occurs. The cells that are transmitted are always pure populations of lymphocytes. They never show any blast cell response in these situations and the population always contains a low proportion of cells with surface Ig. So there is selection, it is specific, it relates to this one class of lymphocytes, and it occurs through normal and not high endothelium.

Gowans: I take your point that there is a normal vascular bed in the kidney which suddenly decides to transmit a pure population of small lymphocytes. But something has changed in that vascular bed to enable this to happen in the renal allograft and the hydronephrotic kidney. What has happened?

Morris: Whatever has happened doesn't allow the transmission of a random selection of blood cells and the vascular bed has not been transformed into one composed of high endothelial vessels. The interesting thing is that the somatic vessels transmit the Ig$^-$ cells, while the blood vessels in the lymph node transmit a very much higher proportion of Ig$^+$ cells. Maybe this difference is due to high endothelial venules.

Born: I was trying to think of an experiment by which the lymphocytes could be immobilized in the act. Perhaps when the process is at its maximum, infusing an agent such as prostaglandin E_1, which inhibits the mobility of lymphocytes, into the arterial supply to the site might immobilize the traffic locally and temporarily. Then the site could be examined histologically and in other ways. There could be complicating effects, e.g. on blood flow, but they would perhaps be restricted to the arterial side whereas the lymphocyte traffic occurs on the venous side. Thus there may be a way of analysing the traffic by catching a large population of lymphocytes in transit.

Hall: We must try and do that kind of experiment to answer Dr Gowans' question of what has happened to these capillaries.

Ford: Isn't the adjuvant granuloma in the sheep a more promising model for this purpose, in that it is associated with a greater increase in lymphocyte traffic (Smith et al 1970) than develops through an area inflamed by contact sensitivity?

Hall: That is a chronic phenomenon, whereas the effect of oxazolone is evanescent and reversible. By a week or 10 days it will be back to normal. The granuloma takes some time to develop and even longer to go away. In the granuloma there are structural changes in the endothelium. I can't say that none occurred after treatment with oxazolone, as the sampling error in electron microscopy is obviously high, but we didn't see any.

Weissman: Is there any way of making sheep specifically unresponsive to oxazolone and seeing whether the increased output of cells has anything to do with antigenic specificity rather than with oxazolone as a chemical?

Hall: That would be a nice one to test. Asherson & Barnes (1973) made mice unresponsive to picryl chloride by injecting them with picryl sulphonic acid, which is more soluble. I tried their regimen on a sheep, which died a disgusting death immediately afterwards. It is the sort of experiment you can do with a mouse but not with a sheep.

Weissman: Ian McConnell's technique is close to that.

Howard: Wouldn't a much simpler way of doing it be to induce a very mild physiological inflammation?

Hall: Would you call the effect of oxazolone acute inflammation?

Howard: In all the experiments it is inevitably associated with some kind of inflammation. Presumably oxazolone causes reddening, for example.

Hall: No, it doesn't. Reddening in a pigmented animal is difficult to assess but shorn sheep have very delicate legs. With the primary application of FDNB or the secondary application of oxazolone you can see the swelling and feel the warmth with no difficulty at all. After the primary application of oxazolone there were absolutely no visible changes whatsoever.

Howard: Does capillary permeability remain normal?

Hall: There was a slight increase in both the flow rate and the protein concentration of the lymph but one didn't see polymorphs coming in, though they swarmed in after FDNB.

Gowans: Which of these fascinating things you have demonstrated has anything to do with contact sensitization?

Hall: I don't think any of them necessarily have anything to do with it. Whenever I challenge a sensitized sheep, all I see is an Arthus reaction. That doesn't mean it wouldn't show delayed hypersensitivity as well.

References

Asherson GL, Barnes RMR 1973 Contact sensitivity in the mouse.X. The use of DNA synthesis *in vivo* to determine the anatomical location of immunological unresponsiveness to picryl chloride. Immunology 25:495-508

Hall JG, Smith ME 1971 Studies on the afferent and efferent lymph of lymph nodes draining the site of application of fluoro-dinitrobenzene (FDNB). Immunology 21:69-79

Smith JB, McIntosh GH, Morris B 1970 The migration of cells through chronically inflamed tissues. J Pathol 100:21-29

Metabolic studies of high-walled endothelium of postcapillary venules in rat lymph nodes

P. ANDREWS, W.L. FORD and R.W. STODDART

Department of Experimental Pathology, University of Manchester Medical School, Stopford Building, Oxford Road, Manchester M13 9PT

Abstract In comparison with other endothelia, high-endothelial (HE) cells in lymph nodes show metabolic specialization, including the ability to incorporate [^{35}S]sulphate. The maximum autoradiographic labelling of HE cells is seen in the draining lymph nodes 15–30 min after injection of [^{35}S]sulphate into the rat footpad and at this stage it is almost confined to the Golgi apparatus.

The conditions of sulphate incorporation have been studied in tissue slices. The kinetics of sulphate uptake (and loss in non-radioactive medium) and the effects of metabolic inhibitors indicate that at least two processes are involved. Only a small amount of the sulphate incorporated by the whole tissue is detectable autoradiographically in HE cells. This sulphate is linked to a macromolecule which may be either an alkali-stable glycoprotein or a glycosaminoglycan other than heparan sulphate. Experiments *in vitro* with labelled sugars have shown that glucosamine, glucuronic acid and mannose were also selectively incorporated in high-endothelial venules (HEV) but galactose, fucose and glucose were poorly localized there.

Partly purified sulphated material from lymph nodes has been injected into the skin of rats. This induced an increased localization of recirculating lymphocytes which was not simply a consequence of inflammation since trypsinized lymphocytes did not localize in excess. The possible functions of the sulphated molecule are discussed.

The structure of postcapillary venules in the paracortical area of lymph nodes is strikingly distinctive in that these vessels are lined with a high-walled endothelium surrounded by an elaborate basement membrane and pericyte lattice (Schulze 1925). One of the functions of these high-endothelial venules (HEV) is the selective transport of lymphocytes from the blood into the extravascular compartment of the lymph node (Gowans & Knight 1964). The precise relationship between the structure and function of these specialized vessels has intrigued many investigators over the past 15 years but has remained an enigma. This paper focuses on the specialized nature of HEV as revealed by ultrastructural, histochemical and metabolic studies, and is

© *Excerpta Medica 1980*
Blood cells and vessel walls: functional interactions
(Ciba Foundation symposium 71) p 211-230

secondarily concerned with the possible relevance of these features to lymphocyte migration.

In physiological conditions HEV are confined to the lymph nodes and the gut-associated lymphoid tissues of mammals (Miller 1969). They also develop in non-lymphoid tissue under certain conditions of chronic inflammation in both mammals (Smith et al 1970) and birds (Miller 1969). It is clear that HEV are not obligatory for lymphocyte diapedesis because lymphocytes migrate selectively from the blood in large numbers into the spleen and bone marrow. Brief exposure of small lymphocytes to trypsin *in vitro* transiently depresses their capacity to leave the blood by crossing HEV in lymph nodes (Woodruff & Gesner 1968), gut-associated lymphoid tissues and sites of immune-mediated inflammation but does not impede their entry into any other sites (Rannie et al 1977).

Studies of HE cells by electron microscopy have emphasized their large notched nuclei with a thick even rim of chromatin. In the cytoplasm both free ribosomes and rough endoplasmic reticulum are plentiful (Schoefl 1972, Wenk et al 1974, Anderson et al 1976). Van Deurs & Ropke (1975) stressed that the endoplasmic reticulum often appears to be closely associated with abundant mitochondria. The Golgi apparatus is usually extensive but it is not evident in the HEV of the newborn mouse, although these are otherwise similar in appearance to HEV in adult mouse lymph nodes. Van Deurs & Ropke found that the Golgi apparatus does not become prominent until four days of age when the blood lymphocyte count is increasing. They observed that the Golgi apparatus might lie on any side of the nucleus but was usually oriented towards lymphocytes in transit across the HEV. This contrasts with our suggestion that the Golgi apparatus is usually nearer the luminal border than the base of the HE cell in the adult rat lymph node (Ford et al 1978).

Modifications of the typical morphology of HEV have been reported in several experimental conditions. Goldschneider & McGregor (1968) observed that HE cells were less plump in neonatally thymectomized rats; their cytoplasm was scanty and less pyroninophilic. The appearance of the HEV was restored to normal 24 h after the transfusion of syngeneic lymphocytes. Similar changes were noted in the HEV of congenitally athymic (*nu/nu*) mice (De Sousa et al 1969), after prolonged administration of prednisone to rats (Miller 1969), and in graft-versus-host disease (Clancy 1973). These observations might be interpreted as showing that either a functioning thymus or traffic of i lymphocytes across the endothelium is a necessary condition for at least some of the morphological features of HEV. However van Deurs & Ropke (1975) noted that the development of HEV is the same in intact, neonatally thymectomized and nude mice. Our preliminary studies of congenitally athymic (*rnu/rnu*) rats have revealed marked variability in the morphology of HE cells between different organs. The i.v. injection of [³H]-uridine-labelled lymphocytes followed by autoradiography of tissues removed 10–30 min after injection showed the normal pattern of localization in HEV of lymph nodes (S. Fossum, M.E. Smith & W.L. Ford, unpublished) confirming Goldschneider & McGregor's claim (1968) that the function of HEV in transporting lymphocytes does not depend on the thymus. It is plausible that the occasional appearance of flatter HE cells after thymectomy is a toxic effect associated with a wasting syndrome (Miller 1969). Thus we doubt that this approach will yield any important clues to the functional significance of HEV structure.

Enzyme histochemical techniques applied to lymph node sections have shown HEV to be particularly rich in non-specific esterase, lactic dehydrogenase and acid phosphatase activities when compared to other endothelial cells. Acid phosphatase is associated with the Golgi apparatus rather than with lysosomes (Smith & Henon 1959, Ropke et al 1972, Anderson et al 1976). The presence of mucopolysaccharide was suggested by metachromatic staining with toluidine blue and positive staining with alcian blue but Ropke et al (1972) concluded that HEV do *not* contain considerable amounts of sulphated mucopolysaccharides because they were able to attribute the metachromasia to a high RNA content and the conditions of alcian blue staining suggested that it was caused by some other material. Since their careful study did not appear to be completely conclusive we tested for the incorporation of [³⁵S]sodium sulphate into HEV and found that it was rapidly taken up by HE cells but not by other endothelia (Ford et al 1978). We now report several ways in which this observation has been pursued.

INJECTION OF [³⁵S]SULPHATE INTO INTACT RATS

Sodium [³⁵S]sulphate (Radiochemical Centre, Amersham; ~100 mCi/mmol) was injected subcutaneously into the footpads of AO rats of either sex at a dose of 1 μCi/g body weight. Popliteal lymph nodes were removed at intervals from 15 min up to 24 h after injection. The lymph nodes were either fixed and processed for autoradiography (for both the electron and light microscopes) or they were homogenized for biochemical analysis. Autoradiography of conventional paraffin sections revealed a remarkably selective localization of labelled material to HE cells. This was most pronounced at 15 min after footpad injection and had begun to wane slightly by 30 min after injection. It was still perceptibly concentrated in HE cells although much fainter by 8 h. At 15 min other endothelial cells and mast cells were unlabelled although radioactive sulphate accumulated over the next few hours in some mast cells, especially those in the subcapsular sinus.

After the i.v. injection of $^{35}SO_4$ selective labelling of HEV was consistently found in the several different lymph nodes examined but other endothelial cells, including those in the spleen, were not labelled. To obtain comparable grain intensity to that measured after footpad injection it was necessary to inject 5 mCi of ^{35}S i.v.

Electron microscopic autoradiography showed that at 30 min after injection $^{35}SO_4$ in HE cells was almost entirely located in the Golgi apparatus and associated vesicles, as previously illustrated (Ford et al 1978). However, by 2 h the localization in the Golgi apparatus had declined. Vesicles throughout the HE cell cytoplasm were sparsely labelled and the nuclear membrane of the HE cell appeared to be labelled (Fig. 1). At no time was any concentration near the plasma membrane of the HE cell discernible and the significance of a few grains over some migrating lymphocytes was uncertain.

The observation that sulphate localization was detectable by autoradiography indicated that it had been incorporated into a substance that was rendered insoluble by fixation of the tissue in glutaraldehyde. An attempt was made to purify this material from a homogenate prepared from lymph nodes removed from rats 2.5 h after the footpad injection of $^{35}SO_4$. The method is summarized in Fig. 2.

The fraction eluted from the ECTEOLA-cellulose column (fraction 5) has been studied in several ways. It contained 30–40% of the radioactivity initially present in the lymph node pellet. Discontinuous polyacrylamide gel electrophoresis of fraction 5 under non-reducing conditions on 7.5% acrylamide gels in sodium dodecyl sulphate (SDS) (Neville 1971) revealed five to seven Coomassie Blue-staining bands, indicating an extensive but by no means complete purification of the lymph node extract. An appreciable proportion of this sample failed to enter both the stacking and running gels. Addition of 2-mercaptoethanol to a portion of the unreduced sample resulted in the migration of all the Coomassie Blue-staining material into the gel, giving rise to an increased banding complexity.

Amino acid analysis of fraction 5 showed a predominance of acidic and hydrophobic amino acids, as would be expected from the extraction procedure. Hydroxyproline was not found by the method of Woessner (1961), indicating that collagen was not a major contaminant. Carbohydrate analysis by gas–liquid chromatography indicated that the fraction contained a number of neutral sugars, including fucose, with trace amounts of hexosamine. There was no evidence for the presence of either sialic acid or hexuronic acids when the methods of Warren (1959) and Bitter & Muir (1962),

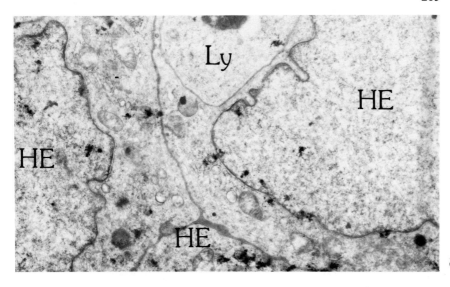

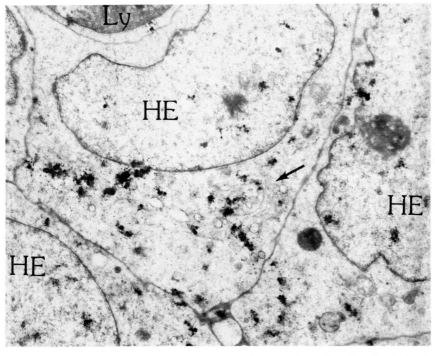

FIG. 1. Electron microscope autoradiographs of part of an HEV from a lymph node 2 h after the injection of $^{35}SO_4$. (Compare with 15 min localization in Plates 2 and 3 of Ford et al 1978.) (a) Light cytoplasmic labelling in three HE cells. The nuclear membrane of the large HE cell on the right appears to be labelled. Ly, lymphocyte. (b) Some radioactivity remains near the Golgi apparatus (arrowed) but most of the radioactivity appears to be associated with cytoplasmic vesicles.

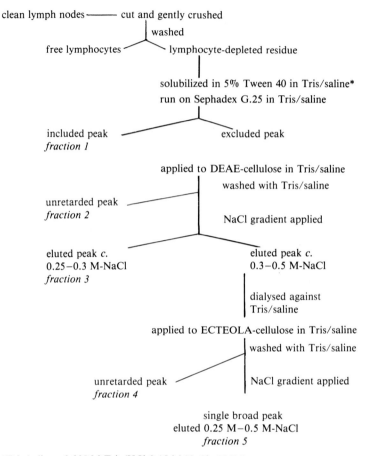

clean lymph nodes ——— cut and gently crushed

washed

free lymphocytes ⟋   ⟍ lymphocyte-depleted residue

solubilized in 5% Tween 40 in Tris/saline*
run on Sephadex G.25 in Tris/saline

included peak   ———   excluded peak
fraction 1

applied to DEAE-cellulose in Tris/saline

washed with Tris/saline

unretarded peak ———
fraction 2

NaCl gradient applied

eluted peak c.   eluted peak c.
0.25–0.3 M-NaCl   0.3–0.5 M-NaCl
fraction 3

dialysed against
Tris/saline

applied to ECTEOLA-cellulose in Tris/saline

washed with Tris/saline

unretarded peak ⟋   NaCl gradient applied
fraction 4

single broad peak
eluted 0.25 M–0.5 M-NaCl
fraction 5

*Tris/saline = 0.005 M-Tris/HCl;0.15 M-NaCl pH 7.4.

Fig. 2. Scheme for purification of ^{35}S-labelled material from lymph nodes. ECTEOLA: epichlorhydrin triethanolamine.

respectively, were used for colorimetric determinations, but since the sulphated material may have been a minor component of fraction 5 this finding is not conclusive.

Analysis of fraction 5 by isopycnic centrifugation in caesium chloride under denaturing conditions revealed that the radiolabelled material migrated with the protein components of the sample, away from the proteoglycan carrier (chondroitin sulphate purified from bovine nasal cartilage). Under non-denaturing conditions the radioactivity was found close to the buoyant density of the glycoprotein carrier (ovalbumin).

When fraction 5 was concentrated by ultrafiltration and applied to a column of Bio-gel A-5 in 0.1% SDS, 0.1M tris/HCl pH 7.4, it resolved, on elution, into two peaks of radioactivity. The larger of these ran just ahead of the excluded volume while the smaller ran slightly in front of the tritiated myosin marker. When fraction 5 was lyophilized and dissolved in 0.01% acetic acid before chromatography on Bio-gel A-5 under the same conditions, a different radioactivity profile was obtained. The major peak was lost, while a new component appeared as a sharp peak running near the included volume. The component running near myosin remained.

Addition of acetic acid to fraction 5, concentrated by ultrafiltration, produced a profile of radioactivity intermediate between the profiles described above; all three of the components already referred to were present and the material running near myosin was substantially increased. These results strongly suggest that the radiolabelled material contained in fraction 5 can exist in various states of aggregation. Nitrous acid caused no loss of radioactivity from fraction 5.

Treatment of fraction 5 with proteolytic enzymes indicated that the radioactive component was significantly susceptible to extensive pronase digestion but only partially susceptible to treatment with papain.

UPTAKE OF [^{35}S]SULPHATE *IN VITRO*

Lymph node slices maintained in a defined medium at 37°C were used to study the uptake of [^{35}S]sulphate into HE cells. This permitted the metabolic requirements of incorporation to be investigated and had the practical advantage of avoiding the wastage of so much ^{35}S that disposal limited experimentation. Phosphate-buffered saline (PBS) supplemented with glucose and sodium sulphate was an adequate medium for the maintenance of tissue slices for up to 4 h, at which time no deterioration of the HE cells was visible histologically. Cervical lymph nodes were removed into PBS from freshly killed AO rats, cleaned of fat and sliced as thinly as possible with a fine pair of scissors. Slices totalling 1.5–6.0 mg (dry weight) were transferred to 0.4 ml of an incubation mixture containing 10 μCi of ^{35}SO$_4$. After a variable period of incubation and 'chasing', slices of lymph node were withdrawn and rinsed three times in cold PBS and once with water. They were transferred to preweighed 1 cm discs of glass fibre paper, on which they were crushed and dehydrated with acetone. After evaporation the discs were reweighed and transferred to a scintillation vial for counting by a modification of the method of Stoddart & Northcote (1967). The acetone extracted less than 5% of the total radioactivity.

Glucose concentrations of from 1 to 50 mM were used at sulphate concentrations of from 1 μM to 10 mM (final concentrations). At any given level of sulphate, increasing the concentration of glucose up to 20 mM promoted the uptake of $^{35}SO_4$, but higher concentrations were inhibitory. Below about 5 mM-glucose the sulphate uptake was slight. At any given glucose level the uptake of sulphate (radioactive plus non-radioactive) was proportional to the sulphate concentration over a wide range (1 μM–5 mM) and departure from linearity was observed only at sulphate concentrations above 5 mM, well above the physiological value. As standard conditions, a glucose concentration of 7 mM and a sulphate concentration of 1.5 mM were used; i.e. close to physiological values in the rat (Fig. 3).

The time course of sulphate incorporation at low glucose concentration (10 mM) presented a curious biphasic appearance (Fig. 3, curve B) with an initial peak of uptake, followed by some loss and then a slower progressive incorporation. At glucose concentrations near to the physiological value this curve was smoother (curve A) with an initial rapid incorporation, followed

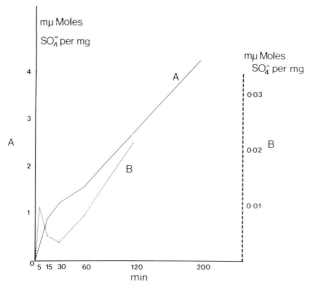

FIG. 3. The uptake of $^{35}SO_4$ into slices of lymph node, incubated in PBS + 1.5 mM-sodium sulphate at two concentrations of glucose (A, 7 mM; B, 10 mM, final concentrations). The 'shoulder' of curve A was consistently found although its size varied. The phase to the end of the shoulder was designated 'uptake I' (0–60 min) and the steadily ascending part of the curve was called 'uptake II' (after 60 min). Curve B is typical of the pattern of incorporation at low glucose levels, where uptake I forms a distinct peak.

by a plateau and then a steady increase of incorporation. Following a large, abrupt increase in glucose concentration, wild oscillations of sulphate uptake were consistently found during the early phase. These results suggested that the initial rapid uptake of sulphate and the later more gradual uptake might represent distinct processes.

When tissue laden with $^{35}SO_4$ was incubated in non-radioactive medium half the radioactivity was lost after 7–8 min at 37°C but subsequently the rate of loss became much slower and the shape of the curve suggested again that more than one process could be involved. On autoradiographic examination of sections of incubated lymph node slices the same pattern of localization was observed as *in vivo*. Although the grain density over HE cells was high, there appeared to be a large discrepancy between the amounts of incorporated radioactivity as measured by scintillation counting and by autoradiography. Moreover, chasing for 15 min in a non-radioactive medium did not alter the pattern or intensity of localization, despite the removal of most of the radioactivity.

We explored this discrepancy by putting samples of incubated lymph nodes through extraction procedures similar to those to which they would be subjected during processing for autoradiographic analysis. At least 95% of the radioactivity was extracted during this procedure, especially by fixation. This extracted material was water-soluble and on gel filtration in PBS on Sephadex-G10 and on Biogel P4 it moved as single peaks of low molecular weight but these did not correspond to free sulphate. In addition the effects of chasing $^{35}SO_4$ in a non-radioactive medium showed that at 25°C the time taken to lose half the radioactivity was 30 min (cf. 7–8 min at 37°C), while at 0°C little radioactivity was lost for more than 90 min. These characteristics of the rapid uptake of sulphate were consistent with its being effected by an endocytotic and probably a diacytotic (transcytotic) mechanism.

The effect of metabolic inhibitors on the incorporation of $^{35}SO_4$ into lymph node slices was tested. The uptake was measured by scintillation counting and in most cases by autoradiography (Table 1). Sodium selenate and sodium fluoride both inhibited the localization of radioactivity in HE cells but showed different effects on the gross incorporation of $^{35}SO_4$ and its localization in mast cells. Mast cells appeared to be more susceptible to the inhibition of oxidative phosphorylation than did HE cells. The effects of selenate showed that the slower, steady uptake of $^{35}SO_4$ (Fig. 3) included not only incorporation into material demonstrable autoradiographically but also some other material that was removed during processing. This might be, for example, a sulphated glycolipid.

A small part of the $^{35}SO_4$ incorporated into tissue slices was not removed by

TABLE 1

Inhibition of [^{35}S]sulphate localization in lymph node slices

Inhibitor	Cell localization		Uptake	
	HE cells	Mast cells	I	II
Selenate (1.5 mM)	−	−	~	↓
Arsenate (10 mM)	+ + +	+ + + +	↓	~
Fluoride (10 mM)	−	+ + + +	↓	~
2,4-Dinitrophenol (1 mM)	+ +	−	~	↑
Azide (10 mM)	+ + +	+ +	n.d.	n.d.
Cyanide (1 mM)	+ + + +	+	~	↓

Uptake I,	initial (non-linear) sulphate incorporation into slices of node.
Uptake II,	slow (linear) sulphate incorporation into slices of node (see Fig. 3).
−,	no selective localization, i.e. complete inhibition.
+ + + +,	no inhibition of selective localization.
↑,	increased uptake.
↓,	decreased uptake.
~,	little or no effect.
n.d.,	not determined.

prolonged incubation in non-radioactive medium or by repeated exchange against 0.1 M-sodium sulphate. Most of this non-exchangeable sulphate was localized in HE cells as assessed by autoradiography. In the column fractionation procedure already described (Fig. 2) the sulphated material produced *in vitro* showed a similar fractionation profile to the material produced *in vivo*. Polyacrylamide gel electrophoresis of fraction 5 produced a similar banding pattern in both cases.

STUDIES OF FIXED SECTIONS OF LYMPH NODES

Several agents were tested for their effect on glutaraldehyde-fixed sections of lymph node slices labelled *in vitro*. Radioactivity within HEV was completely stable to a range of pH values from 3 to 10 but some loss took place above 10. Exposure to nitrous acid was without effect, indicating that radioactivity could not have entered an *N*-sulphate group, as in heparan sulphate. Radioactivity was unaffected by alkaline borohydride and so could not have been incorporated into a saccharide attached to polypeptide by an *O*-

glycoside link. When similar experiments were done with methanol-fixed material, whereas the stability of the molecule to alkali was unaltered at pH 3, in the presence of 0.1% acetic acid most of the radioactivity could be leached out. Glutaraldehyde fixation rendered the radioactive material resistant to acetic acid, suggesting that acetic acid had removed a molecule that included a polypeptide sequence. This sensitivity to acetic acid is also shown by the partially purified material from lymph nodes (fraction 5). Neither chloroform–methanol (1:1 by vol.) nor 1% aqueous mercaptoethanol removed radioactivity from methanol-fixed material. Apparently the bond to sulphate is stable and if that sulphate is attached to a saccharide which is in turn linked to polypeptide, the linkage to peptide must be N-glycosidic or some other form of alkali-stable bond.

THE INCORPORATION OF LABELLED SUGARS INTO HIGH-ENDOTHELIAL VENULES

Because it seemed likely that sulphate could be added to a glycosylated molecule, the incorporation of six labelled (^{3}H or ^{14}C) sugars into HE cells was investigated autoradiographically (Table 2). The effects of supplementing glucose in promoting or inhibiting the localization of sugars suggested that HE cells and mast cells differ in their ability to interconvert sugars. There is no evidence that these sugars were incorporated in the form in which they were supplied, or that they entered the same molecule as sulphate. However, the striking localization of glucosamine, mannose and glucuronic acid in HE cells demonstrates metabolic specialization of these cells in a novel way.

In conclusion, the nature of the sulphated material in HE cells remains an open question and there is a possibility that the molecule may not conform to conventional definitions of either glycoprotein or glycosaminoglycan.

FUNCTIONAL STUDIES OF THE PARTIALLY PURIFIED SULPHATED MATERIAL (FRACTION 5)

In diapedesis, lymphocytes first adhere to the surface of HE cells and then retract their microvilli (van Ewijk 1980). The possibility that the sulphated compound extracted from lymph nodes mediates one of these processes was tested as follows. Lymphocytes from thoracic duct lymph, blood or lymph nodes were incubated with fraction 5, centrifuged and resuspended in medium RPMI 1640 with 10% fetal calf serum and incubated at 37°C for a further 30 min before being fixed for scanning electron microscopy. The incomplete loss and stunting of surface microvilli were consistently noted. Brief exposure to

TABLE 2

Sugar incorporation into lymph nodes

	Cell localization			Comments
Sugar	HE cells	Mast cells	Lympho-cytes	
Glucose (CFB.96)	(+)	+ + + +	−	Mast cells label after 10 min incubation
Mannose (TRA.364)	+ + +	(+)	−	HEV label after 10 min incubation, only in absence of glucose
Galactose (CFB.132)	+ +	+ +	−	Mast cells label only in absence of glucose. HEV label in its presence or absence after 20-60 min
Glucosamine (TRK.398)	+ + +	+ +	(+)	HEV label in 20 min, more rapidly than mast cells. Glucose required
Glucuronic acid (CFB.136)	+ + +	+ + +	−	HEV label in 10 min; mast cells in 30 min. Glucose required.
Fucose (TRK.477)	(+)	−	(+ +)	No labelling of most HEV, with or without glucose

Intensity of localized labelling is scored + to + + + +; −, no label above background. Parentheses indicate that only a few cells were labelled (<5%) at the level indicated, as opposed to a majority.

trypsin alone produced rather different changes in surface appearance, and the combined effect of trypsin and fraction 5 in succession was to render the surface of most lymphocytes almost completely smooth but a minority retained tenuous microvilli (Fig. 4).

Fraction 5 (5–10 μg in 0.1 ml Dulbecco's solution with calcium and magnesium [DAB]) was injected into the flank skin of rats 2 h before the i.v. injection of 2×10^8 syngeneic thoracic duct lymphocytes labelled with ^{51}Cr. The recipients were killed 2, 4 or 6 h later and after whole-body perfusion (Rannie & Donald 1977) the skin sites were excised for gamma counting. Compared to uninjected skin the material always induced a three- to sevenfold increase in radioactivity. However, control sites injected with DAB or chondroitin sulphate never had more than twice the radioactivity of treated skin (Table 3). Other rats which had received intradermal injections of the same materials were given 2×10^8 trypsinized ^{51}Cr-labelled lymphocytes i.v.

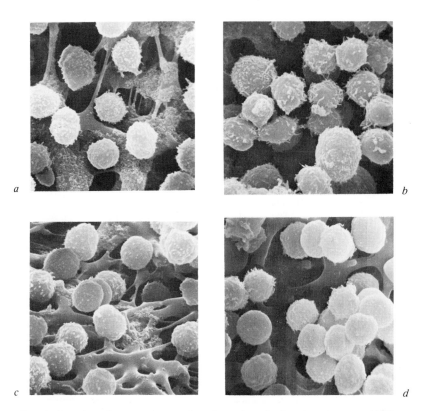

Fig. 4. Scanning electron micrographs of rat thoracic duct lymphocytes. (a) Untreated, (b) trypsin-treated (0.02 mg/ml for 10 min), (c) treated with fraction 5, (d) treated with trypsin followed by fraction 5. See text for description.

As these cells did not localize in the sites of injection of fraction 5 in increased numbers until the effect of trypsin had begun to wane, it was concluded that the material was not simply attracting more lymphocytes because of an inflammatory reaction (Table 3). Rannie et al (1977) found that trypsinized lymphocytes were not inhibited from entering a site of non-immune inflammation.

The increased lymphocyte localization at skin sites has been confirmed in further experiments with different time intervals between the intradermal injection of fraction 5 and i.v. lymphocyte injection, but it is not yet possible to say whether this is caused by an increased rate of migration from the blood or to prolonged retention in the tissue of lymphocytes that have migrated from the blood at physiological rates. Intradermal injections of each of the

TABLE 3

Localization of i.v.-injected lymphocytes in skin sites

Time of sampling (hours)	Treatment of lymphocytes	Radioactivity (^{51}Cr) c.p.m. g^{-1}		Ratio[2] of injected skin c.p.m. g^{-1} / uninjected skin c.p.m. g^{-1}		
		Lymph nodes	Uninjected skin	ECTEOLA Fraction 5	DAB	Chrondroitin sulphate
2	None	42 000	81	3.5	1.6	1.5
	Trypsin	1 400	119	2.1	2.0	1.9
4	None	49 400	66	4.7	2.2	2.0
	Trypsin	2 800	112	1.8	1.5	1.5
6	None	132 000	54	6.3	2.1	2.3
	Trypsin	15 400	73	4.1	1.8	2.7

[a]^{51}Cr-labelled lymphocytes injected i.v. 2 h after skin injections. Time given refers to interval between i.v. injection and killing recipient.
[b]Each ratio is the mean of two observations.

other fractions obtained during the purification procedure (Fig. 2) sometimes produced inflammatory reactions but none except fraction 5 increased the localization of untreated lymphocytes in preference to trypsinized lymphocytes.

GENERAL CONCLUSIONS

The object of these experiments was to explore the nature and function of a sulphated compound found in HE cells of lymph nodes. Since the compound is precisely localized to the Golgi apparatus 30 min after injection it probably represents sulphation of a molecule synthesized within the HE cell itself. The early waning of autoradiographic localization indicates that the material is either secreted from the HE cell or is metabolized rapidly.

The experiments on lymph node slices *in vitro* showed that several processes are involved in sulphate uptake into the tissue as measured by scintillation counting. Most of the early rapid uptake into an exchangeable pool probably represents endocytosis and is of uncertain relevance to the incorporation into HEV as detected by autoradiography. On the other hand the slow progressive incorporation of sulphate into a non-exchangeable form probably includes the material seen in HEV by autoradiography. Selenate completely inhibited the autoradiographic localization in HEV and reduced the non-exchangeable incorporation by half, suggesting that the uptake of sulphate by other cells also contributed to the slow incorporation of activity.

A crucial question is whether the radioactive material in fraction 5 is identical to the sulphated substance seen by autoradiography in HEV. There is no rigorous evidence that these molecules are the same but all the tests to which they have been subjected indicate their similarity. The susceptibility of both substances to acetic acid is highly suggestive. We shall pursue this by testing whether an antibody raised against fraction 5 binds selectively to HEV in frozen sections.

We have tried to test the notion that the sulphated molecule synthesized in HEV is secreted into the bloodstream, where it acts as a signal to lymphocytes first to adhere and second to retract their microvilli. This is rather similar to the suggestion of Wenk et al (1974) that the Golgi apparatus of HE cells produces material which is incorporated into the HE cell coat and could thus influence lymphocytes in the blood. Both their notion and ours explicitly assume that, compared to other leucocytes, recirculating lymphocytes must be uniquely responsive to this 'pheromone'.

No evidence was found that the sulphated molecule was incorporated in the cell membrane or coat, but partially purified material (fraction 5) did fulfil two predictions arising from our hypothesis. These findings are extremely tenuous for two reasons: (1) the uncertain relationship between the ^{35}S-labelled material in fraction 5 and the material detected by autoradiography in HE cells and (2) the possible biological activity of other components of fraction 5. Several other functions could be speculatively ascribed to a molecule synthesized only by HE cells: (1) it may enable lymphocytes which have attached by other means to HE cells to move freely between the cells and eventually detach themselves; (2) it could reseal the gaps made by migrating lymphocytes and so maintain the integrity of the endothelium; (3) it might set up a chemotactic gradient to promote the migration of lymphocytes through the basement membrane and onwards through the labyrinth of the node; (4) it might prevent other leucocytes from adhering to HE cells, and (5) the release of the substance might meter the size of the recirculating lymphocyte pool.

All these possibilities seem less plausible than the signal hypothesis. The role of the sulphated compound is worth pursuing since elucidation of the mechanism of lymphocyte diapedesis across HEV might open up new approaches to the study of other instances of selective leucocyte migration.

ACKNOWLEDGEMENTS

The work was largely supported by MRC programme grant G972/455/B. It is a pleasure to thank Mrs Tamar Aslan for technical assistance and members of the Department of Medical Biochemistry, especially Drs D.W. Milsom and J.C. Anderson and Professor J.E. Scott, for their advice.

References

Anderson ND, Anderson AO, Wyllie RG 1976 Specialized structure and metabolic activities of high endothelial venules in rat lymphatic tissues. Immunology 31:455-473

Bitter T, Muir HM 1962 A modified uronic acid carbozole reaction. Anal Biochem 4:330-334

Clancy J 1973 Non-specific inhibition of adult thoracic duct lymphocyte migration in neonatal graft-versus-host disease. Lab Invest 29:387-397

de Sousa MAB, Parrott DMV, Pantelouris EM 1969 The lymphoid tissues in mice with congenital aplasia of the thymus. Clin Exp Immunol 4:637-644

Ford WL, Smith ME, Andrews P 1978 Possible clues to the mechanism underlying the selective migration of lymphocytes from the blood. In: Curtis ASG (ed) Cell–cell recognition. Cambridge University Press, Cambridge, p 359-392

Goldschneider I, McGregor DD 1968 Migration of lymphocytes and thymocytes in the rat. I. The route of migration from blood to spleen and lymph nodes. J Exp Med 127:155-168

Gowans, JL, Knight EJ 1964 The route of recirculation of lymphocytes in the rat. Proc R Soc Lond B Biol Sci 159:257-282

Miller JJ 1969 Studies of the phylogeny and ontogeny of the specialized lymphatic tissue venules. Lab Invest 21:484-490

Neville DM 1971 Plasma membrane protein sub-unit composition. J Biol Chem 216:6335-6338

Rannie GH, Donald KJ 1977 Estimation of the migration of thoracic duct lymphocytes to non-lymphoid tissues. Cell Tissue Kinet 10:523-541

Rannie GH, Smith ME, Ford WL 1977 Lymphocyte migration into cell-mediated immune lesions is inhibited by trypsin. Nature (Lond) 267:520-522

Ropke C, Jorgensen O, Claësson MH 1972 Histochemical studies of high-endothelial venules of lymph nodes and Peyer's patches in the mouse. Z Zellforsch Mikrosk Anat 131:287-297

Schoefl GI 1972 The migration of lymphocytes across the vascular endothelium in lymphoid tissue. A reexamination. J Exp Med 136:568-588

Schulze W 1925 Untersuchungen über die Capillaren und Post-capillaren Venen lymphatischer Organe. Z Anat Entwicklungsgesch 76:421-462

Smith C, Henon BK 1959 Histological and histochemical study of high endothelium of post-capillary veins of the lymph node. Anat Rec 135:207-211

Smith JB, McIntosh GH, Morris B 1970 The migration of cells through chronically inflamed tissues. J Pathol 100:21-29

Stoddart RW, Northcote DH 1967 Separation and measurement of microgram amounts of radio-active polysaccharides in metabolic experiments. Biochem J 105:61-63

van Deurs B, Ropke C 1975 The postnatal development of high-endothelial venules in lymph nodes of mice. Anat Rec 181:659-677

van Ewijk W 1980 Immunoelectron-microscopic characterization of lymphoid microenvironments in the lymph node and thymus. In this volume, p 21-33

Warren L 1959 The thiobarbituric acid assay of sialic acids. J Biol Chem 234: 1971-1975

Wenk EJ, Orlic D, Reith EJ, Rhodin JAG 1974 The ultrastructure of mouse lymph node venules and the passage of lymphocytes across their walls. J Ultrastruct Res 47:214-241

Woessner JG 1961 The determination of hydroxyproline in tissues and protein samples containing small proportions of the amino acid. Arch Biochem Biophys 93:400-447

Woodruff J, Gesner BM 1968 Lymphocytes: circulation altered by trypsin. Science (Wash DC) 161:176-178

Discussion

McConnell: Does the labelled fraction 5 bind specifically to lymphocytes and have you raised an antibody to it?

Andrews: We have so far failed to detect binding of ^{35}S-labelled fraction 5 to lymphocytes. However, if the extraction procedure (Fig. 2, 216) is halted after the initial solubilization with Tween-40 a heavily labelled, surface-membrane-enriched fraction is obtained. I have incubated lymphocytes with such a membrane preparation on the assumption that the material presented in such a manner may be in a more physiological form than when solubilized. Liquid scintillation counting indicated that the label was associated with the lymphocytes after centrifugation through a calf serum gradient and washing. Such experiments do not determine whether lymphocytes bound the sulphated molecule or a separate structure on the same membrane vesicle.

In answer to your second question, we are attempting to raise an antiserum to fraction 5.

Zigmond: Have you tried tunicamycin to see if it blocks your preparation?

Ford: No; but we will try a wider range of inhibitory agents on lymph node slices. The best way to quantify inhibition is to count autoradiographic grains over HEV; therefore these are rather laborious experiments.

Weissman: In the pulse experiments, as the sulphate marker makes its way out through the cell membrane does it have any preferential localization on any part of the cell membrane?

Ford: We fervently hoped we would find localization of the sulphated material on the cell membrane but there is no evidence of it. The material which we think is taken up into lymph node slices by endocytosis leaves the cells very rapidly. Fortunately the more interesting macromolecular material has a rather longer half-life within the high endothelial cell, although this is probably less than an hour.

Gryglewski: Could you measure the number of high endothelial cells, the rate of incorporation of radioactive sulphate, or the amount of extracted glycoproteins from lymph nodes stimulated with a large dose of sheep erythrocytes?

Ford: Over the three days after the stimulation of the popliteal lymph node with sheep erythrocytes the incorporation of $^{35}SO_4$ into macromolecular material doubled while the weight of the node trebled. There seems to be little change in the frequency of HEV per unit area of paracortex but this has not been precisely measured. Division of high endothelial cells or labelling after [^{3}H]thymidine injection was only occasionally seen as noted by Anderson et al (1975).

Vane: Do these high-walled endothelial cells contain Weibel–Palade bodies?

Ford: Anderson et al (1976) failed to find them.

Gowans: What are these bodies?

Simionescu: They are elongated structures, about 0.1 μm thick and up to 3 μm long, and they consist of several tubules about 15 nm thick embedded in a dense matrix (Weibel & Palade 1964). They have been found in the endothelium of arteries and veins, much less frequently in arterioles and venules, and very rarely in capillary endothelium. Their role is not known, so far.

Weiss: Have you looked at other high endothelial or other blood vessels to see whether there is such a concentration of sulphated material there?

Ford: After an i.v. injection of 5 mCi of $^{35}SO_4$ no other endothelial cells, including those in the spleen, perceptibly incorporated radioactivity.

Andrews: We gave direct injections into the splenic artery.

Zigmond: I am still concerned that the uptake may be pinocytotic rather than synthetic. It is too bad that the Golgi has acid phosphatase in it. Is there any other way that you could see whether your label really is in a digestive vacuole rather than a synthetic vesicle?

Ford: The early rapid uptake of $^{35}SO_4$ into tissue slices may be largely by pinocytosis but the smaller amount of radioactivity that is seen to be concentrated in HEV by autoradiography represents incorporation into a macromolecule for a number of reasons described in our paper. The most suggestive evidence is the selective inhibition of the concentration in HEV by fluoride and selenate.

Woodruff: When you use intravenous injections of ^{35}S is the material present in the lamina propria of the intestine?

Andrews: After i.v. injection of $^{35}SO_4$ there appeared to be some incorporation of sulphate throughout the gut wall. Both the intestinal epithelium and lamina propria were diffusely labelled but the significance of this is uncertain.

Woodruff: If that is the case then it would be of interest to know whether the ^{35}S-labelled molecules isolated from the gut also affect migration of labelled transfused lymphocytes. Did you treat the material with trypsin?

Andrews: No. However the labelled material in fraction 5 was sensitive to pronase digestion and partially sensitive to papain, but resistant to pepsin degradation.

Woodruff: It would be very nice if you could see it on the surface. I wonder whether the specific activity of the ^{35}S-labelled molecules could be increased to make the assay more sensitive.

Howard: When I first heard about the amazing discovery of the accumulation of sulphate in high-walled endothelium I assumed that you would use it as a marker for the isolation of high-walled endothelial cells, since it appears to be totally specific. Obviously, thanks to Stamper & Woodruff's work (1976), everybody has been sticking lymphocytes onto those cells *in situ*. But if you were to use hot sulphate simply as a marker before doing an enzymic extraction of the whole endothelium—perhaps with collagenase perfusion *in vivo,* as done by John Humphrey and his colleagues (this symposium)—to bring everything apart to start with, you would be able to do 'rosettes' the other way round. You could stick free high-walled endothelial cells to lymphocytes.

Andrews: I have used *in vitro* collagenase digestion of lymph node slices in order to isolate HEV cells, with some success. Using highly purified collagenase in the presence of $Na_2{}^{35}SO_4$ it is possible to produce a single cell suspension containing a number of large, pale-staining labelled cells. We had planned to attempt purification of these labelled cells either by density gradient centrifugation or by specifically binding the HEV cells to an immobilized lymphocyte monolayer.

Howard: Disappearance of sulphate from the cells is dependent on metabolic activity, so once the marker is in can you keep it stable with cyanide or azide or cold so that you can handle the cells to complete purification?

Andrews: That seems feasible but we haven't tried.

Weissman: Jonathan Howard's suggestion may be the best way to try to isolate large numbers of cells. What proportion of the cells that you got out after collagenase showed this kind of marker? And have you considered using some of the oncogenic viruses in an attempt to transform populations of endothelial cells from lymph nodes and then looked for cell lines that have the same properties?

Andrews: Before we had discovered the sulphate concentration in HEV, we simply looked at the morphology of the cells released into suspension after digestion of lymph nodes. Identifying HEV cells was impossible, although occasionally a large cell was seen in association with a few lymphocytes, which we supposed were HEV cell/lymphocyte rosettes. The advent of sulphate labelling of HEV cells allowed us to detect isolated HEV cells as long as the digestion conditions were carefully controlled. In an average digestion perhaps 10% of the large cells released are labelled to varying degrees, although less than 1% of these were as heavily labelled as those seen *in vivo*.

I had not considered direct transformation of the HEV cells themselves, but it has been suggested that we should try to fuse these cells with a suitable myeloma cell.

Butcher: Have you determined what effect, if any, this partially purified sulphated material has on the ability of lymphocytes to interact with HEV? For instance, have you examined its effect using your system of perfusing lymphocytes through the isolated mesenteric lymph node chain of the rat (Sedgley & Ford 1976)?

Ford: A practical difficulty with testing fraction 5 in a perfusion system is that it strongly agglutinates lymphocytes and might produce embolization.

Owen: The morphological features of the HEV might be a result of lymphocyte migration rather than a cause of it. Have you manipulated the system so that you don't have lymphocyte migration but still have HEV?

Ford: We have observed that HEV in the lymph nodes of athymic nude rats and in heavily irradiated rats concentrate $^{35}SO_4$. Although this has not been quantified it suggests that sulphate concentration is not completely dependent on a normal traffic of lymphocytes across HEV.

References

Anderson ND, Anderson AO, Wyllie RG 1975 Microvascular changes in lymph nodes draining skin allografts. Am J Pathol 81:131

Anderson ND, Anderson AO, Wyllie RG 1976 Specialized structure and metabolic activities of high endothelial venules in rat lymphatic tissues. Immunology 31:455-473

Sedgley M, Ford WL 1976 The migration of lymphocytes across specialized vascular endothelium. I. The entry of lymphocytes into the isolated mesenteric lymph node of the rat. Cell Tissue Kinet 9:231-242

Stamper HB, Woodruff JJ 1976 Lymphocyte homing into lymph nodes: in vitro demonstration of the selective affinity of recirculating lymphocytes for high endothelial venules. J Exp Med 144:828-831

Weibel ER, Palade GE 1964 New cytoplasmic components in arterial endothelia. J Cell Biol 23:101-112

General discussion II

Gowans: I wonder whether anyone would like to speculate further in this discussion about the physiological or immunological functions of lymphocyte traffic *in vivo,* particularly bringing in some of the phenomena we considered in earlier sessions? Thus, what part, if any, does the 'shutdown' of recirculation through a lymph node play during the induction of an immune response? What is the immunological significance of the 'feeding' of proximal nodes with cells from nodes placed more distally along the lymphatic chain? It was hinted that events are initiated in the distal nodes which allowed the proximal node to generate killer cells, but I am unclear about the facts and about the interpretation.

Much of the detailed analysis of the cellular aspects of immune responses has been carried out *in vitro.* Can we identify inefficiencies in these systems which may be due to their isolation from the dramatic traffic of lymphocytes that links the lymphoid system *in vivo?* One possibility here is that the recirculation of suppressor T lymphocytes *in vivo* may regulate immune responses and provide a constraint that is lacking *in vivo.*

The idea that recirculation allows selection of precursor cells *in vivo* loses some interest if, as has been claimed, virgin B cells are sessile within the lymph nodes and do not recirculate. However, both B and T memory cells recirculate and selection of T cells has been elegantly demonstrated. Is there more to be said here?

Ford: The evidence that memory cells are recruited from the recirculating lymphocyte pool after initiation of a secondary response is comprehensive and undisputed. In contrast the evidence that virgin B cells do not recirculate is limited to one or two antigens (Strober 1975). If it were postulated that the *sole* function of lymphocyte recirculation was the dissemination of immunological memory then the need for such a dynamic redistribution of antigen-sensitive cells is not obvious.

With regard to the comparison between *in vitro* and *in vivo* studies of cellular interactions in immune responses, I feel that while *in vitro* studies have of course clarified the roles of the cell types involved they are of limited value in elucidating the function of lymphocyte migration.

Howard: The recirculating suppressor is interesting. There is a profound state of immunological tolerance, which frequently happens when a parental strain animal is injected with F1 bone marrow at birth. The recipient is tolerant and will accept skin from an allogeneic donor. Many experiments have suggested that in such a case the peripheral pool is depleted of that set of cells which normally mounts the immune response against the tolerizing transplantation antigen. Since that time the idea has appeared that immune responses are in fact regulated by a different method and that there is a state of immunological silence which is due to the positive activity of a set of cells which close immune responses down. In other words, tolerance is not merely clone deletion but rather, or perhaps also, the presence of something else which is immunoregulatory in a negative sense.

Bruce Roser (Roser & Dorsch 1979) has done some remarkable experiments on the thoracic duct lymphocytes of animals which are deeply tolerant by all normal criteria and accept fully allogeneic skin grafts indefinitely. Roser and his colleagues have shown that in the thoracic duct of these animals there is a T cell which has the ability to confer on another animal a somewhat similar state of tolerance. One relevant fact is that this cell, which is clearly a negative immunoregulatory cell, is rapidly recirculating. It moves from blood to lymph at the same tempo, in so far as it has been analysed, as a 'normal' T cell. The second fact is that it is apparently antigen-sensitive so a re-definition of clone deletion is required. If there is clone deletion in these tolerant animals it can't be complete, it seems, because the suppressor appears to recognize antigens. In other words the suppressor for its activity seems to have to see the tolerated antigen, therefore the suppressor is not itself deleted.

I was a little bit cautious when I said that this cell can confer a similar tolerance on the recipient because the recipient itself becomes less tolerant than the donor. The recipient can accept a skin graft permanently but only because it is already partially immunosuppressed. It appears as if the suppressor cell isn't able to operate on a mature system but only on a system which is itself immunologically damaged. The existence of such a suppressor cell in the circulation may be quite compatible with deletion of all the positively reactive cells that normally occupy the recirculating pool. This suppressor cell may have a different role entirely—the shutting down of the immature cell population which is about to become a recirculating cell. The surprising thing is that the cells should be present in the recirculating pool itself. Silvers and colleagues (1975) have shown that the thymus is constantly putting pressure on the maintenance of tolerance and that the effects of this have to be negated. One could argue that the role of the recirculating suppressor cell is exactly this—negating the effects of a thymus which is constantly trying to export new immunologically competent precursor cells.

Gowans: What proportion of the recirculating pool consists of suppressor cells?

Weissman: If you believe that suppressor cells are Lyt 1, 2, 3 or 2–3, and in addition have I–J on the surface, it must be a very tiny population of cells. It is very difficult to identify many I–J-positive cells by standard cytotoxic assays. The Lyt story doesn't distinguish suppressors from killers or precursors of suppressors and killers. When people do the experiments carefully, both the precursors of suppressors and the precursors of killers are Lyt 1, 2 and 3-positive. That makes up 50% of the T cells from suspensions of spleen and lymph nodes.

Gowans: So the normal consequence of immunization is tolerance, not immunity?

Howard: In view of the possible role of MHC antigens on thymus epithelial cells in determining the T cell repertoire, it may be quite unreasonable to suppose that neonatally induced tolerance to major transplantation antigens is an experimental version of normal self-tolerance. Heaven knows what these recirculating suppressor cells are.

Gowans: Immune responses *in vivo* are self-limiting and various mechanisms have been invoked to explain this. I imagine that suppressor cells serve this general end.

Humphrey: Can you be sure? In one instance when antibody with a known idiotype—actually a myeloma protein from MOPC 315 which binds DNP—was complexed with a DNP-conjugated antigen (DNP-KLH) and quite small amounts (10 μg) were injected intravenously into mice, this stimulated the efficient production of B-memory cells specific for the idiotype of the myeloma protein. When more of the myeloma protein was later administered, large amounts of anti-idiotype were produced, even though the idiotype in question is normally present in the Ig of the mice (Klaus 1978). The implication is that once antibody with a given idiotype is stimulated and becomes complexed with antigen and localized in germinal centres, anti-idiotype is automatically produced and the response becomes self-limiting. Of course in a normal antibody response, which is polyclonal and where the idiotypes are unknown, it would not be possible to detect such a mechanism.

Hall: In a sheep in which the antigenic stimulus has been localized deliberately to one node and the humoral and the cellular effluent has been drained away, how is the immune response switched off if it depends on any sort of feedback mechanism, network or otherwise?

Gowans: Either the antigen disappears so the drive disappears, or suppressor cells are produced within the node.

Hall: But then they would have to be upstream of the plasma cells.

Weissman: They wouldn't just stay in the node.

Davies: Is that true where there is continued antigenic stimulus or is it only true where something like FDNB has been used, which is probably going to last for a short time, or where sheep red cells have been used? If something like *Leishmania tropica* that will stop there for 200 or 300 days is put in the skin, will the node automatically switch off or would it keep going?

Hall: It would keep going because there would be continued antigenic bombardment.

Davies: One of the limitations is that it stops because there is no more antigen.

Hall: That is a straightforward situation that I can understand. Why, therefore, does one have to invoke very elaborate switch-off mechanisms?

Davies: In the mouse if you keep adding oxazolone the switch-off still occurs when ostensibly antigen is no longer limitŧng. I would like to see this done with *Leishmania tropica* but nobody seems to have done that experiment. There one would expect to get continuous stimulation, but oxazolone switches off even if one provides more and more antigen.

Hall: More and more antigen causes more and more of a response; in our sheep unit the oxazolone bill mounts and the response is still going on.

Davies: That makes it quite clear that the mouse isn't a small sheep!

Morris: In the sheep the secondary response is not the same in two lymph nodes if one node has experienced the antigen directly while another one has acquired an immunological memory in virtue of the fact that cells have been liberated from an antigenically stimulated node somewhere else in the body. Certain events occur within the node that receives the antigenic challenge that ensure that in a secondary situation the node responds more promptly than and differently from a node which has acquired antigenic memory indirectly without the previous experience of antigen.

Would somebody care to comment on the selection of antigen-sensitive cells within the recirculating pool? We now know that cells are capable of transferring, for instance, an immune response to sheep red blood cells when they are transferred into semi-allogeneic recipients (McCullagh 1975). One can uncover the presence of these cells in what has previously been said to be a deficient population.

Howard: I initially dealt with transplantation tolerance where the evidence for the deletion of normally competent mature alloreactive cells in peripheral lymphoid tissues is very good.

Morris: McCullagh took thoracic duct lymph from these animals and couldn't promote an immune response in an irradiated syngeneic recipient. When he put the same material into an F1 hybrid he got a perfectly good response.

Howard: The explanation is probably that he obtained complete tolerance of T cells and no tolerance at all of B cells.

Williams: He demonstrated suppressor cells, didn't he? If the 'tolerant' cells were mixed with normal cells, they suppressed the ability of normal cells to respond.

Morris: Exactly.

Howard: But McCullagh then put the thing into an allogeneic system, and all the normal rules of immune induction were abrogated. You can switch on anything with an allogeneic effect: immune responses to non-immunogenic antigens, to antigens under immune response gene control, to autoantigens, and so on.

Morris: But sensitized cells had to be responsible for the reaction as there is no antigen in the allogeneic recipient. There are comments to be made on the selection of antigen-specific cells out of the recirculating population by contact with antigen in lymph nodes. This is the proposition that has been advanced by several workers.

Ford: An old idea is that recirculation is a device to minimize the chances that all lymphocytes in the body will be tolerized. When tolerance induction was naively attributed to an excessive concentration of antigen this notion could be simply formulated: the lymphocytes entering a lymph node immediately after a large dose of antigen would be rendered tolerant, but when the effective concentration of antigen fell new arrivals would encounter an immunogenic concentration and would be activated. It might be that this idea could be modified in the light of new information on the conditions for inducing immunity and tolerance.

Davies: Is it completely axiomatic that we know what the lymphoid system exists for in vertebrates? It seems to me that you are asking a particular question about an aspect of the lymphoid system without having a good general idea of what the biological significance of the system itself is. You could argue that it is concerned with defence mechanisms. But what does the poor octopus do? Whatever it does, it seems to do well. Is the question you are posing about the lymphoid system sensible without at least some attempt to comment on the anterior problem?

Gowans: If you do not regard the lymphoid system as a defence mechanism what else do you suppose it might be doing?

Davies: You are usually looking at a particular aspect of the lymphoid system. As long as you are convinced that it is mediating a defence mechanism all sorts of problems of Darwinian selection are apparently solved. Yet most of the antigenic stimuli that people have worked with are completely remote from any of the realities that Bede Morris seems to be concerned with. How many of us here regularly work on infectious agents as antigenic stimuli? How many of us work on immune responses which are concerned with defence in any meaningful circumstances? Our ground rules may be totally irrelevant to nearly all real defence mechanisms. In any case I intuitively don't believe that it is a defence mechanism.

McConnell: Bill Ford said that rapid recirculation is not needed for the dissemination of memory but perhaps a rapid one is needed for the recognition of antigens. If antigen enters the system it has to be recognized fairly rapidly. Hay & Hobbs (1977) calculated that in five days 60% of the total recirculating pool will pass through a node in the blood but of that fraction only 15% will actually enter the node. Obviously it depends on how many antigen-reactive cells present in that 15% are specific for a given antigen but to maximize the specific interactions a rapid recirculation would appear to be necessary.

Ford: Another possibility is that lymphocyte migration facilitates the cooperation between T and B lymphocytes in antibody responses by enabling infrequent antigen-sensitive cells of each type to come together. This would overcome the problem of sessile B lymphocytes in the primary response because it is only necessary for one of the partners to be mobile. Perhaps in a primary response a stream of T cells flows past immobile B cells. In secondary responses the situation could be reversed—the T cells stay put after recognizing antigen while B cells stream past it.

Weissman: The report (Strober 1972) that virgin B cells don't recirculate doesn't mean that they aren't circulating and appearing in the blood at a high rate. The bone marrow produces so many cells per day that must be B cells that one can argue that short-lived virgin B cells are produced at a high rate and disappear if they don't meet their antigen. There has to be an entry site where they can meet antigen and interact with other cells. Once they get into the lymphoid tissues there are T and B cell domains which tend to keep the cells apart. That is a difficulty.

The most interesting experiment on that is Bill Ford's work (1975) on trying to induce an immune response in a perfused spleen. The cells most necessary for that response are cells in the perfusion circuit rather than cells in the spleen. We haven't followed up on the pathway of antigen entry into spleen or lymph node to ask whether the recirculation pathway is a sieve or a net to which antigen-specific lymphocytes bind and where they meet each other. If we take those experiments seriously, it is the cells in the circulation that are going to have to meet up with antigen in their entry site. That fits with Bill's idea that what we are working with is a system of collaboration and that it is the most efficient system for getting together cells that are normally held apart in lymphoid tissues.

Ford: How plausible is the idea that cooperation between B and T lymphocytes occurs within germinal centres, as has been suggested on the basis of your own observations (Gutman & Weissman 1972)?

Weissman: I would suggest that after antigen activation some important cells can move to germinal centres. I made a mistake though. I thought that they would move within that lymph node. They may be antigen-activated in one lymph node and then become participants in germinal centres one node down the line.

Weiss: I agree with Tony Davies that we have to understand the functions of the lymphocyte and that saying this system deals with defence is not enough. We are breaking it down now, talking about meeting antigens and things of that sort, and this is perhaps going to be productive. The point about T and B cells not being kept apart even though they are kept in different places may be a very good way of hanging things together. T and B cells don't have huge interfaces with one another but there may be T and B cell zones with a large interface and in addition to that the T and B cells within their separate zones are rolling about. At those interfaces of T and B cells antigens and macrophages typically collect. The arrangements appear to make selection efficient.

Gowans: How important is the phenomenon of shutdown?

Morris: If people believe that T and B cell domains are precisely defined in terms of traffic areas in lymph nodes they have to understand that this is so only if the route of entry of cells into the lymph node is specified. The way cells distribute themselves in a lymph node when they arrive by the lymph stream is very different from the way they distribute themselves when they come from the blood. That seems to me a crucial point. You are not going to understand the immune response if you believe that these cells have definitive zones of migration and remain within those zones, with no contacts being made between those cells. It is a very flexible situation and my guess is that one of the reasons for recirculation is that it gives cells an opportunity to dissociate themselves from a particular environment from time to time. If they remain fixed there in a continuum they would be destined to transform into end cells of a particular kind. To get out of that bind they dissociate themselves from that environment and change their destiny.

McConnell: There is an element of shutdown that may be relevant here. Shutdown might provide a mechanism whereby random mixing of T and B cells can occur.

Hall: I take a contrary view. I think shutdown occurs either because polymorphs (or their products) are induced to accumulate or because complement is activated. But it is essentially irrelevant, even though it nearly always happens after an antigenic challenge, particularly in the sheep which has pre-existing, complement-binding antibodies to a great many things. There is no doubt that with oxazolone perfectly typical immune responses can occur without shutdown. I don't think that shutdown is a necessary prerequisite for immune responses in lymph nodes.

McConnell: I don't disagree but it may depend on the sort of immune response you are thinking of. For a T and B cell collaborative response it would be interesting to know whether shutdown was essential for the relevant T—B interactions to occur. We are currently testing this to see whether an anti-NIP response occurs in nodes challenged with NIP-PPD in the presence of indomethacin which prevents shutdown.

Morris: There is no argument. Shutdown is not necessary to invoke an immune response. If you put the animal's own lymphocytes into the next node you get a shutdown. That is what I was trying to suggest. Other regulatory activities go on in nodes which inhibit the outflow of cells.

Howard: We have heard quite a lot here about a potentially very large traffic of lymphocytes through non-lymphoid sites—the liver, skin and kidney, for example. We all knew that lymphocytes were there but on the whole we thought of the peripheral lymphoid tissues (spleen and lymph nodes) as being where immune function is induced. A good reason for believing that is that so much of the immune response is executed in solid lymphoid tissues. Also the main dynamic fluxes of cells are through the spleen and lymph nodes. Antigen is concentrated there and immune responses are mediated there. One can draw an animal as being a sort of centripetal structure of afferent lymphatics leading to lymph nodes and immunologically competent cells. It is almost as if the whole of the animal's antigenic environment is confined by its immunological apparatus.

I have been interested for a long time in the induction and maintenance of tolerance to things which are outside that catchment area. Billingham & Silvers (1962) long ago pointed out that the epidermis itself was an immunologically privileged site. A pigmented cell carrying with its pigment the indication that it came from an animal with other major transplantation antigens could reside permanently in the epidermis of a guinea-pig or a mouse without being recognized by the immune system. Does the traffic of lymphocytes through the non-lymphoid tissue constitute a screening mechanism? Are lymphocytes in effective contact with autoantigens when they are migrating through non-lymphoid sites?

One experiment raises that question very vividly. Andrews (1974) did an immune surveillance experiment on the induction of methylcholanthrene tumours in the skin of mice. When this potent carcinogen is painted on the skin of the mice there is first a period of inflammation, then hyperplastic growth of the dermal layers, formation of papillomata, and finally, after several months, tumours growing in the skin. These tumours are exceedingly immunogenic. If inactivated cells from such a tumour in an inbred strain of mice are injected into a normal mouse of the same strain, the normal animal will be primed against the tumour, yet the original mouse dies of the tumour. Why doesn't surveillance work in the original mouse? If the lymphocytes that are coursing through these skin sites are potentially seeing antigen—in this case autoantigen, although it has something to do with methylcholanthrene—why don't they reject the tumour?

Andrews' experiment was remarkable. He picked a piece of skin off the methylcholanthrene-painted site of an animal in the inductive phase, when there were no manifestations of the tumour, and then put it back again. He 'aerated' it, as someone said. There was a rapid massive infiltration of lymphocytes into this autograft, the kind of infiltration we associate with allografting. The beginnings of malignant growth were reversed and no tumour developed at that site.

That tells us two things. First, the very early tumours were indeed immunogenic, which you might have questioned. You might have said that antigenicity could only follow full differentiation of the tumour. Secondly, the animal was competent to recognize this antigenicity, but only when there was inflammation at this site, damage, and possibly even release of antigen.

In this rather roundabout way I am simply asking whether the periphery, meaning that part of the animal which is not involved in the organized lymphoid tissue, is normally recognizably antigenic to lymphocytes.

Davies: How is that demonstrated to be an immunological experiment?

Howard: It doesn't happen in animals which have lost their T cell systems.

Davies: Lymphocytes might be concerned in some way with normal tissue regulation. The problem is that nude mice seem to grow moderately well.

Howard: That is not the point. Do you accept the evidence that the integrity of the T cell system is required for the Andrews effect to be manifested as evidence for its immunological basis?

Davies: It might be a hormonal problem. It may be completely irrelevant. Epithelia are well known to be responsive to a wide variety of steroid hormones.   You have adopted one particular explanation and I am not satisfied that it is unique.

McConnell: Peter Alexander clearly demonstrated specific retention of cells at sites of high antigen concentrations in a tumour system.   If a methyl-cholanthrene-induced tumour was present on one side of a rat, plasma cells proliferated in the draining node.   When Alexander transplanted a small piece of the tumour to the other side it grew—that is, the specific cells involved in rejection were 'trapped' on the original stimulated node.   This was shown by the fact that if the node was excised the second tumour was rapidly rejected.

Howard: My whole point is that 'rejection' depends on release of antigen, but in order to get immunogenicity you have to have antigenic material in the right place.   The right place, as I understand it, is a lymph node.   If you have antigenic material in the wrong place you run a serious risk of not getting an immune response.   This is the same point so elegantly made by Barker & Billingham (1968) and Tilney & Gowans (1971), who showed that skin sites artificially deprived of their lymphatic drainage were immunologically privileged.   That is why the allografted kidney is a key organ: where is that immune response induced?   Is it direct lymphocyte–kidney contact, is there a macrophage mediator, are there primitive induction structures which resemble the things in lymph nodes?   Are these exceptions or does lymphocyte induction normally happen peripherally?

Humphrey: Frey & Wenk (1957) showed quite a long time ago that chemical sensitizing agents applied to the skin of guinea-pigs failed to sensitize them if the lymphatic drainage was cut, even though the blood supply to the treated skin remained intact.

Owens: A much simpler explanation is that the success of the tumour graft depends on a balance between the immune response to the tumour and the growth rate of the tumour.   When you remove the tumour and then replace it you severely limit its growth, to the advantage of the immune system.

Zigmond: You get an inflammatory response too.

Weissman: There are many experiments where the removal of lymphatic drainage prevents sensitization.   For example, allogeneic skin transplanted to an alymphatic site is not rejected, and does not sensitize the host.   Intravenous injection of allo-sensitized cells into these hosts leads to effector cell ingress and destruction of the graft.   The only exception to a lack of peripheral sensitization we are hearing of now is the renal homograft.   Maybe there is something special about that.

Morris: The renal homograft is a classical instance of peripheral sensiti-zation, in the restricted sense that the regional node is not involved.

References

Andrews EJ 1974 Failure of immune surveillance against chemically induced in situ tumors in mice. J Natl Cancer Inst 52:729-732

Barker CF, Billingham RE 1968 The role of afferent lymphatics in the rejection of skin homografts. J Exp Med 128:197

Billingham RE, Silvers WK 1962 Studies on cheek pouch skin homografts in the Syrian hamster. In: Transplantation. Churchill, London (Ciba Found Symp), p 90-108

Ford WL 1975 Lymphocyte migration and immune responses. Prog Allergy 19:1-59

Frey JR, Wenk P 1957 Experimental studies on the pathogenesis of contact eczema in guinea pigs. Int Arch Allergy Appl Immunol 11:81

Gutman GA, Weissman IL 1972 Lymphoid tissue architecture. Experimental analysis of the origin and distribution of T-cells and B-cells. Immunology 23:465-479

Hay JB, Hobbs BB 1977 The flow of blood to lymph nodes and its relation to lymphocyte traffic and the immune response. J Exp Med 145:31-44

Klaus GGB 1978 Antigen-antibody complexes elicit anti-idiotype antibodies to self idiotypes. Nature (Lond) 272:265-266

McCullagh P 1975 The mechanism of unresponsiveness in lymphocytes specifically selected by antigen in normal rats. Aust J Exp Biol Med Sci 53:175-185

Roser BJ, Dorsch S 1979 The cellular basis of transplantation tolerance in the rat. Immunol Rev, in press

Silvers WK, Elkins WL, Quimby F 1975 Cellular basis of tolerance in neonatally induced mouse chimeras. J Exp Med 142:1312-1315

Strober S 1972 Initiation of antibody responses by different classes of lymphocytes. V. Fundamental changes in the physiological characteristics of virgin thymus-independent ('B') lymphocytes and 'B' memory cells. J Exp Med 136:851-871

Strober S 1975 Immune function, cell surface characteristics, and maturation of B cell subpopulations. Transplant Rev 24:84

Tilney NL, Gowans JL 1971 The sensitization of rats by allografts transplanted to pedicles of skin. J Exp Med 133:951

Adherence of lymphocytes to the high endothelium of lymph nodes *in vitro*

JUDITH J. WOODRUFF and BARRY J. KUTTNER

Department of Microbiology and Immunology, State University of New York, Downstate Medical Center, Brooklyn, New York 11203

Abstract   Recirculating lymphocytes emigrate from the blood into lymph nodes by crossing high-endothelial cell venules (HEV). The site of entry in the node is specific and no migration occurs through other vessels, suggesting that lymphocytes may have surface receptors for recognizing the specialized endothelium. This problem has been investigated using a system whereby lymphocyte interaction with HEV can be monitored *in vitro*. Rat thoracic duct lymphocytes adhere selectively to HEV when the cells are deposited over glutaraldehyde-fixed sections of lymph node. TDL also adhere to unfixed HEV but the extent of binding is markedly enhanced when the sections are pretreated with bifunctional reagents such as glutaraldehyde or the diimidoester, dimethyl suberimidate. Lymphocytes from lymph nodes and spleen are also capable of attaching to HEV but thymus and bone marrow are deficient in cells with this capability. Both Ig$^+$ and Ig$^-$ thoracic duct cells bind to high-endothelial venules. This reactivity is not species-specific and both mouse and human lymphocytes bind to rat HEV.

Adherence of lymphocytes to HEV requires metabolically intact cells, is dependent on calcium but not magnesium, and is inhibited by treating lymphocytes with trypsin. Binding is sensitive to inhibition by cytochalasin B whereas colchicine causes no reduction in adherence. The results suggest that HEV-binding sites are present on recirculating lymphocytes but that formation of a stable complex with the endothelium depends on contractile forces generated by cytoplasmic microfilaments in the lymphocyte.

Recirculating lymphocytes migrate continuously between blood and lymph, moving by a route which takes them through various lymphoid organs in the body. Though they circulate freely in the bloodstream and have opportunity for interaction elsewhere they are not promiscuous but faithfully seek out those blood vessels which provide entry into these tissues. One of the most striking features of lymphocyte recirculation is that the cells migrate from blood into lymph nodes via a system of specialized microvascular structures.

© *Excerpta Medica 1980*
Blood cells and vessel walls: functional interactions
(Ciba Foundation symposium 71) p 243-263

The cells enter only by crossing high-endothelial cell venules (HEV) and not by migrating through other vascular endothelia (Gowans 1957, 1959, Gowans & Knight 1964). The basis of this phenomenon is not known but it must involve a highly selective mechanism of recognition.

HEV are present in lymph nodes, Peyer's patches and tonsils, and have been observed in several mammalian species, including the human species (Miller 1969). Typically the specialized endothelium is formed by polygonal cells linked together by discontinuous junctional complexes. Ultra-structurally, most high-endothelial cells have abundant cytoplasm and contain a prominent Golgi apparatus, vesicles, and free and clustered ribosomes and mitochondria (Anderson & Anderson 1976). Non-specific esterase activity is readily detected in the cytoplasm of high-endothelial cells but not in cells of other endothelia (Smith & Henon 1959). Nothing is known of the relationship of these peculiar metabolic and structural features of HEV endothelium to their function in lymphocyte homing.

Studies from this laboratory have been aimed at determining whether recirculating lymphocytes possess surface receptors for HEV recognition. Initially we adopted the approach originated by Gesner of using enzymes to modify surface constituents of radiolabelled lymphocytes; after brief *in vitro* treatment the cells were injected intravenously and their pattern of distribution was determined by isotopic counting techniques or by autoradiography. Such experiments showed that the homing of lymphocytes into lymph nodes is reduced by glycosidases, neuraminidase or trypsin (Gesner & Ginsburg 1964, Woodruff & Gesner 1968, 1969).

Subsequent work revealed that a variety of reagents capable of reacting with surface determinants of lymphocytes are able to induce membrane changes which alter the normal pattern of lymphocyte migration in the body (Zatz et al 1972a, Woodruff & Woodruff 1974, Schlesinger & Israel 1974, Durkin et al 1975, Freitas & DeSousa 1977). Although it was reasonable to conclude from these findings that surface membrane components of lympho-cytes play a role in determining their distribution in the body, precise interpretation of the data was not possible. Firstly, inhibition of lymph node homing might be caused by the rapid removal of donor lymphocytes from the circulation (i.e. in the liver after treatment with glycosidases or neura-minidase) before the cells had the opportunity of reaching the nodes. Secondly, there was no way to distinguish effects which might be due to de-struction of the putative homing receptor from those arising secondarily through non-specific and/or conformational changes in the membrane or metabolism of the lymphocyte.

An approach was therefore sought which would permit direct visualization of the interaction between lymphocytes and the specialized endothelial cells. This was accomplished through the use of lymph node sections. The endothelium of HEV is exposed in the section and lymphocytes deposited over the tissue can be assayed for their capacity to bind to this structure. The results of such studies have shown that recirculating lymphocytes do indeed possess surface membrane constituents which enable them to adhere to the specialized endothelium (Stamper & Woodruff 1976).

Our first experiments with this system were done with inbred Wistar-Furth rats but no difference in results was observed when we used Sprague-Dawley rats, a closed but inbred stock, and the findings to be discussed have largely been obtained using Sprague-Dawley animals.

ADHERENCE OF LYMPHOCYTES TO HEV OF LYMPH NODES *IN VITRO*

The basic procedure has been described in detail elsewhere (Stamper & Woodruff 1977). A brief outline is given here in order to present certain modifications which have increased the sensitivity of the assay and eliminated the variability noted in our earlier reports.

The system we use employs 8-μm thick frozen sections of cervical lymph nodes. These are fixed with 3% glutaraldehyde in 0.1 M-cacodylate for 10 min at 4 °C, thoroughly rinsed with buffer and then incubated for 10 min with 0.2 M-lysine. Since glutaraldehyde readily reacts with the amino groups of lysine this treatment eliminates free aldehyde residues which might be present in the tissue section. This step ensures that lymphocyte binding occurs via interaction with a constituent of the endothelium and not with the fixative itself. Thereafter the sections are rinsed with medium (RPMI 1640), overlaid with 0.2 ml of a washed suspension of thoracic duct cells, and placed on a rotating table operating at 80 rev./min. Incubation is for 30 min at 7 °C. The sections are then washed in phosphate-buffered saline in order to remove non-adherent lymphocytes, briefly fixed and finally stained with methyl green-thionin.

Lymphocytes adhere to the sections in a highly regular and reproducible fashion. They are localized over the endothelium of the HEV and stand out clearly as small round cells above the plane of focus of the tissue itself. Because the lymphocytes are intact they stain more intensely than the transected parenchymal lymphoid cells within the tissue section (Fig. 1). In optimal conditions about 75–90% of the HEV in each section contain two or more adherent lymphocytes; these are scored as 'positive' HEV. The number of cells which attach to each vessel varies but is usually proportional to the cross-sectional area of endothelium. About 25–50% of the HEV have more

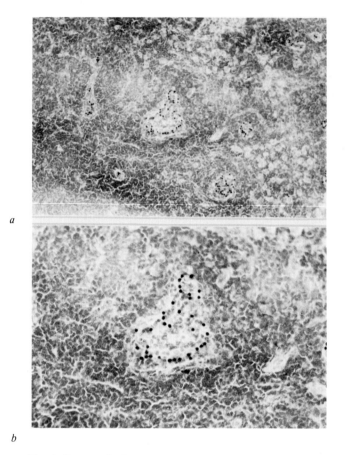

a

b

Fig. 1. Pattern of adherence of overlaid rat thoracic duct lymphocytes (TDL) to lymph node sections. TDL are darkly stained and above the plane of focus of the tissue section. Magnification (a) × 100 and (b) × 200.

than five adherent cells, with some containing as many as 60–80. The extent of HEV binding is also proportional to the concentration of the overlaid lymphocytes, maximal numbers being achieved with 5 to 15×10^6 cells/ml (Fig. 2). Adherence occurs rapidly and is essentially complete in 15–30 min; serum is not required (Stamper & Woodruff 1977).

SPECIFICITY OF THE *IN VITRO* REACTION AND CORRELATION WITH LYMPH NODE HOMING *IN VIVO*

Overlaid lymphocytes are rarely found in subcapsular sinuses and they do not adhere to capillaries, lymphatic channels or non-specialized venules in the cortex or medulla. There is some binding to other cell types in the section, but 'non-specific' adherence ordinarily occurs at a very low level and the density of overlaid lymphocytes attached to HEV is at least 100-fold greater per unit area than that found at non-HEV sites. There is no uniformity in the pattern of adherence of lymphocytes to non-HEV sites. The results of the experiment presented in Fig. 2 show that nearly all the cells (>85%) which adhere to the section bind to the specialized endothelium even though this comprises only about 1 to 2% of the total area of the tissue.

Further studies provided evidence that *in vitro* HEV reactivity is a property of recirculating lymphocytes and not a trait of lymphoid cells incapable of homing into lymph nodes *in vivo*. Thus thoracic duct lymph, lymph nodes and spleen contain substantial numbers of recirculating lymphocytes and comparable levels of HEV binding are observed in sections overlaid with cells obtained from these compartments. In contrast, thymus and bone marrow are deficient in cells with this capability; the extent of HEV binding in sections overlaid with these cell populations is 10% or less of that observed with thoracic duct cells (Fig. 3). Comparable findings have been made in studies

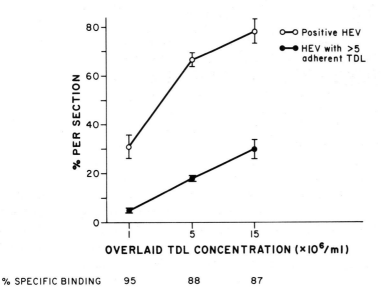

FIG. 2. Pattern of adherence of rat thoracic duct lymphocytes (TDL) to lymph node sections. Alternate sections from one lymph node were overlaid in sequence with TDL at the indicated concentration. Specific binding is the percentage of cells attached to the section which are localized at HEV. Mean ± SEM of four sections/group.

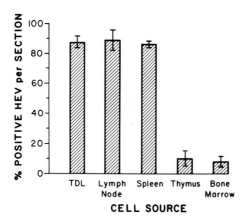

FIG. 3. Tissue distribution of rat lymphocytes capable of adhering to HEV in lymph node sections. Concentration of overlaid cells was 10×10^6/ml. Mean ± SEM of four sections/group.

with mouse lymphocytes and lymph node sections (Stamper & Woodruff 1976). Thus, the *in vitro* system exhibits the dual specificity with regard to blood vessel and lymphocyte type which has been found to characterize the lymphocyte–lymph node homing interactions that occur under physiological conditions.

Our interpretation of these results is that the recirculating lymphocyte has on its surface membrane a constituent which mediates interaction with high endothelial cells. The low levels of binding obtained with bone marrow and thymus cells suggest that acquisition of this surface membrane property is a differentiation event which accompanies the appearance of other functional activities that become manifested as precursors mature into immuno-competent lymphocytes.

It is known that thoracic duct lymphocytes contain a mixture of T and B cells (Howard et al 1972) and adoptive transfer studies have shown that both classes of lymphocytes enter the lymph nodes by crossing HEV (Gutman & Weissman 1973). We have studied the *in vitro* HEV-binding activities of T and B cells using the nylon wool column technique (Hodes et al 1974) to fractionate thoracic duct cells into predominantly Ig-negative (T) and Ig-positive (B) populations. The results shown in Fig. 4 indicate that both T and B cells are capable of binding to HEV *in vitro*. Cell dose-response relationships are the same, suggesting that these populations do not differ significantly in their HEV-binding properties. The results imply that lymphocyte surface determinants which mediate adherence are not among those constituents which are exclusive components of either population. It seems unlikely

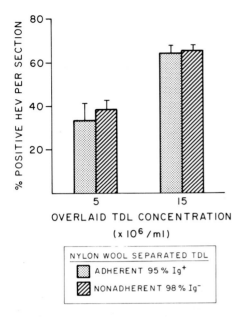

FIG. 4. Comparison of HEV-binding activity of nylon wool-adherent (B) and non-adherent (T) populations. Ig$^+$ thoracic duct cells (B cells) detected using FITC-conjugated rabbit anti-rat Ig deaggregated by ultracentrifugation at $100\,000 \times$ g. Mean $\pm$ SEM of four sections/group.

therefore that the surface determinants that have been identified as specific for T cells or B cells would be responsible for the binding reaction.

Another aspect of this problem concerns the role of surface molecules which are products of the major histocompatibility complex. There has been speculation about the function of these components in the homing of lymphocytes to lymph nodes, since infusion of syngeneic lymphocytes pretreated with anti-H-2 serum (Schlesinger 1976) or injection of histoincompatible lymphocytes results in reduced accumulation of donor cells in lymph nodes of recipients (Zatz et al 1972b). It was therefore of interest to compare the affinity of lymphocytes for syngeneic, allogeneic and xenogeneic HEV. Experiments with spleen cells and lymph node sections from CBA and C57BL/6 mice revealed no difference in HEV-binding activity between these strains (data not shown). Further, lymphocytes from mouse and rat spleen are capable of binding to HEV in rat and mouse lymph node sections, respectively. Usually, binding to xenogeneic HEV approximates that found in syngeneic combinations, although in some experiments with mouse spleen lymphocytes evidence of species preference was obtained (Table 1, expt. 2). Additional studies have shown that human blood lymphocytes are also able to

TABLE 1

Lymphocyte binding to sections of xenogeneic high-endothelial venules (HEV)

Lymphocyte source	HEV source	% Positive HEV/section			
		Experiment	1	2	3	4
Rat spleen	Rat	75	25	-	-
Mouse spleen	Rat	61	24	-	-
Rat spleen	Mouse	72	33	-	-
Mouse spleen	Mouse	69	76	-	-
Human blood	Rat	68	33	21	39

Rat or mouse lymphocytes overlaid at 5×10^6/ml and human lymphocytes at 30×10^6/ml; values shown are means of four sections per group.

adhere to the endothelium of rat HEV. The results in Table 1 show the HEV-binding activity exhibited by peripheral blood lymphocytes obtained from three human donors. Although the results are more variable and the extent of adherence is lower than that obtained with murine cells, human lymphocytes adhere in a highly specific manner with about 90% of the cells attached to the section localized to HEV. Thus the properties of lymphocytes which mediate their affinity for the specialized endothelium are not species-restricted.

PROPERTIES OF THORACIC DUCT CELLS ADHERING TO HEV *IN VITRO*

Since thoracic duct lymphocytes adhere to glutaraldehyde-fixed HEV, binding does not require viable endothelial cells and they must, therefore, be passive participants in the *in vitro* reaction. In contrast, glutaraldehyde treatment of thoracic duct cells abolishes their capacity to bind to HEV, suggesting that lymphocytes play an active role (Stamper & Woodruff 1977). This has been confirmed by the results of experiments examining the effects of reagents which impair metabolism and the function of certain cytoplasmic structures of lymphocytes (Woodruff et al 1977). Thus, sodium iodoacetate, an irreversible inhibitor of glycolysis, and sodium azide, a reversible inhibitor of oxidative phosphorylation, each cause dramatic reduction in binding with 85–90% inhibition occurring with 30 mM-iodoacetate or 10 mM-azide. There is no evidence that cell mortality could account for these results and the effects of sodium azide are largely reversible when the drug is washed out.

It therefore appears that binding is not a passive consequence of the interaction of lymphocyte surface receptors with HEV but rather that the formation of a stable complex requires activation of certain cytoplasmic components of lymphocytes. This is supported by the finding that lymphocyte attachment is very sensitive to inhibition by cytochalasin B, a drug which causes dysfunction of actin-like cytoplasmic microfilaments which form a network of contractile elements beneath the plasma membrane (Wessells et al 1971). At 10 μg of cytochalasin B/ml binding is almost completely eliminated and even at 0.1 μg/ml it is significantly reduced (30%). The effects are not due to inhibition of glucose uptake, a process which this drug can suppress, since extracellular glucose is not required for normal lymphocyte binding nor for the inhibitory effects exerted by cytochalasin B. In addition binding activity is restored when the drug is washed out. Colchicine, which causes irreversible disaggregation of microtubular subunits, has no effect on the attachment of lymphocytes to HEV.

Microfilaments could play a role in the adherence of lymphocytes through an effect on surface receptor movement (Schreiner & Unanue 1976). Interaction with HEV may stimulate lymphocyte contractile forces which are needed to modulate the position of surface receptors, or in some other way affect membrane movement, thereby providing multiple binding complexes along the surface in contact with the endothelium. In lymphocytes treated with cytochalasin B the necessary modulation of the surface membrane would not occur and adherence would be prevented.

It is also possible that contractile forces may influence binding through an effect on surface membrane configuration—such as the number and shape of microvilli—particularly since studies in animals have revealed that lymphocytes in the HEV lumen make contact with the endothelium via these projections (van Ewijk et al 1975, Anderson & Anderson 1976).

Another aspect of the binding reaction is the requirement for calcium but not magnesium ions. Calcium might be needed for the formation of the bond linking lymphocytes and endothelial cells, perhaps by acting as a cationic bridge. We could not, however, elute adherent lymphocytes which had attached to HEV in the presence of calcium by reincubating the section in medium containing EGTA. In view of the evidence that microfilaments are calcium-responsive structures (Schreiner & Unanue 1976), calcium chelators may exert their effects by inhibiting contractile protein function involved in the early stages of lymphocyte-HEV adherence. A summary of these findings is presented in Table 2.

It should also be noted that thoracic duct cells treated with trypsin lose their capacity to bind to HEV *in vitro* but neuraminidase causes no reduction in adherence. Our findings are in accord with observations made by Ford et al

TABLE 2

Factors affecting binding of rat thoracic duct lymphocytes to high-endothelial venules

Treatment of lymphocytes	% inhibition of binding
Sodium azide	85
Cytochalasin B	95
Colchicine	0
EDTA	100
EDTA + Ca^{2+}	0
EDTA + Mg^{2+}	85
EGTA	100
EGTA + Ca^{2+}	0
Trypsin	90
Neuraminidase	0

Concentrations of reagents used were: 1 mM-EDTA and EGTA: 2 mM-Ca^{2+} and Mg^{2+}: 10 mM-sodium azide: 10 μg/ml cytochalasin B, colchicine and trypsin: 25 units/ml neuraminidase. Details as in Woodruff et al (1977).

(1976) on the effects of these enzymes on the behaviour of lymphocytes infused directly into isolated mesenteric lymph nodes.

There are three features of the effects of trypsin which are particularly notable. The first is that only lymph node-homing properties of thoracic duct cells appear to be affected. The capacity of the cells to migrate into the spleens of intravenously transfused recipients is not inhibited and the cells show no tendency to accumulate in atypical sites (Woodruff & Gesner 1968). The second is that the cells are very sensitive to trypsin; for example in vitro HEV binding is inhibited by treatment with as little as 10 μg trypsin/ml (Table 2). Although an effect on metabolic activities cannot be excluded, use of the enzyme at this concentration has not been found to increase cyclic AMP levels of peripheral lymphocytes (Shneyour et al 1976). Third, trypsinization of lymphocytes does not inhibit their capacity to adhere to non-specialized endothelial cells in vitro (De Bono 1976).

The simplest proposition is that trypsin produces changes in lymphocyte membrane function which prevent recognition of HEV and attachment of the cells to the surface of the specialized endothelium. Evidence supporting this view has been obtained in recent experiments in which lymph node sections were pretreated with supernatants of trypsinized thoracic duct cells (10^9 cells/ml incubated at 37 °C for 5 min with 20 μg trypsin/ml and 50 μg DNase/ml followed by 200 μg soybean trypsin inhibitor/ml). This significantly reduced the HEV-binding activity of normal lymphocytes. The result is consistent with the view that trypsin-sensitive surface proteins may play a role in lymphocyte adherence to HEV.

EFFECTS OF FIXING LYMPH NODE SECTIONS ON HEV-BINDING OF THORACIC DUCT CELLS

When the idea of investigating the capacity of thoracic duct cells to bind to HEV in lymph node sections was first conceived it was anticipated that unfixed tissue would provide the most suitable targets for analysis of this phenomenon. Numerous attempts to demonstrate binding with unfixed sections were unsuccessful, largely due to the fact that in preparations air-dried for 5 to 10 min the endothelium of HEV adheres poorly to the slide and loss occurs during the incubation and washing steps. This problem has been overcome by placing frozen sections on thoroughly cleaned glass slides coated with poly-L-lysine or on cleaned Thermanox slides. Morphological integrity of the endothelium is preserved and the overlaid cells bind selectively to these structures. However, results obtained with unfixed sections and tissue fixed with glutaraldehyde as described here exhibit marked differences in reproducibility and sensitivity. With fixed tissue excellent binding is consistently observed when the sections are overlaid with cells at a concentration of 5×10^6 cells/ml. With unfixed tissue the extent of binding is lower (20 to 50% positive HEV/section) and specific adherence is observed in only about 50% of the experiments; the optimal lymphocyte concentration is 30×10^6/ml. In the experiment shown in Fig. 5, cells at 10^6/ml overlaid onto fixed sections gave significantly higher levels of binding than that produced by overlaying unfixed sections with lymphocytes at a concentration of 30×10^6 cells/ml.

We next examined the effects of other fixatives on the binding reaction. The results summarized in Table 3 show that adherence to HEV in sections treated with methanol or formaldehyde is poor, whereas high binding levels are observed with glutaraldehyde or the diimidoester dimethyl suberimidate (DMS)-fixed sections. Both glutaraldehyde and DMS have two reactive groups capable of interacting with and forming bridges between the side-chains of amino acids in proteins (Wold 1972). In contrast, methanol is not a cross-linking agent and although formaldehyde may be capable of cross-linking, it is not bifunctional and therefore its effectiveness in this capacity is limited. Thus the results suggest that the binding of lymphocytes to HEV *in vitro* is facilitated under conditions where the endothelial binding site would be likely to be present in an oligomeric or aggregated form. *In vivo* the

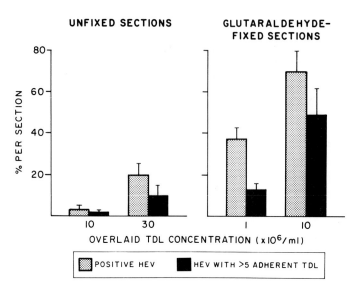

FIG. 5. Comparison of rat thoracic duct lymphocytes (TDL) binding to fixed and unfixed HEV. Sections were placed on poly-L-lysine-coated slides and air-dried at 4 °C for 4 h; alternate sections were then fixed with glutaraldehyde. Mean ± SEM of four sections/group.

endothelial cells would be expected to be capable of modifying the arrangement of these binding sites in order to accommodate lymphocytes and promote their adherence.

CONCLUSIONS

Direct membrane interaction between lymphocytes and endothelium is the first step in a sequence of events which result in the passage of lymphocytes from the bloodstream into the parenchyma of lymph nodes. The mechanism underlying the affinity between these cell populations is unknown but the data presented demonstrate that recirculating lymphocytes display surface membrane constituents which enable them to bind selectively to HEV. The precise location of the endothelial binding site is not known and we can only speculate that adherence *in vitro* is via attachment to membrane components of HEV exposed in the lymph node section. Higher resolution techniques (scanning and transmission electron microscopy) will be necessary to prove that binding is to the cell surface and not to the cytoplasmic space. Nevertheless, the evidence warrants the assumption that moieties on the cell surface participate in a process of recognition and specific adhesion, and that these molecules play a major role in the selectivity of lymphocyte–endothelial inter-

TABLE 3

Effect of different fixatives on binding of rat thoracic duct lymphocytes to high-endothelial venules (HEV) in sections of lymph nodes

Fixative		% positive HEV/section
Glutaraldehyde (3×10^6 cells/ml)	$\overset{\text{O}}{\overset{\|}{\text{H}-\text{C}}}-(\text{CH}_2)_3-\overset{\text{O}}{\overset{\|}{\text{C}}}-\text{H}$	60-80
Methanol (15×10^6 cells/ml)	$\text{H}-\underset{\underset{\text{H}}{\|}}{\overset{\overset{\text{H}}{\|}}{\text{C}}}-\text{OH}$	5-15
Formaldehyde (15×10^6 cells/ml)	$\overset{\text{O}}{\overset{\|}{\text{H}-\text{C}}}-\text{H}$	0-5
Dimethyl suberimidate (15×10^6 cells/ml)	$\text{H}_3\text{CO}-\underset{}{\overset{^-\text{Cl}^+\text{H}_2\text{N}}{\overset{\|}{\text{C}}}}-(\text{CH}_2)_6-\underset{}{\overset{\text{NH}_2^+\text{Cl}^-}{\overset{\|}{\text{C}}}}-\text{OCH}_3$	50-70

Treatment of sections with glutaraldehyde, methanol or formaldehyde was for 10 min at 4°C; fixation with dimethyl suberimidate by the method of Hassell & Hand (1974). Concentration of overlaid lymphocytes in parentheses.

action and consequently exert a decisive influence on the route of lymphocyte migration through the body.

Receptors on lymphocytes which mediate HEV binding might be for complementary sites on high-endothelial cells, or the endothelium and lymphocytes may have in common receptors for a cross-linking structure produced by either endothelial cells or lymphocytes, or by both. Alternatively a soluble factor or factors, released into the lumen of HEV, may affect surface and metabolic properties of lymphocytes, inducing changes which stimulate them to adhere to other cells in their environment—in this case the specialized endothelium.

ACKNOWLEDGEMENTS

This work was supported by Grant AI 10080 from the National Institutes of Health. Barry J. Kuttner is a predoctoral trainee supported by Immunology Training Grant AI 07131 from the National Institutes of Health. The excellent technical assistance of Robert Rasmussen is gratefully acknowledged.

References

Anderson AO, Anderson ND 1976 Lymphocyte emigration from high endothelial venules in rat lymph nodes. Immunology 31:731-748

De Bono D 1976 Endothelial-lymphocyte interaction *in vitro*. I. Adherence of nonallergised lymphocytes. Cell Immunol 26:78-88

Durkin HG, Caproale L, Thorbecke GJ 1975 Migratory patterns of B lymphocytes. I. Fate of cells from central and peripheral lymphoid organs in the rabbit and its selective alteration by anti-immunoglobulin. Cell Immunol 16:285-300

Ford WL, Sedgley M, Sparshott SM, Smith ME 1976 The migration of lymphocytes across specialized vascular endothelium. II. The contrasting consequences of treating lymphocytes with trypsin or neuraminidase. Cell Tissue Kinet 9:351-361

Freitas AA, DeSousa MAB 1977 Control mechanism of lymphocyte traffic. A study of the action of two sulphated polysaccharides on the distribution of ^{51}Cr and ^{3}H adenosine-labelled mouse lymph node cells. Cell Immunol 31:62-77

Gesner BM, Ginsburg V 1964 Effect of glycosidases on the fate of transfused lymphocytes. Proc Natl Acad Sci USA 52:750-755

Gowans JL 1957 The effect of the continuous re-infusion of lymph and lymphocytes on the output of lymphocytes from the thoracic duct of unanaesthetized rats. Br J Exp Pathol 38:67-78

Gowans JL 1959 The recirculation of lymphocytes from blood to lymph in the rat. J Physiol (Lond) 146:54-68

Gowans JL, Knight EJ 1964 The route of recirculation of lymphocytes in the rat. Proc R Soc Lond B Biol Sci 159:257-282

Gutman GA, Weissman IL 1973 Homing properties of thymus-independent follicular lympho-cytes. Transplantation 16:621-629

Hassell M, Hand AR 1974 Tissue fixation with diimidoesters as an alternative to aldehydes. I. Comparison of cross-linking and ultrastructure obtained with dimethylsuberimidate and glutaraldehyde. J Histochem Cytochem 22:223-229

Hodes RJ, Handwerger BS, Terry WD 1974 Synergy between subpopulations of mouse spleen cells in the *in vitro* generation of cell-mediated cytotoxicity. Evidence for the involvement of a non-T cell. J Exp Med 140:1646-1659

Howard JC, Hunt SV, Gowans JL 1972 Identification of marrow-derived and thymus derived small lymphocytes in the lymphoid tissue and thoracic duct lymph of normal rats. J Exp Med 135:200-219

Miller JJ 1969 Studies of the phylogeny and ontogeny of the specialized lymphatic tissue venules. Lab Invest 21:484-490

Schlesinger M 1976 Cell surface receptors and lymphocyte migration. Immunol Commun 5:775-793

Schlesinger M, Israel E 1974 The effect of lectins on the migration of lymphocytes *in vivo*. Cell Immunol 14:66-79

Schneyour A, Patt Y, Trainin N 1976 Trypsin-induced increase in intracellular cyclic AMP of lymphocytes. J Immunol 117:2143-2149

Schreiner GF, Unanue ER 1976 Membrane and cytoplasmic changes in B lymphocytes induced by ligand-surface immunoglobulin interaction. Adv Immunol 24:37-165

Smith C, Henon BK 1959 Histological and histochemical study of high endothelium of post capillary veins of the lymph node. Anat Rec 135:207-213

Stamper HB, Woodruff JJ 1976 Lymphocyte homing into lymph nodes: *in vitro* demonstration of the selective affinity of recirculating lymphocytes for high-endothelial venules. J Exp Med 144:828-833

Stamper HB, Woodruff JJ 1977 An *in vitro* model of lymphocyte homing. I. Characterization of the interaction between thoracic duct lymphocytes and specialized high endothelial venules of lymph nodes. J Immunol 119:772-780 van Ewijk W, Brons HHC, Rozing J 1975 Scanning electron microscopy of homing and recirculating lymphocyte populations. Cell Immunol 19:245-261

Wessells MK, Spooner BS, Ash JF, Bradley MO, Ludena MA, Taylor EL, Wrenn JT, Yamada KM 1971 Microfilaments in cellular and developmental processes. Science (Wash DC) 171:135-143

Wold F 1972 Bifunctional reagents. Methods Enzymol 25:623-651

Woodruff J, Gesner BM 1968 Lymphocytes: circulation altered by trypsin. Science (Wash DC) 161:176-178

Woodruff JJ, Gesner BM 1969 The effect of neuraminidase on the fate of transfused lymphocytes. J Exp Med 129:551-567

Woodruff JJ, Woodruff JF 1974 Virus-induced alterations of lymphoid tissues, IV: The effect of Newcastle disease virus on the fate of radiolabeled thoracic duct lymphocytes. Cell Immunol 10:78-85

Woodruff JJ, Katz IM, Lucas LE, Stamper HB 1977 An *in vitro* model of lymphocyte homing. II. Membrane and cytoplasmic events involved in lymphocyte adherence to specialized high endothelial venules of lymph nodes. J Immunol 119:1603-1610

Zatz MM, Gingrich R, Lance EM 1972a The effect of histocompatability antigens on lymphocyte migration in the mouse. Immunology 23:665-675

Zatz MM, Goldstein AL, Blumenfield CO, White A 1972b Regulation of normal and leukemic lymphocyte transformation and recirculation by sodium periodate oxidation and sodium borohydride reduction. Nat New Biol 240:252-255

Discussion

Born: Platelet to platelet adhesion depends on fibrinogen acting (with calcium) as a bridge. Platelets are apparently able to adhere to each other before they change shape and produce extrusions on their surfaces. An interesting difference with your lymphocytes seems to be, therefore, that their adhesion requires microvilli. As I understood the evidence, there are always microvilli on the cells which emigrate. The microvilli may initiate cell-to-cell contact by the Bangham mechanism.

You are using washed sections, Dr Woodruff, so the observed adhesion is likely to depend on some direct receptor–receptor interaction between the surfaces of the two types of cell. Would it be reasonable to try other inhibitors? I believe there is evidence that lymphocytes do not retain microvilli if the cyclic AMP concentration is raised.

Woodruff: Treatment of lymphocytes with drugs to alter their cyclic AMP or GMP levels did not affect binding to high endothelial cells.

Hall: Your preparation works better if you fix the section with glutaraldehyde. If you fix lymphocytes with glutaraldehyde does this inhibit or enhance their adhesion?

Woodruff: Glutaraldehyde-fixed lymphocytes do not bind, dead lymphocytes don't bind, and lymphocytes treated with sodium azide don't bind.

Hall: Can you elute lymphocytes once they are bound? We calculate bogus extraction efficiencies from experiments *in vivo,* saying that 10% of lymphocytes going to the node are extracted from the blood and transmitted to the lymph. Is this a random selection, though? If you eluted the lymphocytes which had bound, you could see whether they would then bind more efficiently than a random population.

Woodruff: We tried to elute the adherent cells by incubating the sections at 37 °C. We did that six times. In three experiments HEV-bound lymphocytes were eluted but in the other experiments there was no elution.

Davies: That was a visually stunning demonstration of binding, Dr Woodruff. Is it specific for the high-endothelial venules or will other vessels do it? And if you put the cells that didn't bind on another section, would you lose all the binding cells with repeated attempts? Is there any heterogeneity in relation to cells in binding?

Woodruff: I wish we could answer that. It would be nice to pour off the fluid and get the cells back to test them again but we can't do that because many cells adhere to the glass. We have not seen binding to other vessels.

Davies: Why did some HEV bind and some not bind?

Woodruff: We include in the total HEV count some venules which are very small; if these are excluded the percentage of positive HEV would approximate 100%. This is related to the finding that the number of lymphocytes which bind to an HEV is proportional to the size of the endothelium exposed in the tissue section.

Davies: The cross-linking possibilities of glutaraldehyde and the other agent you used could signify two things. Either cross-linking on the surface brings about some kind of concentration of receptors on the surface, as you suggested, or there are unsaturated linking possibilities. So it is difficult to see why you shouldn't get binding over the whole tissue section.

Woodruff: That is why we routinely treat the sections with lysine which readily reacts with free aldehyde groups which might be present in tissue sections fixed with glutaraldehyde.

Humphrey: Do you see anything like this in the spleen at any sites to which lymphocytes adhere? And have you treated lymphocytes with dilute solutions of iron salts, say 10^{-5} M-ferric citrate, which Maria de Sousa (1978) says prevents lymphocyte migration from the blood to the spleen or lymph nodes?

Woodruff: We do not see lymphocyte binding to any particular area of the spleen, and I have not done any experiments with iron.

Humphrey: Her hypothesis is that there are receptors for transferrin, for example, and the iron saturates the surface of the lymphocyte, making it unable to attach. I just wondered whether this sort of phenomenon, if it is significant at all, came into the sort of attachment that you are looking at.

Ford: One of the most significant inhibitory agents may be sodium azide. Your results fit with our experiments in which lymphocytes were treated *in vitro* with sodium azide before i.v. injection into recipients. Migration into lymph nodes was inhibited while migration into the spleen was unaffected. Autoradiographic analysis suggested that the adherence of treated lymphocytes to HEV was reduced; the inhibition could not be explained by more sluggish movement across the vessel wall (Ford et al 1978). The failure of azide-treated lymphocytes to stick to HEV cannot be accounted for by loss of microvilli because in fact microvilli become more prominent.

Williams: Does that happen at 7 °C?

Ford: I don't know. We did our experiments at 37 °C and they suggest that the presence of both microvilli and a hypothetical receptor is not sufficient for lymphocyte binding to high endothelial cells.

Woodruff: Temperature does affect microvilli. At 37 °C there are many microvilli but they are long and filamentous. At 4 °C the number is reduced but those that are present are short and stubby. Such microvilli might provide better contact between lymphocytes and HEV. That may be why the assay works in the cold.

Ford: Are there any microvilli on the azide-treated cells in your conditions?

Woodruff: The only study that I know of is the one by Lin et al (1973) showing the changes in microvilli with temperature.

Weiss: Thoracic duct lymphocytes have an affinity for HEV but there are also T and B cell zones in the lymph node. The thoracic duct lymphocytes didn't seem to have any affinity for the T cell zone, for example. Is that correct?

Woodruff: All lymphocytes enter the nodes by crossing the HEV (Sprent 1973, Gutman & Weissman 1973). Once they are in the substance of the node they segregate out. I don't have any evidence that if you put a B cell on a lymph node section it would bind better to the HEV close to germinal centres than to HEV in the deep cortex.

Weiss: I wasn't thinking of that. In your sections of lymph nodes whose surface is flooded with living cells the cells attach to the same door that they attach to and pass through in an intact lymph node. You are dealing with a collection of lymphocytes that are largely T cells and these cells have access to and are concentrated in T cell zones in living lymph nodes. The interesting thing is that they don't adhere there in your sections. What I am thinking of are the approaches that could be used to get specific adherence of these cells not only to the postcapillary venules but also to those specialized areas beyond, where they are selectively held in life. Presumably once the cells pass through the postcapillary venules they lose their microvilli. If the numbers of microvilli are reduced would the T cells adhere to T cells? Or perhaps the cells pick something up as they pass through the postcapillary venule that permits them to adhere further on.

Weissman: Whether microvilli are the attachment points and what happens at raised temperatures are interesting points if there is a concentration of receptors at the end of microvilli and not at other places. But we all seem so far from identifying the receptor molecules by using specific probes such as antibodies that it is not a very fruitful discussion. It may be that you can collapse microvilli and still see receptor microzones.

Williams: But is there really a membrane on these sections? We are talking about this sophisticated stuff but I would have thought the section would have gone straight through the cell.

Morris: That is the crucial part of it all. How do you know you are looking at membrane-directed and not cytoplasm-directed migration?

Woodruff: I said that I didn't know. One needs high resolution techniques for that.

van Ewijk: Dr Butcher and I looked with the scanning electron microscope at frozen sections incubated with lymphoid cells. It was possible to recognize the cell surface of HEV in these sections.

Butcher: I think two main points emerged from those observations of lymphocytes binding to HEV in frozen sections. The first is that the adherent lymphocytes have numerous microvilli and, just as *in vivo* (van Ewijk, this volume), many of these villi are in contact with the HEV in the tissue section. Secondly, as Willem van Ewijk mentioned, the cells of the frozen section itself maintain distorted but recognizable cellular outlines, as if the cryostat blade sheared around them rather than simply cutting through the cytoplasm of every cell. Thus it may be that cell membranes are more highly represented in these tissue sections than one would expect. It remains quite possible, therefore, that adherent lymphocytes are actually binding to HEV membranes rather than to cytoplasmic elements. This question remains unanswered in my mind.

Weissman: But even if there is no membrane there, it wouldn't be terribly important. A particular ligand produced by a particular cell can be produced inside the cytoplasm and expressed on the cell membrane. I don't think that is crucial to any argument. What is specific is what Judith Woodruff showed. It sticks to those cells and not to other cells.

Williams: Can you make the thin section, then cut the other way and see whether there is a lipid bilayer?

Butcher: Willem van Ewijk and I have discussed this possibility, and because of the technical difficulties involved, as well as the considerations just mentioned by Irv Weissman, we have decided to postpone it indefinitely.

Woodruff: That experiment needs to be done before we know precisely whether there is contact.

Gowans: Are you all happy now about the problem of whether it matters whether there is a membrane there?

Ford: I think the question of whether lymphocytes are binding to cytoplasm matters very much. The cytoplasm is not a place where one would expect to find receptors that bind cells together.

Davies: Oestrogen receptors are thought to be cytoplasmic.

Ford: The oestrogen receptors are not actually holding cells together against shearing forces.

McConnell: Don't plasma cells have 'receptors' within the cytoplasm?

Gowans: You do get specific binding on these sections. What surprises me is that that technique works at all. I think this is an important technical advance.

Williams: Another point about technique, Professor Woodruff, is that you are correlating the glutaraldehyde effect with fixation but perhaps lymphocytes bind after fixation by adsorbing to the lysine which you use to block excess aldehydes (membranes bind very well to polylysine). You are using higher glutaraldehyde concentrations than we do for studying binding to cell surface antigens. We fix at room temperature, which makes comparison somewhat difficult, but we can totally fix cells by incubating them in 0.15% glutaraldehyde for 5 min at room temperature, while you used 3% for 10 min at 7 °C. We also know that we knock out some antigens if we go much above a concentration of 0.15%. If you drop the glutaraldehyde concentration to a level which is minimal for fixation, do lymphocytes still bind to the sections?

Woodruff: In the original work we used PBS as a buffer for glutaraldehyde. In the work I have just described, in which we increased the sensitivity of the system, we are using cacodylate. I haven't cut down the glutaraldehyde concentration so I can't answer your question. Certainly glutaraldehyde treatment of sections is not an absolute requirement for lymphocyte–HEV binding.

Williams: There was a big difference between fixed and unfixed sections.

Woodruff: There is a difference in degree. Clearly the cells bind to HEV in unfixed sections. Also the rat lymph node section doesn't adhere very well to the glass when it is not fixed. That is one of the reasons why we tried glutaraldehyde in the first place.

Williams: Have you ever fixed a fibroblast monolayer and put cells on that?

Woodruff: No. I have looked at a number of other tissues and saw no binding to the liver, heart, spleen, thymus or submaxillary glands. There is binding, however, to the brain sections but it is restricted to the myelinated regions; non-myelinated axons and nerve cells don't bind.

McConnell: You had a nice system for trying to analyse what cell surface antigens might be involved in this system. Have you tried treating the lymphocytes or the HEV with F(ab')$_2$ or F(ab') antibody to see whether the lymphocyte-HE cell interaction can be inhibited by antibodies to cells? One could refine this approach for particular antigenic determinants.

Woodruff: We have begun experiments like that.

Ford: Your technique could be exploited to test whether the sulphated material plays a role in the adhesion of lymphocytes to HEV, Professor Woodruff. For example when a section includes only the basal part of a high endothelial cell no sulphate concentration is seen. It might be possible to determine whether lymphocytes adhere to this part of the cell.

van Ewijk: We have studied the denaturing effect of the concentration of glutaraldehyde in studies of antibody binding to glutaraldehyde-fixed cells. With concentrations of glutaraldehyde above 0.1% we see a severe drop in the antibody binding. At concentrations higher than 1% we see non-specific binding of antibody. So it is important to control the glutaraldehyde concentration during fixation. The type of glutaraldehyde is important too: different brands give different effects at all sorts of concentrations.

Gowans: What antibodies were binding to what?

van Ewijk: These were rabbit anti-mouse immunoglobulin antibodies, so we were staining B cells preferentially. We analysed antibody binding quantitatively with a fluorescence-activated cell sorter and a radioimmunoassay.

Woodruff: So you are looking at cell surface immunoglobulin?

van Ewijk: Yes.

Williams: The extent of loss of antigenic activity on fixation varies with different antigens. To achieve fixation 0.1% glutaraldehyde is needed, but for one antigen that we studied a twofold increase in this resulted in complete loss of antigenic activity. However, that doesn't happen in all cases. Do you fix at room temperature?

van Ewijk: No, in the cold. With respect to other antigens: 0.1% glutaraldehyde is not only optimal for cell-surface-bound immunoglobulin but also for H-2 molecules.

Woodruff: But lymphocytes treated with glutaraldehyde won't bind, so there is agreement on that point.

The point about the different preparations of glutaraldehyde is a very critical one; with some commercial preparations we observe very poor or erratic HEV binding.

Gowans: Have you tried lower concentrations of glutaraldehyde on lymphocytes?

Woodruff: Yes, we used concentrations substantially lower than that used to treat the sections.

Humphrey: Surely the ε-amino group will bind to glutaraldehyde, so the end result is to replace a terminal amino group with an amphoteric group. To leave the charge unchanged you would need to treat with a diamine after the glutaraldehyde.

Williams: You are making a correlation with fixing. All I am saying is that Dr Woodruff is using glutaraldehyde at levels vastly higher than are needed for fixing. So is the correlation with fixing or with something else?

Woodruff: That is why I showed you the data with methanol and formaldehyde, which are excellent fixatives with respect to morphology. Dimethyl suberimidate is a very poor fixative, judging from the morphology, but we get very good HEV binding with it. One can get excellent morphology and HEV binding by briefly treating the sections with methanol and then with dimethyl suberimidate. The effect is therefore not specific to glutaraldehyde.

References

de Sousa M 1978 Lymphoid cell positioning: a new proposal for the mechanism of control of lymphoid cell migration. In: Curtis ASG (ed) Cell–cell recognition. Cambridge University Press, Cambridge (Soc Exp Biol Symp 32) p 393-449

Ford WL, Smith ME, Andrews P 1978 Possible clues to the mechanism underlying the selective migration of lymphocytes from the blood. In: Curtis ASG (ed) Cell–cell recognition. Cambridge University Press, Cambridge (Soc Exp Biol Symp 32) p 359-392

Gutman GA, Weissman IL 1973 Homing properties of thymus-independent follicular lymphocytes. Transplantation (Baltimore) 16:612-629

Lin PS, Wallach DFH, Tsai S 1973 Temperature-induced variations in the surface topology of cultured lymphocytes are revealed by scanning electron microscopy. Proc Natl Acad Sci USA 70:2492

Sprent J 1973 Migration of T and B lymphocytes in the mouse. I. Migratory properties. Cell Immunol 7:10-39

Cellular, genetic, and evolutionary aspects of lymphocyte interactions with high-endothelial venules

EUGENE C. BUTCHER and IRVING L. WEISSMAN

Department of Pathology, Laboratory of Experimental Oncology, Stanford University School of Medicine, Stanford, California 94305

Abstract   Lymphocytes leave the blood by adhering to and migrating through the specialized endothelium of postcapillary high-endothelial venules (HEV).   The study of lymphocyte–HEV interaction shows promise of delineating, at cellular and molecular levels, the general mechanisms involved in cell–cell interactions. Evidence from quantitative *in vivo* and *in vitro* assays of lymphocyte adherence to mouse HEV suggests that: (1) The ability to bind HEV is equivalent in and predominantly limited to mature lymphocyte populations. (2) Lymphocyte–HEV interactions determine the organ specificity of lymphocyte migration (i.e. the preferential migration of intestinal and peripheral lymphocytes through gut and peripheral nodes respectively).   It is proposed that selective recognition is mediated by interaction of lymphocyte surface receptors with organ- or region-specific endothelial cell surface ligands. (3) Murine thymic lymphomas express characteristic HEV-binding properties including (in some instances) a high degree of organ specificity. (4) Lymphocytes exhibit strain differences in binding to HEV that are susceptible to genetic analysis.   One major difference between BALB/c and C57BL mesenteric node lymphocytes, predominantly affecting adherence to mesenteric and peripheral node HEV, is attributable to a single region on chromosome 7. (5) Surface structures mediating lymphocyte–HEV adherence have been randomly but continuously modified during evolution despite morphological (and presumably functional) conservation of the interaction in all mammalian species.

Mature small lymphocytes recirculate from blood to lymph.   As described by Gowans & Knight (1964), lymphocytes enter the lymphatic system from the blood by migrating through the specialized endothelium of postcapillary high-endothelial venules (HEV), located predominantly in the thymus-dependent regions of lymph nodes and Peyer's patches.   The ability to recognize and adhere specifically to the endothelium of HEV is a specialized property of mature lymphocytes.   We have been interested in comparing the ability of various normal and neoplastic lymphocyte populations to bind to HEV in the

© *Excerpta Medica 1980*
Blood cells and vessel walls: functional interactions
(Ciba Foundation symposium 71) p 265-286

mouse, using a modification of the Stamper-Woodruff frozen section assay system (Stamper & Woodruff, 1976, 1977, Woodruff & Kuttner, this volume), which we have previously shown to be an accurate assay of *in vivo* lymphocyte–HEV adherence (Butcher et al 1979). We now report and discuss experiments designed to examine the degree of cellular and organ specificity of the adherence of lymphocyte populations to syngeneic HEV, the genetics of strain variation in lymphocyte–HEV binding properties, and the extent of species-specific differences in lymphocyte surface structures mediating the interaction with HEV.

THE IN VITRO FROZEN SECTION ASSAY

A suspension of lymphocytes in Hanks' balanced salt solution containing 1% bovine serum albumin is incubated at 7 °C for 30–35 min on freshly cut, unfixed frozen sections of mouse mesenteric or peripheral lymph nodes or Peyer's patches. The medium is removed and the section is fixed with 1% glutaraldehyde in cold phosphate-buffered saline. After fixation, non-adherent cells are gently rinsed off. To allow quantitative comparisons, an internal standard population of fluorescent lymphocytes (labelled by 15 min incubation with 60 μg/ml fluorescein isothiocyanate, FITC, at 37 °C) is mixed with each population before incubation. A reference sample, usually of syngeneic mesenteric node lymphocytes, is included. Data from multiple sections are pooled for each sample. The ratio of sample to standard cells in the incubation mixtures (R_I) and adherent to HEV (R_{HEV}) is determined. A specific adherence ratio R_{HEV}/R_I, is calculated for each sample ($SAR_{A,B,C...}$). To permit direct comparison of the adherence of two sample populations, A and B, their relative adherence ratio (RAR) is calculated:

$$RAR \text{ of A to B} = (SAR_A - SAR_N)/(SAR_B - SAR_N)$$

where SAR_N is the background SAR of a negative control population of erythrocytes or formalin-fixed lymphocytes. (SAR_N is about 1% of the SAR of mesenteric node lymphocytes.) Standard deviations of all ratios are calculated using the delta method.

LYMPHOCYTE ADHERENCE TO HEV IN SYNGENEIC MESENTERIC NODES

Sprent (1973) has demonstrated that most if not all B and T lymphocytes in secondary lymphoid organs of the mouse (lymph nodes, Peyer's patches, and spleen) are mobilized into the thoracic duct lymph during prolonged thoracic duct drainage. Most of the cells appearing in the thoracic duct are destined to re-enter the lymphatic system via HEV from the blood, perhaps after a transit period in the spleen. Thus, most mature lymphocytes must have, at some stage in their existence, the capacity to recognize and migrate through HEV.

Analysis of the binding of various syngeneic lymphocyte populations to mesenteric lymph node HEV *in vitro* supports the concept that the ability to bind HEV is shared equally by most mature lymphocytes. Fig. 1 summarizes results from several experiments comparing the ability of various lymphocyte populations to adhere to syngeneic mesenteric node HEV. Cells and mesenteric node sections were prepared from 6–8-week-old (BALB/c × C57BL/6J)F$_1$ hybrid mice. The HEV-binding abilities of B and T mesenteric node lymphocytes were determined by two independent methods, the results of which are pooled in Fig. 1: (1) immunofluorescent staining of lymphocytes adherent to HEV after a standard *in vitro* incubation, using FITC-conjugated rabbit anti-mouse immunoglobulin (B cells), or rabbit anti-mouse T cell serum followed by an FITC-conjugated goat anti-rabbit immunoglobulin second stage; and (2) comparison of B lymphocytes from the mesenteric node of syngeneic 'B' mice (thymectomized, irradiated, and reconstituted with fetal liver cells) with normal mesenteric node lymphocytes. Each of the relatively

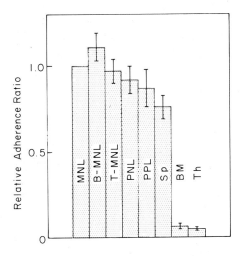

Fɪɢ. 1. *In vitro* adherence of normal lymphocyte populations to HEV in frozen sections of syngeneic mesenteric lymph node. Data expressed as the relative adherence ratio of sample to mesenteric node lymphocytes. Mesenteric node lymphocytes (MNL); B (B-MNL) and T (T-MNL) mesenteric node lymphocytes; peripheral (axillary and brachial) node lymphocytes (PNL); spleen cells (Sp); bone marrow cells (BM); and thymocytes (Th). Bars represent standard errors.

pure populations of mature lymphocytes tested—peripheral (axillary and brachial) node, Peyer's patch and mesenteric node (T or B) cells—bind mesenteric node HEV equally well. In addition, nucleated spleen and bone marrow cells bind in proportion to their content of mature lymphocytes (roughly 75% and 6% as well as mesenteric node lymphocytes, respectively). These results suggest that the capacity to recognize and bind to mesenteric node HEV may be common to all lymphocytes.

Most available evidence indicates that the thymus does not contain mature recirculating lymphocytes (Cantor & Weissman 1976), yet Zatz & Lance (1970) have shown that a small subpopulation of thymocytes is capable of homing to lymph nodes after intravenous injection. Consistent with these findings, the *in vitro* binding of thymocytes to HEV, although poor, is significantly above background (RAR = 0.05 ± 0.01). A significant proportion of this minor lymph node-seeking, HEV-binding thymocyte subpopulation may consist of cortisone-resistant cells. These cells represent about 5% of the total thymocyte population, are located predominantly in the thymic medulla, demonstrate many characteristics of mature peripheral T lymphocytes (e.g. size, surface antigens and mitogen sensitivity [Cantor & Weissman 1976]), are enriched in lymph node-seeking cells (Blomgren & Andersson 1972), and bind better than the total thymocyte population to mesenteric node HEV *in vitro* (RAR to mesenteric node lymphocytes = 0.24). Nearly mature cells on the verge of leaving the thymus represent a second class of thymocytes that is probably capable of interacting with HEV, since we have observed thymic migrants in the mesenteric lymph node within 30 min of intrathymic injection of fluorescent label (fluorescein isothiocyanate) (Scollay et al 1979). The relative contribution of these two possible HEV-binding populations to the observed adherence of thymocytes to HEV is unknown, and the possibility that other subpopulations are involved is not excluded.

ORGAN SPECIFICITY OF LYMPHOCYTE–HEV ADHERENCE

It is well known that certain lymphocyte populations selectively home to or recirculate through specific organs or regions of the body. Lymphocytes and immunoblasts derived from the gut migrate preferentially to Peyer's patches and to the gut wall lamina propria, whereas lymphocytes and immunoblasts from peripheral lymph nodes return preferentially to peripheral lymph nodes (see Gowans & Knight 1964, Griscelli et al 1969, Guy-Grand et al 1974, 1978, McWilliams et al 1975, Scollay et al 1976, Cahill et al 1977). The basis of this organ-specific lymphocyte migration has been the subject of speculation. Although localized antigen can effect the eventual localization of antigen-specific antibody-forming cells (Husband & Gowans 1978) or memory B cells (Ponzio et al 1976, Rowley et al 1972), it is now generally accepted that antigen is not a factor in the early stages of organ-selective homing (see Halstead & Hall 1972, Parrott & Ferguson 1974, Guy-Grand et al 1974, Husband & Gowans 1978). We present results demonstrating that selective lymphocyte migration can be mediated at the level of lymphocyte interaction with organ-specific postcapillary venules, and we suggest the existence of at least two highly specific and separate migration pathways: through the peripheral lymph nodes and through the gut-associated HEV of the Peyer's patches.

Organ specificity of the interaction of normal lymphocyte populations with Peyer's patch and peripheral node HEV

The relative binding preference of peripheral (axillary and brachial) node, mesenteric node and Peyer's patch lymphocytes for HEV in peripheral (axillary and brachial) nodes and Peyer's patches was determined both *in vitro,* using the frozen section assay, and *in vivo.* In the latter experiment ^{51}Cr-labelled sample cells (100 μCi/ml ^{51}Cr, 37 °C, 1 hour) were injected via the tail vein into three or four splenectomized syngeneic recipients. The animals were killed 15 min later (a time when most lymphocytes are still confined to HEV: see Butcher et al 1979), and the counts (c.p.m.) localizing in peripheral lymph nodes, mesenteric node, and per milligram of Peyer's patch were determined. As a measure of the *in vivo* or *in vitro* organ preference of the sample populations, a preference index (PI) was calculated as follows, using the relative localization of mesenteric node lymphocytes (MNL) as an arbitrary standard:

In vivo: PI (of sample A vs. MNL for peripheral node over Peyer's patch) =

$$\frac{\text{Average ratio (c.p.m. in peripheral nodes/c.p.m. per mg Peyer's patch) in recipients of A}}{\text{Average ratio (c.p.m. in peripheral nodes/c.p.m. per mg Peyer's patch) in recipients of MNL}}$$

In vitro: PI (of sample A vs. MNL for peripheral nodes over Peyer's patch HEV) =

$$\frac{\text{SAR}_A \text{ on peripheral node/SAR}_A \text{ on Peyer's patch}}{\text{SAR}_{MNL} \text{ on peripheral node/SAR}_{MNL} \text{ on Peyer's patch}}$$

where sample A in these experiments is either Peyer's patch or peripheral node lymphocytes. The *in vivo* and *in vitro* results (Table 1) are in substantial agreement and demonstrate a significant preference of mesenteric and peripheral node lymphocytes for peripheral node over Peyer's patch HEV

TABLE 1

Organ preference of HEV-binding by normal lymphocyte populations. Relative preference of sample cells for syngeneic peripheral node over Peyer's patch HEV, determined *in vitro* (frozen section assay) and *in vivo* (15 min after i.v. injection, ^{51}Cr label)

| Cell source | *In vitro* | | | | *In vivo* | |
| | *(BALB/c × C57BL/ 6J)F1* | | *AKR/Cum* | | *(BALB/c × C57BL/ 6J)F1* | |
	Preference Index[a]	*P value*[b]	*Preference Index*	*P value*	*Preference Index*	*P value*
Mesenteric node	Unity	-	Unity	-	Unity	-
Peripheral nodes (axillary and brachial)	0.84 ± 0.11[c]	N.S.[d]	1.58 ± 0.36	N.S.	1.43 ± 0.24	<0.05
Peyer's patches	0.55 ± 0.06	<0.001	0.34 ± 0.07	<0.001	0.48 ± 0.13	<0.001

[a]An index of the relative adherence of sample cells to peripheral (axillary and brachial) node over Peyer's patch HEV, as compared to the same ratio for mesenteric node lymphocytes (see text).
[b]Significance level of difference between sample preference index and the preference index of mesenteric node lymphocytes (i.e. unity). Two-tailed test.
[c]Standard error.
[d]Not significant.

(preference index 0.8–1.6), as compared with Peyer's patch lymphocytes (preference index 0.3–0.6).

This organ preference of lymphocyte adherence could be mediated by mechanisms of selection expressed only on Peyer's patch HEV (with all lymphocytes binding equally to peripheral node HEV), or only on peripheral node HEV (with non-selective binding to Peyer's patch HEV), or by lymphocyte selection at both Peyer's patch and peripheral node HEV. The results in Table 2 suggest that HEV in each organ may be capable of selective lymphocyte recognition; peripheral and mesenteric node lymphocytes bind in significantly greater proportion to peripheral node HEV than do Peyer's patch lymphocytes ($P<0.001$), and conversely Peyer's patch lymphocytes bind in greater proportion to Peyer's patch HEV ($P<0.03$).

The quantitative organ preference of lymphocyte-HEV binding demonstrated by these experiments implies significant differences between peripheral node and Peyer's patch lymphocyte populations, and between peripheral node and Peyer's patch HEV, but does not reveal the degree of

TABLE 2

Relative adherence of peripheral node, mesenteric node and Peyer's patch lymphocytes to peripheral node and Peyer's patch HEV *in vitro*.

Cell source	Adherence to peripheral node[a] HEV		Adherence to Peyer's patch HEV	
	Relative adherence ratio	P value[b]	Relative adherence ratio	P value
Mesenteric node	Unity	-	Unity	-
Peripheral node[a]	0.93 ± 0.08[c]	N.S.[d]	1.05 ± 0.09	N.S.
Peyer's patches	0.68 ± 0.06	<0.001	1.22 ± 0.11	<0.03

[a]Axillary and brachial.
[b]Significance level of difference between the relative adherence ratio of sample cells and that of mesenteric node lymphocytes (i.e. unity). One-tailed test.
[c]Standard error.
[d]Not significant.

selectivity possible in the interaction of individual lymphocytes with HEV. Thus we chose to extend this study using clonal lymphocyte populations (thymic lymphomas).

Binding of thymic lymphomas to HEV

Since murine lymphomas probably represent the outgrowth of solitary transformed precursor cells, each lymphoma might be expected to consist of a population of cells homogeneous in their capacity to adhere selectively to HEV. Thus, we anticipated that study of the interaction of murine lymphomas with HEV might resolve whether organ preferential homing can be absolute at the level of each cell, or whether the difference between cells is simply a relative organ bias.

We examined the HEV-binding properties of cells from the thymus and/or lymph nodes of C57BL/Ka mice with advanced thymic lymphomas induced by radiation leukaemia virus—a virus first isolated from a radiation-induced lymphoma, but which now acts independently of radiation. The RARs of sample cells to syngeneic mesenteric node lymphocytes were determined on BALB/c × C57BL/6J F_1 sections of Peyer's patches and of axillary and brachial lymph nodes. Representative results, obtained from four individual animals, have been selected for discussion (Fig. 2). Several points should be emphasized:

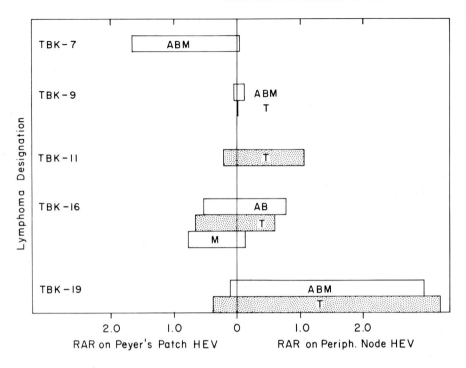

FIG. 2. *In vitro* adherence of thymic lymphoma cells to HEV in peripheral (axillary and brachial) nodes and Peyer's patches. Lymphoma cells obtained from pooled axillary and brachial nodes (AB), pooled axillary, brachial and mesenteric nodes (ABM), or thymus (T). RAR, relative adherence ratio. The adherence of normal syngeneic C57BL/Ka mesenteric node lymphocytes to HEV in either organ is defined as unity.

(1) A remarkable degree of specificity is possible in the interaction of lymphocytes with HEV of Peyer's patches and peripheral lymph nodes. Lymphoma TBK-19 binds almost exclusively to peripheral node HEV (preference index 27, see Fig. 3), whereas lymphoma TBK-7 adheres with equal selectivity to Peyer's patch HEV (preference index 1/30). These preference indices may actually underestimate the specificity of interaction, since the cell populations may have been contaminated with a small number of normal lymphocytes.

(2) Expression of receptor for HEV, although a characteristic of mature lymphocytes, does not necessarily lead to emigration from the thymus. Thus, although TBK-11 cells demonstrated a greater ability to bind to peripheral node HEV than normal mesenteric node lymphocytes, the lymphoma remained entirely limited to the thymus; there was no gross or microscopic evidence of tumour in the lymph nodes, Peyer's patches, spleen or liver of the

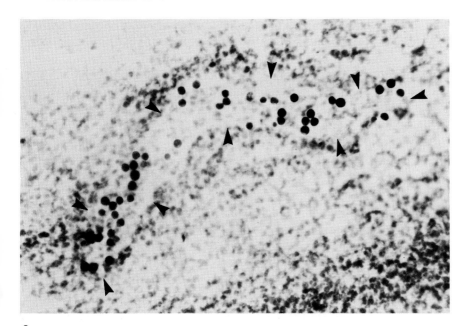

FIG. 3. Organ specificity of adherence of thymic lymphoma TBK-19 to HEV *in vitro*. × about 200. (a) Numerous lymphoma cells (dark, round, large cells) binding to HEV in frozen section of peripheral node. (b) Few cells, predominantly smaller lymphocytes, adherent to HEV in nearby section of Peyer's patch. Arrows delineate HEV basement membrane.

animal in which it arose, even though the thymus was massively enlarged.

(3) In most lymphomas, surface expression of specific HEV receptor(s) (or their apparent absence, as in TBK-9) is determined in the thymus and is maintained in tumour metastases regardless of local environment. Thus, lymphoma cells removed from the thymus, from the peripheral or mesenteric lymph nodes, or from Peyer's patches (data not shown) of a given animal usually demonstrated similar HEV binding behaviour. (It also appears that HEV binding properties may reflect metastatic properties of a lymphoma; the animal bearing TBK-7 had lymphomatous involvement of the T-cell region of its Peyer's patches, whereas in the mouse bearing TBK-19 the Peyer's patches appeared to be uninvolved.)

(4) One animal (bearing TBK-16) offered an apparent exception to the clonal stability of organ preference. TBK-16 lymphoma cells from the thymus and peripheral nodes bound nearly equally well to peripheral node and Peyer's patch HEV (preference indices 0.9 and 1.5), while cells from the mesenteric node demonstrated significant preference for HEV of Peyer's patches (preference index 1/6). This lymphoma may represent malignant expansion of (a) two transformed clones, one recognizing peripheral node HEV and the other Peyer's patch HEV, which have segregated unequally between the lymphoid organs; or (b) a single clone of cells expressing a receptor or receptors capable of recognizing both types of endothelia, but which expresses Peyer's patch HEV receptors preferentially in the mesenteric node microenvironment. Cloning experiments are planned to determine whether the HEV specificity of lymphomas can in fact be altered, since clonal maintenance and stability of specific surface receptors for HEV could have significant implications for the understanding and prediction of the *in vivo* behaviour of lymphomas.

Discussion

These experiments with normal and neoplastic lymphocyte populations suggest that lymphocytes may be programmed for specific migration and recirculation pathways by virtue of the surface expression of receptors for organ- or region-specific endothelial cell surface ligands. We have demonstrated the existence of two distinct and highly selective lymphocyte receptor–HEV ligand combinations, involving Peyer's patches and peripheral node HEV. Region-specific lymphocyte–HEV interaction may occur in other tissues as well.

Any model of the programming and induction of region-specific receptors for HEV during lymphocyte maturation must explain the ability of Peyer's patches and peripheral node lymphocyte populations to bind in significant numbers to HEV in both organs. Two extreme models are suggested: (1) HEV specificity is already expressed on virgin lymphocytes leaving the thymus or bone marrow, and is clonally maintained on their progeny. A significant number of these lymphocytes bear receptors specific for HEV in individual organs, but a major fraction of the cells in each organ are capable of interacting with all HEV, either via a general HEV receptor, or via the surface expression of several organ-specific receptors. This model is consistent with the observed HEV-binding properties of thymic lymphomas. (2) Virgin lymphocytes express a receptor or receptors of general interaction ability. Antigenic stimulation and blastogenesis result in the expression of region-specific receptors, perhaps selectively induced by the local microenvironment at the site of stimulation. This model explains the general observation that blast cells demonstrate a greater degree of homing specificity than unfractionated lymphocyte populations (see Griscelli et al 1969, Guy-Grand et al 1974, 1978, McWilliams et al 1975), and is similar to that proposed by Guy-Grand et al (1978) for the generation of gut-specific homing properties in T lymphocytes stimulated to divide in the Peyer's patch microenvironment.

GENETIC ANALYSIS OF A DIFFERENCE IN THE HEV-BINDING CHARACTERISTICS OF BALB/c AND C57BL MESENTERIC NODE LYMPHOCYTES

In experiments designed to detect genetic variability in the HEV-binding properties of lymphocytes from various strains of mice, we observed two major differences in the binding characteristics of mesenteric node lymphocytes from BALB/c and C57BL/6J mice. First, BALB/c mesenteric node lymphocytes bind to mesenteric node HEV in greater proportions than do C57BL/6J mesenteric node lymphocytes (RAR of BALB/c to C57BL = 1.5 ± 0.2, SEM, 6 experiments). Second, BALB/c mesenteric node lymphocytes consistently demonstrate a relative preference for peripheral (axillary and brachial) node HEV over Peyer's patch HEV, by comparison with C57BL mesenteric node lymphocytes (preference index of BALB/c vs. C57BL for peripheral node over Peyer's patch HEV = 2.6 ± 0.4, SEM, five experiments); in most experiments, this effect is due primarily to differences in adherence to peripheral node HEV. (In most of these experiments, and all of those described below, BALB/c $\times$ C57BL/6J F_1 hybrids were used as the source of HEV.)

We did several experiments in an attempt to find a genetic linkage of this strain difference or differences. The first (Table 3) showed that the increased adherence of BALB/c over C57BL mesenteric node lymphocytes to mesenteric node HEV is not determined by the *H-2* locus. A further set of experiments used seven independent recombinant inbred lines which have different contributions of BALB/c and C57BL genetic material, as described by Bailey (1971). The value of these recombinant inbred lines lies in the fact that the distribution of any new characteristic among the seven strains can be compared with the distribution of known BALB/c and C57BL genetic information, to give an indication of possible linkages.

As shown in Table 4, the HEV-binding properties of mesenteric node

TABLE 3

Differences in adherence of BALB/c and C57BL mesenteric node lymphocytes to mesenteric node HEV are not determined by *H-2*

Strain of origin of mesenteric node lymphocytes	H-2 genotype	Relative adherence ratio
C57BL/6J	$H\text{-}2^b/H\text{-}2^b$	Unity
(BALB/c × C57BL/6J)F$_1$	$H\text{-}2^d/H\text{-}2^b$	1.22 ± 0.16
BALB/c	$H\text{-}2^d/H\text{-}2^d$	2.43 ± 0.33
BALB/c.H-2^b	$H\text{-}2^b/H\text{-}2^b$	2.10 ± 0.31

TABLE 4

HEV-binding properties of mesenteric node lymphocytes from seven BALB/c × C57BL/6 recombinant inbred strains.

Strain	Relative adherence ratio (to mesenteric node HEV)	Preference Index[a]	Assigned trait (probable)
C × BD	0.71[b]	0.71[b]	C57BL
C × BE	0.54	0.62	C57BL
C × BG	1.10	1.12	BALB/c
C × BH	0.54	0.60	C57BL
C × BI	1.03	1.18	BALB/c
C × BJ	0.57	0.73	C57BL
C × BK	0.85	0.88	C57BL(?)
C57BL/6	0.70	0.56	C57BL
BALB/c	1.10	1.52	BALB/c
(BALB/c × C57BL/6)F$_1$	Unity	Unity	Heterozygous

[a]An index of relative adherence to peripheral (axillary and brachial) node over Peyer's patch HEV (see text).
[b]Each value is the mean of two independent determinations.

lymphocytes from these recombinant inbred strains did not separate cleanly into the two parental types, suggesting that the HEV-binding differences between BALB/c and C57BL mesenteric node lymphocytes are determined by more than one genetic locus. However, the two major effects we described above for BALB/c lymphocytes—increased binding to mesenteric node HEV and a relative preference for peripheral node over Peyer's patch HEV, compared with C57BL mesenteric node lymphocytes—segregated together, and were expressed predominantly by cells from two of the seven strains, C × BG and C × BI. Comparison of the strain distribution of this BALB/c 'trait' with that of known BALB/c and C57BL genetic material suggested possible linkage with histocompatibility loci *H-18* and *H-20* on chromosome 4, or *H-1* on chromosome 7.

In a third set of experiments we used congenic mice bearing the above BALB/c histocompatibility loci on C57BL/6 or C57BL/10 genetic backgrounds. As shown in Table 5, mesenteric node lymphocytes of B10.C (41N)Sn mice (a strain carrying BALB/c information at the closely linked loci *H-1, Hbb* and *c,* on a C57BL/10 background) demonstrated the BALB/c HEV-binding trait. Mesenteric node lymphocytes of the other congenic mice tested (including another strain with the C57BL/10 genetic background) behaved like C57BL mesenteric node lymphocytes.

TABLE 5

Expression of BALB/c and C57BL HEV-binding properties by mesenteric node lymphocytes of congenic strains. The BALB/c trait is linked to *H-1, Hbb* and *c.*

Strain	Donor loci carried	Donor strain	Inbred partner (genetic backgr.)	Preference Index[a] (vs.C57BL/ 6 MNL)	P value[b] versus: BALB/c	C57BL/- 6	HEV-binding trait
C57BL/6	-	-	-	Unity	<0.001	-	C57BL
BALB/c	-	-	-	2.6 ± 0.4[c]	-	<0.001	BALB/c
B6.H-18[c]	*H-18*	BALB/c	C57BL/6	1.1 ± 0.2	<0.001	N.S.[d]	C57BL
B6.H-20[c]	*H-20*	BALB/c	C57BL/6	1.0 ± 0.3	<0.002	N.S.	C57BL
B10.C(4lN)Sn	*H-1, Hbb, c*	BALB/c	C57BL/10	2.1 ± 0.3	N.S.	<0.001	BALB/c
B10.S (9R)	*H-2*	A.SW and B10.A	C57BL/10	1.0 ± 0.3	<0.002	N.S.	C57BL

[a] An index of the relative adherence to peripheral (axillary and brachial) node over Peyer's patch HEV (see text).
[b] Two-tailed test for significance of difference from BALB/c or C57/BL preference index.

[c] Standard error.
[d] Not significant.

We conclude that a major difference in the HEV-binding properties of BALB/c and C57BL mesenteric node lymphocytes is determined by a locus of or loci on chromosome 7, closely linked to *H-1, Hbb* and the albino coat colour locus, *c*. Since the C57BL/10 and B10.C(41N)Sn lines differ only in this region, it will now be possible to study the properties that the region determines in isolation from other differences demonstrated by lymphocytes from the two parent strains.

EVIDENCE OF EVOLUTIONARY DRIFT IN SURFACE STRUCTURES MEDIATING LYMPHOCYTE–HEV ADHERENCE

In studies designed to examine the species-specificity of lymphocyte–HEV interactions we observed that the ability of spleen cells from different vertebrate species to bind to mouse mesenteric node HEV decreased exponentially with increasing evolutionary separation of the lymphocyte donor from the mouse host (Fig. 3). A similar evolutionary decrease in adherence to mouse HEV was observed for mesenteric node lymphocytes from several species, and, in all the cases that we examined, adherence ability was predominantly confined to mature lymphocyte populations—rat and rabbit thymocytes and chicken bursal lymphocytes bound poorly in relation to their mature counterparts (Butcher et al 1979).

The observed relationship between ability to adhere to HEV and evolutionary distance approximates a negative exponential function. It is intriguing to consider this relationship in the light of studies of the molecular evolution of proteins. Dickerson (1971) has reviewed data demonstrating that several proteins (cytochrome *c,* haemoglobin, and others) have incorporated amino acid substitutions at characteristic and fairly constant rates during evolution, even though they have continued to serve the same function in diverging lineages. A linear rate of amino acid substitution implies an *exponential* decline in the fraction of mutable amino acids that a protein shares with its progenitor or with diverging proteins. Such an exponential decline in shared amino acids could explain a roughly parallel decline in interaction compatibility of the proteins in diverging species. We suggest that the observed exponential decline in lymphocyte–HEV adherence compatibility may be the functional result of a linear rate of evolutionary mutation in the components —presumably amino acids—of the cell surface structures mediating the interaction. Although selection would ensure the continuing compatibility of 'drifting' lymphocyte and HEV surface receptors within a given species, no selective pressure would prevent the progressive development of incompatibility of interacting structures in diverging lineages.

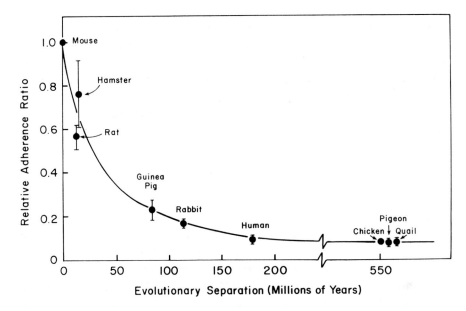

Fig. 4. Adherence of spleen cells from several vertebrate species to mouse mesenteric node HEV *(in vitro)*. Exponential decline in binding ability with increasing separation of the lymphocyte donor from the mouse host.

Such a hypothesis of evolutionary drift in macromolecular interactions, when combined with the known process of gene duplication (Ohno 1970), suggests a model for the generation of multiple region-specific lymphocyte-HEV interactions (e.g. peripheral node and Peyer's patch specific interactions) from a primitive lymphocyte–endothelial cell recognition phenomenon. The products of duplicated genes, although initially identical, might drift apart at about the same rate as that demonstrated between species. Although abrupt changes in receptor structure could not be tolerated, and would result in loss or suppression of the altered gene, minor receptor variations would be generated until differential expression of duplicated lymphocyte and HEV surface receptors on subpopulations of lymphocytes and vessels conferred a selective advantage. Subsequent selective pressure could then rapidly generate a high degree of specificity in the duplicated lymphocyte receptor–HEV ligand combinations.

CONCLUSIONS

In this paper we have summarized studies of interactions between lymphocytes and high-endothelial venules at cellular, organ, strain and species levels. We have confirmed that adherence to HEV is a specific property of mature lymphocyte populations, and have demonstrated the existence of mechanisms permitting nearly absolute organ restriction of the interaction. We have described and mapped to a single chromosome a genetically defined strain difference in lymphocyte–HEV adherence and have presented evidence that surface structures mediating the binding of lymphocytes to HEV have undergone random but progressive evolutionary alteration. Further cellular studies may elucidate the maturational sequence and functional importance of the expression of organ-specific HEV receptors on lymphocytes and the extent and significance of functional genetic polymorphism in this cell–cell recognition interaction. Biochemical studies of membrane proteins from clonal lymphocyte populations of defined HEV specificity, and of the genetically defined BALB/c chromosome 7 gene products, may eventually reveal the molecular basis of the specific recognition and adherence.

We have identified several murine lymphomas capable of organ-restricted adherence to HEV, and others apparently lacking the ability to interact well with any HEV. Human lymphomas may express similar restrictions, thus offering a possible explanation for the clinical presentation and behaviour of certain lymphoid malignancies of characteristic and limited organ distribution. If so, knowledge of the HEV-interactional ability and specificity of lymphomas or leukaemias in individual patients may prove to be of prognostic and therapeutic importance.

ACKNOWLEDGEMENTS

We thank Shelley Jacobs and Claire Wolf for excellent technical and secretarial assistance and Dr Roland Scollay for continuing help, advice and support.

References

Bailey DW 1971 Recombinant-inbred strains: an aid to finding identity, linkage, and function of histocompatibility and other genes. Transplantation 11:325-327

Blomgren H, Andersson B 1972 Recirculating lymphocytes in the mouse thymus are part of the relatively cortisone resistant cell population. Clin Exp Immunol 10:297-303

Butcher E, Scollay R, Weissman I 1979 Lymphocyte-high endothelial venule interactions: examination of species specificity. In: Müller-Ruchholtz W, Müller-Hermelink HK Function and structure of the immune system. Plenum, New York (Adv Exp Med Biol 114), p 65-72

Cahill RNP, Poskitt DC, Frost H, Trnka Z 1977 Two distinct pools of recirculating T lympho-cytes: migratory characteristics of nodal and intestinal T lymphocytes. J Exp Med 145:420-428

Cantor H, Weissman I 1976 Development and function of subpopulations of thymocytes and T lymphocytes. Prog Allergy 20:1-64

Dickerson RE 1971 The structure of cytochrome *c* and the rates of molecular evolution. J Mol Evol 1:26-111

Gowans JL, Knight EJ 1964 The route of re-circulation of lymphocytes in the rat. Proc R Soc Lond B Biol Sci 159:257-282

Griscelli C, Vassalli P, McCluskey RT 1969 The distribution of large dividing lymph node cells in syngeneic recipient rats after intravenous injection. J Exp Med 130:1427-1451

Guy-Grand D, Griscelli C, Vassalli P 1974 The gut-associated lymphoid system: nature and properties of the large dividing cells. Eur J Immunol 4:435-443

Guy-Grand D, Griscelli C, Vassalli P 1978 The mouse gut T lymphocyte, a novel type of T cell. J Exp Med 148:1661-1667

Halstead TE, Hall JG 1972 The homing of lymph-borne immunoblasts to the small gut of neonatal rats. Transplantation 14:339

Husband AJ, Gowans JL 1978 The origin and antigen-dependent distribution of IgA-containing cells in the intestine. J Exp Med 148:1146-1160

McWilliams M, Phillips-Quagliata JM, Lamm ME 1975 Characteristics of mesenteric lymph node cells homing to gut-associated lymphoid tissue in syngeneic mice. J Immunol 115:54-58

Ohno S 1970 Evolution by gene duplication. Springer, New York

Parrott DMV, Ferguson A 1974 Selective migration of lymphocytes within the mouse small intestine. Immunology 26:571

Ponzio NM, Chapman JM, Thorbecke GJ 1976 Effect of antigen on localization of immuno-logically specific B cells. Adv Exp Med Biol 73B:65-75

Rowley DA, Gowans JL, Atkins RC, Ford RC, Smith WL 1972 The specific selection of recirculating lymphocytes by antigen in normal and preimmunized rats. J Exp Med 136:499

Scollay R, Hopkins J, Hall J 1976 Possible role of surface Ig in non-random recirculation of small lymphocytes. Nature (Lond) 260:528-529

Scollay R, Butcher E, Weissman I 1979 Thymus cell migration: quantitative aspects of cellular traffic from the thymus to the periphery in mice. Eur J Immunol, in press

Sprent J 1973 Circulating T and B lymphocytes of the mouse. I. Migratory properties. Cell Immunol 7:10-39

Stamper HB Jr, Woodruff JJ 1976 Lymphocyte homing into lymph nodes: *in vitro* demon-stration of the selective affinity of recirculating lymphocytes for high endothelial venules. J Exp Med 144:828-833

Stamper HB Jr, Woodruff JJ 1977 An *in vitro* model of lymphocyte homing. I. Charac-terization of the interaction between thoracic duct lymphocytes and specialized high-endothelial venules of lymph nodes. J Immunol 119:772

Woodruff JJ, Kuttner BJ 1979 *In vitro* adherence of lymphocytes to the high endothelium of lymph nodes. In this volume, p 243-257

Zatz MM, Lance EM 1970 The distribution of chromium 51-labelled lymphoid cells in the mouse. Cell Immunol 1:3-17

Discussion

Weissman: In the mouse most lymphomas start in the thymus and metasta-size to involve peripheral lymph nodes. They go to the T cell areas almost exclusively. In humans Hodgkin's disease apparently has a potential for systemic involvement, selectively, of lymphoid organs. When tumour-bearing local lymph nodes are removed by surgery or sterilized by irradiation there is very often selective spleen involvement but not involvement of other peripheral nodes outside the radiation field. One explanation for this behaviour is specific homing of tumour cells to the spleen.

Fidler & Nicolson (1976) have selected melanoma cell lines in animals which metastasize specifically to bone, or lung, or brain, or liver. Although they have not yet tested whether these metastases localize due to selective homing or by selective retention in the tissues, our analysis would predict binding to blood vessels.

Butcher: There are several examples of human lymphoid disease which demonstrate specific localizations.

Gowans: Your genetic analysis was very interesting. What was the maximum difference between the good responder and the bad responder?

Butcher: On the average it was 1.5-fold.

Marcus: In your initial assays you used erythrocytes as a control. Have you tried other cells, such as macrophages, granulocytes or platelets, both washed and unwashed?

Butcher: Macrophages and platelets are sticky cells; I would prefer not to use as a negative control a population of cells which would stick to everything. It is hard to know what the perfect negative control would be.

Marcus: I just thought that perhaps another cell type, in addition to the erythrocyte, would serve as a useful control.

Butcher: A formalin-fixed lymphocyte comes fairly close to being ideal, and we have observed no differences in background adherence of erythrocytes and fixed lymphocytes.

Marcus: If you start with a low concentration of fetal lymphocytes and increase the quantity, is there a concentration at which no binding occurs? Also, does one lymphocyte bind to another?

Butcher: All my quantitative analyses are done in terms of ratios, not in terms of absolute numbers of cells bound. However, I saw no qualitative differences in binding when I used cell densities from just over a monolayer to four to-five times a monolayer. The lymphocytes do not appear to stack up on top of each other like platelets.

Marcus: It is difficult to know whether the binding seen represents a metabolic property of the lymphocyte or some property of the cell surface that can be modified in one way or another. If binding takes place naturally at 7 °C, it is difficult to conceive of it as a metabolic process.

Butcher: Judith Woodruff has shown that it doesn't occur as well at 1 °C. Some metabolic function must be involved. Another explanation of why *in vitro* interaction does not occur well at 37 °C is that it may be reversible at that temperature.

Marcus: Is the phenomenon you are observing totally irreversible?

Butcher: As Judith Woodruff mentioned, when you raise the temperature in some cases the cells come off.

Born: Could it be that lymphocytes are nicely motile at 37 °C but that there is a time factor as well, so that they never really get the chance to adhere? Perhaps motility counteracts adhesiveness.

Butcher: The binding reaction occurs very rapidly. We can demonstrate binding as early as 2.5 min after we layer the lymphocyte suspension on the section. The number of cells binding increases rapidly for the first 5 or 10 min, and then remains relatively constant, I don't know how fast lymphocytes move, but it seems unlikely that their motility could be a significant factor on this time scale.

Born: I was merely trying to think of an explanation for the artifact that is brought out by the difference in temperature.

Weissman: Some of these lymphoma cells have other properties with other surface receptors (such as receptors for viruses) that we can measure directly and quantitatively (McGrath & Weissman 1979). At 37 °C the cells shed those receptors fairly rapidly and resynthesize them just as rapidly. If this is true for HEV receptors one has a nice explanation of how a cell binding to an endothelial cell can crawl through by shedding bound receptors and then expressing new ones at the leading edge of the lymphocyte in the HEV.

Woodruff: In the strain difference experiments, were you using lymph node sections from one strain and then testing a variety of other lymphocytes on that single strain?

Butcher: Most of the experiments analysing the genetic difference between C57BL and BALB/c lymphocytes (Tables 3, 4 and 5) were done using BALB/c × C57BL F_1 hybrids as HEV donors. We have also done a series of experiments (E.C. Butcher, R.G. Scollay and I.L. Weissman, unpublished work), to determine whether lymphocytes are restricted in their ability to bind to allogeneic HEV. We have looked at several strain combinations, and we have never observed a preference for syngeneic over allogeneic HEV. That is, the relative binding ability of lymphocytes from two strains of mice is independent of the strain of the HEV on which they are tested.

Woodruff: We have done similar experiments and found that it is important to assay lymphocyte binding to allogeneic and syngeneic HEV simultaneously. In experiments where we used lymph nodes from BALB/c mice, BALB/c lymphocytes gave higher binding levels than that observed using C57BL lymphocytes. However BALB/c cells also showed higher binding levels on C57BL HEV.

Butcher: Thus, just as in the cases we have examined the HEV strain doesn't make a difference. The BALB/c lymphocytes bind better than C57BL lymphocytes on both types of HEV.

Woodruff: Yes, but that is in experiments where comparisons are made simultaneously. Such a finding is not necessarily reproducible. The age of the animals and infection with viruses or other agents, even though asymptomatic, may influence the results. That is why we always do the criss-cross experiments. There may be minor differences which are more apparent than real.

Butcher: Some of the results I showed you were pooled data from multiple replicate experiments. In our hands, the statistical variation between experiments, even using different animals, is similar to that within each experiment. Thus we do not find the day-to-day variation or individual animal variations which you are describing, although we have found some minor differences in lymphocyte populations from animals of different ages and so forth.

Woodruff: This may be because we are using different systems to quantitate the binding reaction. And of course when one uses different species without performing criss-cross experiments, interpretation of the results becomes very difficult.

Butcher: I was glad to see that you had done the rat/mouse criss-cross. We have not yet been able to do such criss-cross analyses *in vitro*. We have confined ourselves to the use of fresh-frozen sections, and since, like yourself, we have found relatively poor binding to unfixed rat sections, our experiments have been limited to mouse tissues.

Howard: It seems extraordinary that the low binding to HEV is dominant in the F1.

Butcher: By manipulating the assay conditions it is possible to make F1 lymphocytes behave more like the high-binding type, although their binding is always intermediate, in actuality. We are now trying to define the conditions which determine the binding properties of F1 cells, in an attempt to understand the basis of the parental binding differences more clearly.

Howard: What evolutionary distance between the high and low alleles approximately equals the difference between a mouse, rat and guinea-pig?

Butcher: One implication of our evolutionary hypothesis is that there probably exists within each species a variety of different alleles of the lymphocyte–HEV interaction molecules which differ to a small but significant extent—not to the extent that they will be incompatible, but probably to the extent where they are detectable. The evolutionary distance separating rats and hamsters from mice is very small. One would expect only one or two amino acid changes in, for instance, haemoglobin during the seven or so million years since these species have diverged (Dickerson 1971). Thus it would not be surprising to find, in various strains of mice, allelic differences as large as those between these very closely related xenogeneic species.

However, in this discussion there has been some assumption that the strain difference we have observed may be due to a difference in a lymphocyte–HEV receptor. I want to emphasize that we have no evidence that that is the case. We may be looking simply at a regulatory function, or a shift in lymphocyte populations, or something like that.

McConnell: Is there any correlation between chick lymphocytes being poor HEV binders and the lack of lymph nodes in this species?

Butcher: I wouldn't think so. Human lymphocytes also bind very poorly to mouse HEV, yet humans have rather well developed lymph nodes. Furthermore, lymphocyte–endothelial cell interactions are not absolutely limited to lymph nodes and may well occur in chickens even in the absence of lymph nodes.

Ford: Although chickens do not have lymph nodes, high-endothelial cells have been described in granulomas in that species (Miller 1969).

Davies: There are genetic differences between the binding capacity and there are clearly differences in standard haematological parameters between strains of mice. What the differences are due to is another matter but does anything correlate with the rather peculiar phenotype that you described?

Butcher: We know of no correlations yet. We haven't even compared the *in vivo* behaviour of BALB/c and C57BL lymphocytes yet. However, we plan to look for *in vivo* parallels to the *in vitro* differences now that we have congenic mice differing only in this trait (Table 5, p 277).

Davies: The density of cells in the thoracic duct lymph, for example, or the peripheral blood lymphocyte count, might be monitored.

There are said to be differences between transformed and non-transformed cells in relation to their capacity for recirculation. Have you compared non-malignant transformed cell populations with untransformed cells? For example, in thoracic duct lymph which is activated there can be a reasonably high proportion of morphologically transformed blast cells, which can be separated out.

Butcher: I haven't examined the binding properties of blast cells yet.

Davies: Where you put C57BL cells onto CBA nodes, does priming make any difference to the adhesion? It is difficult to see why it should, but immunological priming might affect your results.

Butcher: I don't know whether priming makes a difference.

Davies: My last question is, are the phenotypic differences that you relate to particular genetic backgrounds maintained in a radiation chimera?

Butcher: That would be an extremely interesting experiment.

Vane: It is obvious that lymphocytes do stick to HEV cells, although they have to be very cold, which worries me. What is the next step? If this is the first step in migration, as I assume you believe, what then drives the lymphocyte to crawl between the cells? Is it a chemotactic stimulant from outside the endothelial cell? Is it a repellent stimulant from the blood? Or is the endothelial cell doing something to the lymphocyte? Quite a bit of work now shows that receptors on a cell surface can move around. Insulin receptors, for example, seem to move around quite easily. Could the receptor in the HEV cell for the lymphocyte be dragging the lymphocyte around the cell with it? Is any of this known?

Woodruff: Lymphocytes are inherently motile, so when they come into contact with endothelial cells passage across the vessel wall becomes possible.

Vane: I don't see how its motility induces the lymphocyte to crawl between cells. It could just crawl along the surface without dipping down.

Gowans: If the motility of a lymphocyte is abolished without its being killed and it binds to the endothelium without penetrating it, can the possibility then be ruled out that the endothelium is exclusively the active partner in the migration through the HEV?

Born: One ought to consider the enzymes on cells' external surfaces—best characterized, I believe, for polymorphonuclear leucocytes. It would be interesting to find out whether there are qualitative and/or quantitative differences between cells which carry microvilli and those which are rounded up.

References

Dickerson RE 1971 The structure of cytochrome *c* and the rates of molecular evolution. J Mol Evol 1:26-111

Fidler IJ, Nicolson GL 1976 Organ selectivity for implantation, survival, and growth of B16 melanoma variant tumor lines. J Natl Cancer Inst 57:1199-1202

McGrath M, Weissman IL 1979 AKR leukemogenesis: identification and biological significance of thymic lymphoma receptors for recombinant AKR MuLV MCF-247. Cell 17:65-75

Miller JJ 1969 Studies of the phylogeny and ontogeny of the specialized lymphatic tissue venules. Lab Invest 21:484-490

Macrophages and the differential migration of lymphocytes

J.H. HUMPHREY

Department of Immunology, Royal Postgraduate Medical School, Hammersmith Hospital, London W12 0HS

Abstract   Two kinds of macrophage are described to which B or T lymphocytes adhere and which occur in anatomical sites in lymphoid tissues where these lymphocytes accumulate in their migration after leaving the blood stream. Macrophages in the marginal zone of the spleen white pulp and in the marginal sinus of lymph nodes are distinguishable from other macrophages by selective uptake of fluorescent labelled uncharged polysaccharides, notably ficoll and hydroxyethyl starch.   Some of the experimental work which led to their identification is described.   These macrophages, when isolated from mouse spleens after collagenase perfusion, are morphologically distinct and characteristically have lymphocytes attached to them.   There is a case for regarding such macrophages as involved in determining the migration pattern of B lymphocytes, but it remains to be proved.

T lymphocytes in lymphoid tissues migrate through areas containing interdigitating cells, with which they make intimate contact and which probably determine their migration pattern.   The work of B.M. Balfour and her colleagues suggests that interdigitating cells are derived from veiled cells, of which Langerhans cells are one form.   Veiled cells have general characteristics of macrophages but are strongly positive for Ia antigen.   Some properties of these interesting cells are described.

Other contributors to this symposium have discussed the pathways taken by lymphocytes in their migration through lymphoid tissues and the importance of the microenvironments in which T and B lymphocytes and antigen associated with macrophages may meet. I think that, despite all the elegant studies presented on the traffic of lymphocytes across high endothelial venules in lymph nodes and across the marginal zone of the white pulp of the spleen, there remain unanswered problems about what causes T and B lymphocytes to move subsequently into the T and B dependent regions. A possible explanation, at least at a superficial level, might be that the lymphocytes show

© *Excerpta Medica 1980*
Blood cells and vessel walls: functional interactions
(Ciba Foundation symposium 71) p 287-298

differential adherence to other cells in these regions which are themselves relatively stationary. In this paper I describe two possible candidates for these other cells, namely marginal zone macrophages in the spleen and marginal sinus macrophages in lymph nodes on the one hand, and interdigitating cells on the other. The former occur in the B cell and the latter in the T cell traffic areas. I shall not discuss the dendritic follicular cells which bind antigen–antibody complexes via C3 in germinal centres and appear to be important for the generation of B-memory lymphocytes (Klaus & Humphrey 1976) since traffic through these areas represents a separate problem.

DIFFERENTIAL UPTAKE OF POLYSACCHARIDES BY MACROPHAGES

The observation that marginal zone macrophages differed from other macrophages in the spleen arose from investigations (J.H. Humphrey, unpublished) designed to extend the finding that hapten-conjugated type 3 pneumococcal polysaccharide (S3) is a potent specific inhibitor of anti-hapten responses (Mitchell et al 1972) to haptens conjugated to other polysaccharides. Conjugates of dinitrophenyl (DNP) lysine were prepared with a variety of uncharged or negatively charged naturally occurring or synthetic polysaccharides, and tested for their capacity to inhibit secondary IgG anti-DNP responses by spleen cells challenged with antigen after passive transfer to irradiated recipient mice. Somewhat unexpectedly, conjugates of acidic polysaccharides (S3, pectin, alginic acid, hyaluronic acid) proved to be much more effective inhibitors than conjugates with a similar amount of DNP made with uncharged polysaccharides (dextrans, levan, ficoll, hydroxyethyl starch). Some illustrative results are shown in Table 1. Although the former were inhibitory at doses of 10 μg or less, similar amounts of the latter were not, and when injected into intact normal mice elicited a rapid burst of splenic IgM anti-DNP plaque-forming cells (PFC) followed by continuous production of IgM PFC at a lower but substantial level (about 25 000 PFC/spleen) lasting many weeks. When the conjugates were made a small amount of tyramine was also attached, in order to permit radio-iodination so that their persistence in the body could be followed after intravenous injection of 10–100 μg. All the conjugates were retained for long periods, with half-lives of 20–60 days (except for hyaluronic acid 5 days), and evidently resisted degradation. They all disappeared fairly rapidly from the bloodstream, but disappearance was not complete and small amounts (0.01–0.1% of the injected dose) were present in the blood up to at least 15 days. The pattern of their gross distribution in various tissues did not change greatly over the period of study (1–12 days) but there were marked differences between one polysaccharide

TABLE 1

Inhibition of secondary anti-DNP responses by conjugates. 7-day anti-DNP responses (total spleen PFC) as % of control

		Dose of conjugate (µg)			
		1	10	100	1000
$DNP_{2.7}S3^{a}$	Direct	12	18		
	Indirect	25	1		
$DNP_{4.6}S3$	Direct	52	45	3	
	Indirect	90	1	1	
DNP_{20}alginate	Direct	18	18	11	
	Indirect	2	3	1	
DNP_{16}pectin	Direct	50	7	3	
	Indirect	21	0	0	
DNP_4Ficoll	Direct	95	135	62	33
	Indirect	112	69	45	15
DNP_4Levan	Direct	74	39	28	
	Indirect	49	18	11	
DNP_5Dextran	Direct	133	170	164	
2000	Indirect	103	138	144	

Spleen cells (15-25 million) pooled from mice primed with DNP-KLH (keyhold limpet haemo-cyanin) were injected intravenously into groups of five irradiated (8.5 Gy) syngeneic mice, together with the stated amount of the material to be tested. On the following day they were boosted with 25 µg soluble DNP-KLH; direct (IgM) and indirect (IgG) anti-DNP plaque-forming cells in their spleens were measured 7 days later. The results are presented as mean percentages of the number of plaque-forming cells measured in the spleens of a similar group of recipients which received no test material.
[a]The degree of conjugation is given as mols DNP/50,000 daltons irrespective of the molecular weight.

and another, especially in relation to the proportion retained in the liver, spleen and skin (plus underlying connective tissue). Although a substantial proportion of all was retained in the liver, a relatively large amount of the neutral polysaccharides (dextran 2000, ficoll [mol. wt. 400 000] and hydroxy-ethyl starch) was retained in the spleen whereas more of the acidic polysaccharides was retained in connective tissues. Autoradiography of selected tissues revealed some unexpected differences in the cellular distribution of the various polysaccharides labelled with ^{125}I. Whereas the acidic polysaccharides were detectable in what are conventionally regarded as tissue macrophages, the neutral polysaccharides mentioned above were strikingly absent from liver Kupffer cells (though in or on parenchymal cells), and were concentrated in the marginal zones of the white pulp of the spleen and the marginal sinuses of lymph nodes but absent from the red pulp and much of the lymph node medulla.

Because these observations revealed interesting differences between macrophages in different situations, but autoradiography required the use of fixed tissues and was too imprecise for detailed study of cells, I thought it worthwhile to prepare fluorescent derivatives of those polysaccharides which demonstrated the extremes of behaviour. When fluorescein isothiocyanate (FITC) or tetramethyl rhodamine isothiocyanate (TRITC) are reacted with S3, ficoll or hydroxyethyl starch to which had been attached a small number of 1,6-diaminohexane groups it was possible to obtain green or red conjugates with two to four fluorescent groups per 100 000 mol. wt., which was sufficient to permit visualization of intracellular materials but insufficient to change significantly their charge or their biological behaviour. Bright intracellular fluorescence could be seen, using epi-illumination, in frozen sections of tissues taken from mice after intravenous injection of 50–100 μg of ficoll or hydroxyethyl starch, or 100–200 μg of S3. The distribution was similar to that seen by autoradiography, but it was now possible to examine not only cells in sections but also cells live in suspension. Because most conventional methods of fixation failed to render the polysaccharides insoluble in aqueous media, fixed preparations had to be mounted dry and duplicate specimens had to be examined when conventional histological stains were employed. (We have recently found that fixation in 1.25% neutral buffered glutaraldehyde allows retention of these polysaccharides.)

TISSUE DISTRIBUTION OF FLUORESCENT POLYSACCHARIDES

S3 could be detected for at least 10 days in liver Kupffer cells, subcutaneous connective tissue histiocytes, bone marrow macrophages, glomerular mesangial cells in the kidney, and lymph node medullary macrophages. In the spleen it was present in all red pulp macrophages and in scattered macrophages in the white pulp, and was detectable in some but not all of the marginal zone macrophages. In mice killed five or more days after injection it was also seen in a fine lacy pattern in germinal centres, presumably as antigen–antibody complexes on the surface of follicle dendritic cells.

Ficoll and hydroxyethyl starch behaved alike and are considered together. They were not seen in liver Kupffer cells, nor in parenchymal cells, although they appeared to be slightly concentrated in bile canaliculi; they were detectable, though much fainter than S3, in connective tissue histiocytes. In the spleen they were highly concentrated in and confined to large cells in the marginal zone of the white pulp, and in lymph nodes to large cells in the marginal sinus extending some way along the septa and up into the hilum. They were notably absent from the much more numerous macrophages of the spleen red pulp and from bone marrow macrophages. A limited number of studies in rats have shown a similar distribution. The polysaccharides were initially present in tiny cytoplasmic droplets, which became larger the longer the interval between injection and examination. These were presumably phagolysosomes in which the materials remained undegradable.

By injecting a mixture of S3 and ficoll or hydroxyethyl starch into mice it was possible clearly to distinguish marginal zone from other macrophages in spleen cell suspensions. Suspensions prepared by conventional teasing and sieving contained very few intact marginal zone macrophages, but sheets of these (and some of the red pulp macrophages) were present in the debris which is normally discarded. Only after perfusing the mice with collagenase or injecting it *in vivo* could I obtain spleen cell suspensions which contained all or most of the marginal zone and other macrophages in viable form. The yield was variable, but 4- to 6-month-old C3H mice averaged about 1.5×10^6 macrophages per spleen, of which 10% or less were marginal zone macrophages.

PROPERTIES OF MARGINAL ZONE MACROPHAGES AND THEIR RELEVANCE TO CELL TRAFFIC

Marginal zone macrophages from spleen cell suspensions, after lysis of erythrocytes with NH_4Cl, were examined in suspension or in cytocentrifuge preparations or after adherence to plastic or glass surfaces. They were generally larger than red pulp macrophages, and many had cytoplasmic extensions even in the original suspension. The most characteristic feature was that many had several lymphocytes adhering to them, both in suspension and after attachment to glass, whereas the red pulp macrophages generally did not. In fact the adherence of lymphocytes made it impossible to prepare enriched suspensions by density gradient centrifugation, since they mostly moved in the lymphocyte-rich layers. However marginal zone macrophages appear to be unaffected by 850 rad (8.5 Gy) whole-body X-irradiation and it is therefore possible to recover them from spleens empty of lymphocytes. They can be enriched relative to other cells in relatively low density layers on Percoll gradients.

Marginal zone macrophages are larger than those in the red pulp, and morphologically distinguishable in Leishman-stained preparations by large open round or oval nuclei and extensive pale blue-grey cytoplasm with long branching processes. Both stain strongly for cytoplasmic acid phosphatase and non-specific esterase. When examined for surface receptors by rosette formation with coated bovine erythrocytes (EA and EAC) the marginal zone macrophages have both Fc and C3 receptors. The attached lymphocytes also formed rosettes, sufficiently to give the impression that they were all B cells; immunofluorescent staining with specific anti-mouse Ig, however, showed no selective attachment of lymphocytes with surface Ig relative to those without it. It would be important to establish whether these macrophages have surface Ia antigen, but I have not so far been able to demonstrate this.

I do not know the origin of marginal zone macrophages or their relationship to other macrophages in spleen and lymph nodes. It is unlikely, for two reasons, that the distinction based on differential uptake of ficoll or hydroxyethyl starch is an artifact due to their gaining first access to these polysaccharides as a consequence of their location in the zone where the arterial blood first percolates past channels lined with macrophages. One reason is that the polysaccharides continue to circulate in the blood for many hours, and the second is that much of the blood supply to the spleen passes directly to the red pulp without arborizing at the marginal zone (and in the lymph nodes blood does not flow through the marginal sinus at all). Marginal zone macrophages do not seem to belong to the category of interdigitating cells described by Veldman (1970) and discussed below, and in any case are in the wrong place. Their relationship to the spleen dendritic cells of Steinman et al (1979) is uncertain, but in adherence properties and surface markers these seem to be different and in fact more like the interdigitating cells discussed below. The justification for drawing attention to marginal zone macrophages in this symposium is that they are situated at sites in the spleen where B lymphocytes appear to turn back on their tracks, and in lymph nodes at sites where B lymphocytes accumulate in the outer cortical layer. Whether this tendency to attach B lymphocytes actually determines the migration pathway of the latter remains to be proved. It may be significant that marginal zone and marginal sinus macrophages are the cells in which antigen and antigen-antibody complexes are first trapped, and it is in their neighbourhood that the earliest signs of B-cell stimulation are observed (e.g. Veldman et al 1978).

VEILED CELLS AND INTERDIGITATING CELLS

The thymus-dependent areas of lymphoid tissues characteristically contain interdigitating cells, clearly described by Veldman (1970) and Veerman (1974), which have typical bizarre nuclei and long processes extending between the recirculating lymphocytes. In his doctoral thesis van Ewijk (1977) pointed out that the interdigitating cells are probably involved both in the homing of T cells and in the presentation of antigen to them. He also drew attention to resemblances between interdigitating cells and Langerhans cells of the skin, both of which are mononuclear phagocytes and have many characteristics in common, but pointed out that their identity remained to be proved. I propose to expand on this theme, but must make it clear that much of the information has been provided by Dr B.M. Balfour, based on the work of herself and her colleagues.

Veiled cells in afferent lymph have been already mentioned in this symposium by Professor Morris and Dr Hall. They are present in afferent lymph draining the skin of all species examined (humans, rabbit, pig, sheep), constituting 8–30% of the total cell population, but are absent from efferent lymph and from the blood. They are large cells, with a horse-shoe-shaped nucleus, and numerous long thin cytoplasmic extensions, resembling veils, which contain no organelles apart from free ribosomes. When examined in the electron microscope these cells have large bundles of microfilaments near the nuclear membrane, and large numbers of small electron-lucent smooth-surfaced vesicles located in the central part of the cell. Almost all have a large vacuole near the surface containing a diffuse precipitate of electron-dense material. With histochemical methods veiled cells give diffuse reactions for acid phosphatase and also contain non-specific esterase, and the plasma membrane reacts strongly for ATPase (Drexhage et al 1979a,b). In rabbit lymph draining normal skin about 1/6 of the veiled cells contain dense bodies or large phagolysosomes. Some 4% also contain Birbeck granules, small rod-shaped organelles revealed by electron microscopy which are characteristic of epidermal Langerhans cells (Kelly et al 1978). Langerhans cells have lately excited considerable interest as strong candidates for the role of presenting antigens introduced into the skin to lymphocytes (Silberberg-Sinakin et al 1976), and have been shown in the guinea pig (Stingl et al 1978a) and man (Stingl et al 1978b) to possess surface Ia antigen. Veiled cells from pig and humans are also all strongly positive when tested for Ia antigen, unlike other macrophages in lymph which stain very weakly or not at all. Veiled cells have also been observed in mesenteric lymph, and in oil-induced rabbit peritoneal exudates they make up 5–10% of the cell population but do not contain Birbeck granules (Macpherson 1979, J. Cvetanov, personal communication). When examined by time-lapse cinephotography veiled cells are seen to be in constant movement, putting out and withdrawing their veils, and appear to be imbibing fluid. They do not adhere well to glass surfaces nor do they usually ingest particulate matter, though they readily form aggregates with other cells. They form weak rosettes with EA and with EAC.

After intracutaneous injections of diphtheria toxoid the proportion of veiled cells containing Birbeck granules in the draining lymph increased, reaching tenfold within two to three days (Hoefsmit et al 1979b). After injection of paratyphoid vaccine the draining lymph node also shows marked changes (Hoefsmit et al 1979a). During the first 24 hours Birbeck granules can be seen in the interdigitating cells, but these cells then vanish and are replaced by cells resembling lymph-borne veiled cells, with dense cytoplasm and many microfilaments, almost certainly derived from the afferent lymph. Within the next four days these cells appear to transform into typical interdigitating cells. The suggestion that the latter are derived from a precursor cell in afferent lymph is supported by the observation that ligation of all afferent lymphatics to a node results in disappearance of its interdigitating cells by six weeks (Hendriks 1978). Furthermore, when veiled cells labelled in vitro with [^{3}H]adenosine are reinfused into the afferent lymphatic most of them become localized in the paracortical area of the regional lymph node (Kelly et al 1978). The evidence that interdigitating cells in the T cell areas of lymph nodes are derived from veiled cells in the afferent lymph, and are steadily being replaced (at an accelerated pace when antigen is injected into the skin), is reinforced by recent observations in the human species that not only are veiled cells in lymph strongly Ia-positive, but so are the interdigitating cells in normal lymph nodes and in the area surrounding the central arteries of spleen white pulp, as well as in the enlarged paracortical areas of lymph nodes in patients with dermatopathic lymphadenopathy (Lampert et al 1979).

Birbeck granules have not been detected in peritoneal exudate macrophages even after antigenic stimulation (J. Cvetanov, personal communication), nor have they been seen in parathymic draining lymph nodes or in the interdigitating cells of the spleen (E.W.A. Kamperdijk, personal communication). Many such granules are present, however, in interdigitating cells in the thymus (Olah et al 1968). The latter, like epidermal Langerhans cells, are in close contact with squamous epithelial cells. These observations could be brought together by supposing that only those veiled cells which have travelled via squamous epithelium can manufacture Birbeck granules or give rise to interdigitating cells containing granules after stimulation by antigens introduced into the skin.

Because veiled cells both possess surface Ia and can carry foreign skin-sensitizing agents such as dinitrofluorobenzene (B.M. Balfour et al, unpublished), their localization in T-cell areas of lymphoid tissues is likely to be important for antigenic stimulation of T lymphocytes. From the point of view of what directs the movement of T lymphocytes, however, the aspect I wish to emphasize is the very intimate contact which these cells achieve, with narrow extensions of the lymphocyte membrane frequently inserted into invaginations of the interdigitating cell membrane (Veerman 1974). Admittedly there is no evidence of specific attachment of T lymphocytes to interdigitating cells *in vitro,* similar to the attachment of lymphocytes to high-endothelial venules demonstrated by Woodruff and by Butcher in this symposium, but if adherence to macrophages is important it presumably involves dynamic interaction between the cells and would require the microenvironment of living lymphoid tissues.

References

Drexhage HA, Mullink R, de Groot J, Clarke J, Balfour BM 1979a Large mononuclear cells resembling Langerhans cells, present in lymph draining from normal skin and skin after the application of the contact sensitizing agent DNFB. Cell Tissue Res in press

Drexhage HA, Lens JW, Cvetanov J, Kamperdijk EWA, Mullink R, Balfour BM 1979b Structure and functional behaviour of veiled cells, resembling Langerhans cells, present in lymph draining from normal skin and after the application of the contact sensitizing agent dinitrofluorobenzene. In: van Furth R (ed) Mononuclear phagocytes. Martinus Nijhoff, The Hague, in press

Hendriks HR 1978 Occlusion of the lymph flow to rat popliteal lymph nodes for protracted periods. Z. Versuchstierkd 20:105-112

Hoefsmit ECM, Balfour BM, Kamperdijk EWA, Cvetanov J 1979a Cells containing Birbeck granules in the lymph and lymph node. In: Müller-Ruchholtz W, Müller-Hermelink HK (eds) Function and structure of the immune system. Plenum Press, New York (Adv Exp Med Biol 114), p 389-394

Hoefsmit ECM, Kamperdijk EWA, Balfour BM 1979b Reticulum cells and macrophages in the immune response. In: van Furth R (ed) Mononuclear phagocytes. Martinus Nijhoff, The Hague, in press

Kamperdijk EWA, Raaymakers EM, de Leeuw JHS, Hoefsmit ECM 1978 Lymph node macrophages and reticulum cells in the immune response. I. The primary response to paratyphoid vaccine. Cell Tissue Res 192:1-23

Kelly RH, Balfour BM, Armstrong JA, Griffith S 1978 Functional anatomy of lymph nodes II. Peripheral lymph-borne mononuclear cells. Anat Rec 190:5-21

Klaus GGB, Humphrey JH 1976 The role of C3 in the generation of B memory cells. Immunology 33:31-40

Lampert IA, Pizzolo G, Thomas A, Janossy G 1979 Immunohistochemical characterization of cells involved in dermatopathic lymphadenopathy. J Pathol, in press

Macpherson GG 1979 In: van Furth R (ed) Mononuclear phagocytes. Martinus Nijhoff, The Hague, in press

Mitchell GF, Humphrey JH, Williamson AR 1972 Inhibition of secondary antihapten responses with the hapten conjugated to type 3 pneumococcal polysaccharide. Eur J Immunol 2:460-467

Olah I, Dunay C, Röhlich P, Toro I 1968 A special type of cell in the medulla of the rat thymus. Acta Biol Acad Sci Hung 19:97-113

Silberberg-Sinakin I, Thorbecke GJ, Baer RL, Rosenthal SA, Berezowsky V 1976 Antigen-bearing Langerhans cells in skin, dermal lymphatics and in lymph nodes. Cell Immunol 25:137-151

Steinman RM, Kaplan G, Witmer MD, Cohn ZA 1979 Identification of a novel cell type in peripheral lymphoid organs of mice. V. Purification of spleen dendritic cells, new surface markers, and maintenance in vitro. J Exp Med 149:1-16

Stingl G, Katz SI, Abelson LD, Mann DL 1978a Immunofluorescent detection of human B
cell alloantigens on S-Ig-positive lymphocytes and epidermal Langerhans cells. J Immunol
120:661-664

Stingl G, Katz SI, Shevach EM, Wolf-Schreiner E, Green I 1978 b Detection of Ia antigens on
Langerhans cells in guinea pig skin. J Immunol 120:570-664

van Ewijk W 1977 Microenvironments of T and B lymphocytes: a light- and electronmicroscopic
study. PhD Thesis, Erasmus University, Rotterdam

Veerman AJP 1974 On the interdigitating cells in the thymus-dependent area of the rat spleen:
a relation between the mononuclear phagocyte system and T-lymphocytes. Cell Tissue Res
148:247-257

Veldman JE 1970 Histophysiology and electron microscopy of the immune response. (PhD
Thesis), Drukkerij Dijkstra-Niemeijer, Groningen, The Netherlands

Veldman JE, Keuning FJ, Molenaar I 1978 Site of initiation of the plasma cell reaction in the
rabbit lymph node. Virchows Arch B Cell Pathol 28:187-202

Discussion

Hall: Isn't there a considerable species difference in the distribution of Birbeck granules and thus of Langerhans cells? They are not found in the macrophages in the peripheral lymph of sheep (M. Birbeck & J. Hall, unpublished observations). Are they normally present in *Homo sapiens*?

Humphrey: I don't know. Dr Lampert and his colleagues are looking into this. The work I described was done mainly on rabbits and pigs, which are both big enough for the afferent lymphatics to peripheral nodes to be cannulated.

Hall: I don't think anyone knows what these granules are. Michael Birbeck described Langerhans cells with their characteristic granules in the skin of patients with vitiligo (Birbeck et al 1961). It is annoying that they don't appear in sheep.

Humphrey: Interdigitating cells are visible in the thymus and when I asked Dr van Ewijk earlier whether there was a difference between those cells and reticular cells it was because I wanted to make sure that we were talking about two different things.

Gowans: Dr van Ewijk showed a lot of macrophages in peripheral lymph and H.W. Steer and G.G. McPherson (in preparation) have also demonstrated significant numbers of macrophages in lymph from intestinal lacteals in the rat. Now John Humphrey tells us that they might become located in particular areas after arrival in lymph nodes. Are there macrophages in sheep lymph?

Morris: Inherent in the proposition that John Humphrey is putting forward is that central and peripheral lymph nodes have very different mechanisms for directing cell traffic. Only the peripheral nodes receive the cells that originate from the veiled cells in lymph, and these cells don't come out of these nodes.

By peripheral nodes I mean the first nodes on the lymphatic chain. The classical situation where there is a heavy traffic of phagocytic cells in the lymph is between the liver and the portal lymph node. In the sheep most of the cells in the peripheral lymph coming from the liver have been Kupffer cells at some stage. In many species if India ink is injected intravenously the portal node eventually becomes black because phagocytic cells that are monitoring the bloodstream subsequently detach and migrate via the peripheral lymph to that node; they never seem to leave it.

Gowans: Where exactly do the macrophages in peripheral lymph go after arrival at the node?

Morris: Kotani et al (1977) have shown that macrophages in peripheral lymph in the rabbit enter germinal centres, for instance.

Humphrey: There are macrophages and macrophages. Veiled cells are said to be different from other macrophages which adhere well to glass or plastic surfaces.

Weissman: Bede Morris said that the veiled cells were in the peripheral afferent lymph but that they don't come out of the lymph node. Since there is such a massive traffic of lymphocytes coming out of the efferent lymph from the lymph node, what kind of marker experiments would lead you to believe that they all got stuck in that peripheral node?

Morris: You can see these cells quite easily and they are not present in central lymph. What occurs in the portal node is a classical example. Peripheral hepatic lymph carries as many as 10^7 veiled cells per hour to the portal lymph node. They must have come from a precursor which could only be the blood monocyte or the Kupffer cells lining the hepatic sinusoids. Very large numbers of veiled macrophages (up to 10^8/h) are carried in the peripheral lymph coming from a renal allograft and these cells could only have come from a precursor in the bloodstream.

Humphrey: Do they leave in the efferent lymph?

Morris: No; when you transfuse the effluent lymph from a renal allograft into another lymph node they don't pass through the node.

Gowans: Professor Ford, would you comment on the marginal zone in the spleen of the rat?

Ford: In mice and rats the initial localization of both T and B cells in the spleen is in the marginal zone. We haven't seen any discrimination between different parts of the marginal zone by B or T cells (Nieuwenhuis & Ford 1976). The striking observations that you have made *in vitro* might be simply because B cells are more sticky to that sort of surface.

Humphrey: Yes, that was the point I was trying to make. What holds B cells back and stops them from going the way T cells go is perhaps their adhesiveness.

Ford: I don't think that the segregation of B and T cells begins until after they have left the marginal zone. B lymphocytes seem to leave the marginal zone at about the same rate as T lymphocytes and go into the periphery of the periarteriolar lymphoid sheaths before segregating (Nieuwenhuis & Ford 1976).

Gowans: Are people making out a case for antigen being picked up by macrophages in the periphery, passing into the nodes in the peripheral lymph and then locating in precisely the right place in the lymph node or spleen? Is there a turnover of such macrophages?

Humphrey: The marginal zone and red pulp macrophages of the spleen, and indeed macrophages in other tissues such as liver, lung and connective tissue are radioresistant in the sense that they persist after—say—850 rad (8.5 Gy) when lymphocytes, blood monocytes and granulocytes have disappeared. I do not know how fast these tissue macrophages turn over, nor whether they all derive from blood monocytes.

If the antigens are indigestible polysaccharides with haptens attached and if they are of the kind which remain in marginal zone macrophages, they elicit a long-continued low-level IgM antibody response except in high doses which cause tolerance. For example, more than two months after injecting 5–50 μg DNP-levan or ficoll into mice I still find about 25 000 anti-DNP plaques in the spleen. This is not true of DNP-proteins which are rapidly degraded in macrophages. However, the use of very slowly degraded antigens is certainly a special case.

Morris: We have used antigens that are extracted by Kupffer cells in the liver. The immune response was then studied in the regional hepatic lymph node. We used influenza virus and other types of antigens which if they are injected intravenously will arrive in the portal lymph node inside these phagocytic cells. An immune response doesn't occur in the portal lymph when the antigen is injected intravenously; but when we put the antigen underneath the capsule of the liver, in much the same way as one would study the immune response in the popliteal node by injecting it under the skin of the leg, we get an excellent immune response. The phagocytic elements that convey antigens to the lymph node in peripheral lymph are not initiators of an immune response.

References

Birbeck MSC, Breathnach AS, Everall GD 1961 An electron microscope study of the basal melanocytes and high-level clear cells (Langerhans cells) in vitiligo. J Invest Dermatol 37:51-64

Kotani M, Okada K, Fujii H, Tsuchiya H, Matsuno K, Ekino S, Fukuda S 1977 Lymph macrophages enter the germinal center of lymph nodes of guinea pigs. Acta Anat 99:391-402

Nieuwenhuis P, Ford WL 1976 Comparative migration of B and T lymphocytes in the rat spleen and lymph nodes. Cell Immunol 23:254-267

Polymorphonuclear leucocyte chemotaxis: detection of the gradient and development of cell polarity

SALLY H. ZIGMOND

Department of Biology, University of Pennsylvania, Philadelphia, PA 19104

Abstract The ability of polymorphonuclear leucocytes to respond to a chemical gradient has been examined by observing their behaviour in response to the peptide *N*-formylnorleucylleucylphenylalanine (f-NorleuLeuPhe). The cells appear to detect the direction of the chemical gradient by sensing differences in the number of their chemotactic receptors that are bound across their dimensions. When moving, a PMN has a polarized form with ruffles or pseudopods at the front and a knob-like tail at the rear. The potential for forming new pseudopods appears to exist in a gradient from anterior to posterior along the cell axis. Rapidly increasing the concentration of peptide can transiently induce the formation of ruffles over most of the cell surface except the tail. The presence of transient reversible responses to increases in the chemotactic factor suggests that with time the leucocyte adapts to the concentration of peptide to which it is exposed. A simple model which describes the cell polarity and adaptation is presented.

DEFINITION AND ASSAY SYSTEMS

Chemotaxis is locomotion oriented along a chemical gradient. Although the phenomenon appears simple and straightforward, its detection and quantitation have proved difficult. The mere presence of an agent with chemotactic activity is insufficient to induce chemotaxis. It requires a concentration gradient sufficiently steep to stimulate a directional cell response. Early attempts to detect polymorphonuclear leucocyte (PMN) chemotaxis in response to soluble chemicals probably failed because the gradients present were too shallow. In other studies agents were claimed to be chemotactic without any demonstration that they were actually able to affect the direction of locomotion (Harris 1954). Increased movement of a population of cells can occur towards a substance that merely stimulates the rate of locomotion (chemokinesis) without affecting the direction of locomotion. The accumulation of cells in regions of high concentrations of a

299

test chemical can also be due to cell trapping rather than chemotaxis. For example, if high concentrations of a substance inhibit or reduce cell movement, that substance can cause cells to accumulate by trapping any that happen to come in contact with it. The movement by the cells would be random, not directed, and thus the accumulation would not be due to chemotaxis.

The combination of these factors has made detection of chemotaxis *in vivo* difficult. Although the *in vivo* observations by Clark & Clark (1930) and Buckley (1963) strongly support the existence of chemotaxis in the inflammatory response, and agents generated at inflammatory sites have been shown to be chemotactic in *in vitro* assays (Snyderman et al 1971), the role of chemotaxis in cell accumulation *in vivo* has not been definitively demonstrated.

In vitro, chemotaxis has been clearly demonstrated with a variety of techniques. Chemotaxis was first detected by microscopic observations (Leber 1888) and direct microscopic observations combined with dark-field or time-lapse photography have continued to be very useful. Boyden (1962) designed a system in which gradients were established across thin Millipore filters and the migration of cells through the pores of the filter was followed. This method has proved useful for detecting new chemotactic agents. Among the agents found to be active in this system are: serum factors (Keller & Sorkin 1966), in particular C5a (Shin et al 1968), cell-derived factors (Phelps 1969), oxidized lipids (Turner et al 1975), digested, denatured or modified proteins (Wilkinson & McKay 1972) and certain peptides, particularly N-formyl-methionyl peptides (Schiffmann et al 1975). As originally described, the Millipore filter method did not adequately differentiate between the chemokinetic and chemotactic effects. Subsequent modifications were introduced to clearly demonstrate chemotaxis (Zigmond & Hirsch 1973). A further limitation of the technique is the instability of the gradient, particularly when the agent is of low molecular weight and therefore has a high diffusion coefficient. In about 30 minutes a gradient of a small molecule the size of acetic acid would be expected to have decayed significantly (to a gradient in which the change in concentration across the cell's dimensions is only 0.2%) (Zigmond 1979). This may explain why chemotaxis to small peptides cannot be detected unless a protein is present (Wilkinson 1976). The peptides are known to bind to albumin, which would be expected to slow their diffusion.

Two other assays of chemotaxis have recently been introduced. An 'under agarose' method is similar in design to an Ouchterlony assay (Nelson et al 1975). The distance that cells placed in one well of an agarose plate migrate towards adjacent wells containing test substances or buffer solutions is measured. As in the Millipore assay system, controls must be included to differentiate directed locomotion from chemokinetic effects (Tono-oka et al 1978). The time of the assay can be adjusted to allow substances with different diffusion coefficients to establish gradients in the proximity of the cell well. A large molecule may need to be placed in the well for several hours before the cells are added (Zigmond 1979).

In a new visual assay, cells are observed in a thin fluid layer, about 10 μm high and 1 mm across (Zigmond 1977), over a bridge that connects two wells containing buffer or the substance to be tested. The orientation of the movement of cells on the bridge is evaluated by determining the direction of locomotion from morphological features of the cells. Because the direction of locomotion is evaluated, interpretation of chemotaxis is not complicated by chemokinetic effects of the test substance. Chemokinetic effects can be evaluated independently. As with the other assay systems, the optimal timing of the response depends on the diffusion coefficient of the substance tested. The visual assay appears to be best suited to materials of low molecular weight. Small peptides will form a linear gradient after 10 to 15 minutes (Zigmond 1979).

GRADIENT DETECTION

The *in vitro* assays have been used to define agents that are chemotactic for PMNs, to determine the environmental requirements of the response, and to begin to dissect its molecular basis.

Leucocytes interact with the chemotactic peptides by means of specific saturable receptors (Aswanikumar et al 1977, Williams et al 1977). Although various peptides compete for binding to this receptor, C5a does not. There also appears to be a separate receptor for a cell-derived chemotactic factor which does not bind either C5a or the peptides (Spilberg & Mehta 1979). Thus, there are several classes of chemotactic receptors. Rabbit cells bind about 10^5 molecules of N-formylnorleucylleucylphenylalanine (f-NorleuLeuPhe) per cell with a dissociation constant, K_d, of 1.5×10^{-9} (Aswanikumar et al 1977). Human cells bind about 2×10^3 molecules of f-MetLeuPhe per cell with a K_d of about 1.3×10^{-8} (Williams et al 1977). In both cases the binding constants of the peptides correspond approximately to the biological activities of the peptides.

From the binding of its receptors a cell receives information about the direction of the gradient. Bacteria exhibiting chemotaxis have been shown to use a 'temporal' means of sensing a gradient (MacNab & Koshland 1972). They compare the concentration of attractant present at different times as they swim. If the concentration is increasing (bacterium swimming up a gradient), the bacterium tends to continue swimming along the same path; if the concentration is decreasing, the bacterium tends to make a turn. There is no evidence that the bacterium knows whether the concentration of attractant on its right or in front is higher or lower than that on its left or behind. But if the bacteria merely alter the frequency of random turns, depending on whether they are moving up or down the gradient, the net movement of the bacteria changes to up the gradient. In contrast, stationary leucocytes placed in a gradient of a chemical attractant can initiate locomotion up the gradient (Zigmond 1974). Furthermore, introduction of a chemotactic agent to one side of a locomoting leucocyte will induce the cell to turn towards the new source (Bessis 1974, Ramsey 1972). These observations indicate that differences in the concentration of chemotactic factors can be discriminated across the cell's dimensions. Thus the leucocyte has vectorial information about the gradient and can differentiate stimulation on its left from that on its right.

If the leucocytes detect a difference in the number of receptors bound across their dimensions, one would expect optimal orientation to a standard gradient—i.e. a threefold or 10-fold gradient across the 1 mm bridge in the visual assay chamber—to occur at concentrations near the dissociation constant of the peptide being tested (Zigmond 1977). At the concentration of the dissociation constant there would be a maximal change in number of receptors occupied for a standard gradient. In fact, optimal orientation occurs at concentrations slightly below the K_d of binding. At the optimal concentration the cell can detect an increase in the concentration across its dimensions of about 0.6% of the mean concentration present.

CELL POLARITY

The information present in the differential binding of receptors must be transformed into an orientation of cell polarity and movement along the axis of the gradient. The initial response of rounded cells exposed to a gradient of peptide is to selectively form pseudopods towards the higher concentration of peptide and to begin translocation in this direction. As the cells move up the gradient the concentration of peptide at the back of one cell is clearly higher than the concentration at the front of the cell behind, yet the two cells appear morphologically similar. Thus, the induction of pseudopod formation by peptides is not directly proportional to the concentration of peptide present.

The development of a polarized cell morphology does not require a chemical gradient. Cells exposed to a homogeneous concentration of chemotactic factor initially form ruffles over most of their surface. After a few minutes they begin translocation, form ruffles only at the front, and develop a knob-like tail at the rear. Although the concentration of peptide present initially caused ruffling over the entire surface, soon there was an inhibition of ruffling except at the cell front (Zigmond & Sullivan 1979).

The inhibition of pseudopod formation appears to take place along an anterior to posterior gradient. The inhibition along the 'sides' of the cell can be overcome by rapidly increasing the concentration of peptide present. After such an increase the upper surface and sides of a cell again form ruffles. A pseudopod can also be induced to form locally from the 'side' of a cell if one brings a chemotactic stimulant up to the side of the cell in a micropipette. The inhibition can also be removed by cutting off the cell front. This was observed to happen naturally in a cell that was being filmed as it moved in a gradient of peptide. The cell, which was ruffling only in the front, became highly elongated and eventually broke into two fragments of about equal size. The rear part of the cell which had not been ruffling began to ruffle from what was now its front. Clearly, there is a reversible means of coordination within the cell.

The inhibition of ruffling or pseudopod formation is most severe at the tail of the cell. Increasing the concentration of peptide can cause ruffles to form over the entire cell surface except the tail. When the cell recommences locomotion it will move roughly in the direction it was going before the peptides were added. When we reverse the direction of a gradient, either by reversing the concentrations in the wells of the visual assay system or by bringing a pipette containing peptide up to the tail of a moving cell, the cell almost never reverses its direction by forming a new pseudopod from the tail. Rather it makes a series of small turns and moves around in a circle until it is reoriented along a new gradient. Nevertheless, the formation of a tail is reversible. If the peptide is removed the cells round up and the tail disappears within a few minutes. After the tail has disappeared, cells restimulated with peptide ruffle all over and they initiate locomotion randomly without bias to their previous direction of locomotion (S.H. Zigmond & H. Levitsky, unpublished work).

The polarized inhibition of pseudopod formation may involve some rearrangement of cellular components. It is possible that the receptor for the chemotactic factor is excluded from the cap-like tail and is present in highest concentrations in the cell front. Alternatively, the polarity could result from the rearrangement of a cell component required for pseudopod formation. For example, cytoskeletal elements may accumulate in the tail in an unresponsive form. It appears that contractile proteins do accumulate in cell caps (Braun et al 1978). However, the simple rearrangement of cell components is not sufficient to explain all our observations. The fact that a cell was seen to break in two and then begin ruffling indicates that all the elements required for pseudopod formation were present in the rear of the cell.

Cell behaviour was also studied after rapid increases in the concentration of peptide. These cells exhibited ruffling over most of their surface but did not translocate over the substrate. Both the generalized ruffling and cessation of locomotion were transient and soon the cells resumed locomotion and normal morphology. A further sudden increase in peptide concentration again induced a transient response. The dose-response relationship of this behaviour followed that of the binding of peptide to the saturable cell receptor. In fact, the duration of the transient inhibition of translocation correlated roughly with the change in number of receptors occupied that would be expected to occur over a particular concentration change. Thus the time from maximum ruffling until the reinitiation of locomotion was about twice as long when the concentration was increased from zero to 1×10^7 M-f-NorleuLeuPhe (a change in receptor occupancy from zero to 83% of the receptors, using a K_d of 2×10^{-8} M) as after an increase from zero to 1×10^{-8} M (when about 33% of the receptors should be occupied) (Zigmond & Sullivan 1979).

The transient nature of the change does not appear to be due to destruction of the peptide since re-perfusion with the peptide does not induce a second response. In studies of binding over the time course of the response (less than 10 min at 23°C) we did not detect any change in the number of receptors available for binding. Whether the transient response is limited by receptor, cytoskeletal or intermediary translocation or availability, there is some control of the response, with the result that the cell appears to adapt to the concentration of peptide it is experiencing.

Adaptation could contribute to the ability of the leucocyte to sense a chemical gradient by controlling the level of some internal signal(s) which determine cell polarity. The level of such a signal would have to reflect the difference in the number of receptors occupied across the cell. However, if this signal returned to baseline levels during long-term exposure to various concentrations of peptide, the cell could be tuned to detect small local or temporal changes in this signal which result from changes in the concentration of peptide. Furthermore, the cell would be expected to respond to differences in the number of receptors occupied across its dimensions, regardless of the mean number of receptors occupied. This is approximately what is observed, as noted above.

MacNab & Koshland (1972) have suggested that adaptation can be considered as consisting of two processes: one process would cause the production of a stimulatory agent or agents, X, the other the inactivation of X. The rate constants of both processes would be proportional to the number of receptors occupied but the activation time after a change in receptor occupancy would differ in the two processes. Increasing the number of receptors bound would rapidly increase the rate of production of X and more slowly increase the rate of inactivation of X. For leucocytes, X can be imagined as an internal signal inducing pseudopod formation. Immediately after an increase in binding, the production of X would increase and pseudopods would be formed; as the rate of inactivation rises slowly the level of X would return to its normal baseline level.

This simple model effectively describes the transient nature of the leucocyte ruffling response. However, if spatial constraints are placed on the distribution of either the inactivation or production of X, the model can describe both the polarized morphology of the leucocyte and its ability to detect concentration gradients. For example, if the inactivator is free to move in the cell, it could become asymmetrically distributed in the cell, with the highest concentration in the tail. This might occur if the inactivator were associated with membrane proteins or cytoskeletal elements that accumulate in the tail. The production of X would remain closely associated with the receptor occupied. Thus, in the adapted state in a homogeneous concentration of peptide X would be produced uniformly over the cell surface, but X inactivation would be greatest in the tail. Since the levels of both production and inactivation of X are proportional to the number of receptors occupied, the mean concentration of X in the cell would be at the baseline level. However, there would be a gradient of X in the cell with the highest concentration in the front. This fits with the behaviour of cells in a homogeneous concentration of peptide. They ruffle primarily from the front and there is a strong inhibition of ruffling from the tail.

In a gradient of peptide, the cell would adapt to the mean concentration it experiences. However, the distribution of the production of X would now reflect the asymmetrical distribution of receptors bound. The distribution of inactivator would again be in increased concentrations in the tail. A cell oriented up a gradient thus would have a steep internal gradient of X caused by rapid formation at the front and rapid destruction at the rear. The orientation and movement of such a cell would be stabilized. For chemotaxis to occur, cells moving at an angle to the axis of the gradient must be stimulated to turn towards the higher concentration of the chemical. An increased binding of receptors on the high-concentration side of the cell could lead to a local increase in X. Since there would be an anterior to posterior gradient of inactivator, the level of production of X required to induce pseudopod formation would be most likely to occur near the front. This fits with the observation that cells tend to make turns of small angles (Nossal & Zigmond 1976). The absolute level of X in the cell is controlled; thus the magnitude of the signal to turn would depend not on the absolute number of receptors bound but only on the difference in the number of receptors bound from one 'side' of the cell to the other. If the front of the cell moving up the gradient were removed, the number of receptors occupied would decrease and both the production and inactivation of X would decrease to maintain a constant mean concentration of X. Again, an asymmetrical distribution of the inactivator, and a relative increase in receptors bound when the concentration of peptide is highest, would allow ruffling to occur at the front.

The specific features of this model can of course be arranged in a number of different ways. For example, the receptors could be distributed asymmetrically, with a high concentration in the front and a very low concentration in the tail. The concept of adaptation involving the production and inactivation of a stimulatory agent and the different spatial distributions of these two processes is helpful in developing a working model for the chemotactic behaviour of leucocytes.

ACKNOWLEDGEMENTS

I thank Hy Levitsky for help with the cell polarity studies and Susan Sullivan for help with studies on peptide binding. The research was supported by NSF Grant No. PCM 77-4442 and NIH Grant No. HL 15835 to the Pennsylvania Muscle Institute.

References

Aswanikumar S, Corcoran B, Schiffmann E, Day AR, Freer RJ, Showell HJ, Becker EL, Pert CB 1977 Demonstration of a receptor on rabbit neutrophils for chemotactic peptides. Biochem Biophys Res Commun 74:810-817

Bessis M 1974 Necrotaxis: chemotaxis towards an injured cell, Antibiot Chemother (Basel) 19: 369-381

Boyden S 1962 Chemotactic effect of mixtures of antibody and antigen on polymorphonuclear leukocytes. J Exp Med 115:453-466

Braun J, Fujiwara K, Pollard TD, Unanue ER 1978 Two distinct mechanisms for distribution of lymphocyte surface macromolecules. J Cell Biol 79:409-418

Buckley IK, 1963 Delayed secondary damage and leukocyte chemotaxis following focal aseptic heat injury in vivo. Exp Mol Pathol 2:402-417

Clark ER, Clark EL 1930 Observations on the macrophages of living amphibian larvae. Am J Anat 46:91-143

Harris H 1954 Role of chemotaxis in inflammation. Physiol Rev 34:529-562

Keller HU, Sorkin E 1966 Studies on chemotaxis. IV. The influence of serum factors on granulocyte locomotion. Immunology 10:409-416

Leber T 1888 Über die Entstehung der Entzündung und die Wirkung der enzündungerregenden Schädlichkeiten. Fortschr Med 4:460

MacNab RM, Koshland DE Jr 1972 Gradient-sensing mechanism in bacterial chemotaxis. Proc Natl Acad Sci USA 69:2509-2512

Nelson RD, Quie PG, Simmons RL 1975 Chemotaxis under agarose: a new and simple method for measuring chemotaxis and spontaneous migration of human polymorphonuclear leukocytes and monocytes. J Immunol 115:1650-1656

Nossal R, Zigmond SH 1976 Chemotropism indices for polymorphonuclear leukocytes. Biophys J 16:1171-1182

Phelps P 1969 Polymorphonuclear leukocyte motility in vitro. III. Possible release of chemotactic substance after phagocytosis of urate crystals by polymorphonuclear leukocytes. Arthritis Rheum 12:197-204

Ramsey WC 1972 Analysis of individual leukocyte behavior during chemotaxis. Exp Cell Res 70:129-139

Schiffmann E, Corcoran BA, Wahl SM 1975 N-formylmethionyl peptides are chemotactic for leukocytes. Proc Natl Acad Sci USA 72:1059-1062

Shin HS, Snyderman R, Friedman E, Mellors A, Mayer MM 1968 Chemotactic and anaphylatoxic fragment cleaved from the fifth component of guinea pig complement. Science (Wash DC) 162:361-363

Snyderman R, Phillips JK, Mergenhagen SE 1971 Biological activity of complement in vivo. J Exp Med 134:1131-1143

Spilberg I, Mehta J 1979 Demonstration of a specific neutrophil receptor for a cell derived chemotactic factor. J Clin Invest 63:85-88

Tono-oka T, Nakayama M, Matsumoto S 1978 Enhanced granulocyte mobility induced by chemotactic factor in the agarose plate. Proc Soc Exp Biol Med 159:75-79

Turner SR, Campbell JA, Lynn WS 1975 Polymorphonuclear leukocyte chemotaxis toward oxidized lipid components of cell membranes. J Exp Med 141:1437-1441

Wilkinson PC 1976 A requirement for albumin as carrier for low molecular weight leukocyte chemotactic factors. Exp Cell Res 103:415-418

Wilkinson PC, McKay IC 1972 The molecular requirements for chemotactic attraction of leukocytes by proteins. Studies of proteins with synthetic side groups. Eur J Immunol 2:570-577

Williams LT, Snyderman R, Pike MC, Lefkowitz RJ 1977 Specific receptor sites for chemotactic peptides on human polymorphonuclear leukocytes. Proc Natl Acad Sci USA 74:1204-1208

Zigmond SH 1974 Mechanisms of sensing chemical gradients by polymorphonuclear leukocytes. Nature (Lond) 249:450-452

Zigmond SH 1977 The ability of polymorphonuclear leukocytes to orient in gradients of chemotactic factors. J Cell Biol 75:606-616

Zigmond SH 1979 Gradients of chemotactic factors in various assay systems. In: van Furth R (ed) Mononuclear phagocytes – functional aspects. Martinus Nijhoff, The Hague, in press

Zigmond SH, Hirsch JG 1973 Leukocyte locomotion and chemotaxis – new methods for evaluation and demonstration of a cell-derived chemotactic factor. J Exp Med 137:387-410

Zigmond SH, Sullivan SJ 1979 Sensory adaptation of leukocytes to chemotactic peptides. J Cell Biol 82:517-527

Discussion

Weissman: With your assay I could imagine that the concentration gradient sets itself up on the coverglass itself and that cells move over surface concentration gradients. Is this a cell-substrate-oriented movement rather than a movement through fluids? That would affect our appreciation of how this might work with endothelial cells.

Zigmond: The response is rapidly reversible. When I reverse the gradient the cells turn around. I can also bring a micropipette up to a cell and get the cell to orient towards the chemotactic factor. If it interacted via a substrate link, that link would have to be similarly reversible. The extension of lamellipodia into the medium on stimulation by the chemotactic factor suggests that the cell responds to fluid phase stimulants and isn't just walking up a gradient of chemotactic factor, though it is very difficult to eliminate that possibility.

Weissman: Have you had a gradient at right angles to the coverslip? Have you ever put a gradient on a coverslip, removed the coverslip and then set it down again where the cells are to see whether they can move on a substrate?

Zigmond: I would argue that it isn't possible to set up a gradient on the coverslip and have the cells orient on it but if the peptides normally act via a substrate link, this link is rapidly reversible. When I reverse the gradient the cells turn around very rapidly.

Weissman: I still think you could do that with the substrate. You could reverse it just about as rapidly if there are enough sites on the glass.

Zigmond: I can't eliminate it. Showing that there is little or no attachment to the substrate would be one way of ruling out this possibility.

Gowans: Chemotactic agents could act directly on the cell by producing sol–gel transitions on the cell surface, or the cell could sense the agent and then, by some internal mechanism quite distinct from the chemical itself, change its own state. You talk about the cell taking readings: does the substance itself have properties that change the cell?

Zigmond: Probably not. Cells in a gradient can be lined up, all pointing in one direction. It seems unlikely that the chemotactic substance has a direct effect on sol–gel transformation since the rear of a cell further up the gradient is in a higher concentration than the front of the cell behind. So a direct concentration-dependent interaction will not account for the observation. There is some fairly subtle coordination within a cell.

Born: The adaptation may be analogous to one found in platelets. When an agent such as adenosine increases cyclic AMP in platelets, that increase disappears much more rapidly than the inhibition of function produced by the agent. Your agent apparently transduces onto some intracellular mechanism (perhaps, from what one knows about polymorphs, to increase cyclic GMP). The situations seem analogous in that the effect persists in part of the cellular machinery.

Vane: Your film showed a wave of chemotaxis going across, with quiescent cells waking up one by one and crawling along the glass. If you put a similar population of quiescent cells into a non-gradient peptide, just an increase in concentration, would they all wake up and then go to sleep again?

Zigmond: They would wake up to be stimulated into locomotion. The locomotion persists while the peptide persists.

Vane: You mentioned that if you squash a red cell the polymorph goes towards it. Does this happen with any cell if you squash it, including a polymorph?

Zigmond: It is not clearly defined. Damaged polymorphs can be chemotactic. Most of the studies by Bessis were on red cells (Bessis & Burté 1965).

Morris: Bessis called the phenomenon necrotaxis.

Born: Some years ago, we tried to quantify chemotaxis *in vivo* (Atherton & Born 1972). Preparations of hamster cheek pouch were superfused with Krebs' solution without or with Hammarsten casein known to be chemotactic *in vitro*. After increasing times we counted the number of polymorphs lying in standard areas of tissue around venules. The casein brought large numbers of polymorphs out into the tissues. This is a puzzling effect because casein is quite a large protein. We were trying to find the mildest means for inducing chemotaxis *in vivo,* and at that time casein was one of the few reasonably well established chemotactic agents. Nowadays one would use something much better defined chemically and much more active, such as the peptides derived from complement.

Vane: We have done similar experiments, with similar results, and we find that very low concentrations of prostacyclin prevent the rolling leucocytes from settling on the venules.

Williams: What are the prospects for isolating receptors in these cells?

Zigmond: I think they are reasonable. People are trying to do that with affinity columns and so on. The affinity is high but the binding is very reversible. The K_d for the f-NorleuLeuPhe peptide is about 10^{-8}.

Williams: It would be going on and off extremely fast, wouldn't it?

Zigmond: Yes, the half-time is about 30 s. It would be nice to make some kind of ligand that bound irreversibly.

Weissman: In the washing experiment is the on/off time just for the peptide or is this also a cell that might be shedding its receptors rather than re-synthesizing them or re-expressing them?

Zigmond: I don't know, except that it works in the cold. The on/off rates are not markedly different in the warm from those in the cold. The on/off rates also allow the K_d to be predicted from the kinetics as well as the equilibrium findings.

Humphrey: You said that fragments which are broken off would still migrate and show increased chemotaxis at 46 °C. Could that finding be used to test whether the receptor is renewable? If the whole cell responds again and again you presumably have renewable receptors.

Zigmond: That is an interesting question. These cells are metabolically rather inert but they are synthesizing a bit of protein. We don't know whether there is a pool of receptors that can be internalized and reinserted or whether the cells are resynthesizing receptors. They can exhibit chemotaxis in the presence of inhibitors of protein synthesis.

Gowans: Can you locate the tritiated peptides autoradiographically and watch their movement and disappearance?

Zigmond: We would love to but there are no free amino groups so we can't fix them in place with glutaraldehyde. They are so rapidly reversible that a rapid wash to get rid of background binding removes the cell binding. I am about to try to couple peptides to a larger molecule with a fluorescent probe. It would be nice to know whether the receptors are homogeneously distributed.

Vane: One way in which the cell could obtain more orientation would be for the receptors within the cell to swim towards the stimulus.

Zigmond: A lot of things could be explained if the receptors were moving towards the chemotactic substance. Chemotaxis still has to occur so the receptors should not be highly localized, but a lot of the behaviour would be explained if the receptors were excluded from the tail.

Gowans: Are the cells always polarized, with a front and back which are immutable?

Zigmond: Once they are moving there is a strong probability that the front will remain the front. When you reverse the gradient the cell walks around in a circle. But once the cell rounds up the polarity may be entirely dissolved. That cell will move in any direction. It may be that rounded cells still have a polarity but we have lost the marker of it.

Davies: Do polymorphs with full bellies have reaction characteristics any different to those that have not phagocytosed?

Zigmond: Old studies show that phagocytosis tends to inhibit the rate of locomotion of neutrophils. Are you thinking about trying to deplete the number of receptors or something like that?

Davies: One of the functions of polymorphs is presumably phagocytosis. As you have demonstrated beautifully, they move in response to these various factors. When they have eaten do they sit down and sleep?

Zigmond: They are more sluggish.

Vane: Higgs et al (1975) put forward a very nice theory that the phagocytosing cell itself releases chemotactic substances, perhaps prostaglandins. This would call in other cells, as long as the food lasted. Have you been able to demonstrate such a thing with your elegant method?

Zigmond: The stimulus in the film I showed was a line of aggregated gammaglobulin to which polymorphs had attached. The line of gammaglobulin itself was not chemotactic. It depended on the polymorphs being attached to that line. So the chemotactic factor was in fact a cell-derived chemotactic factor. Spilberg et al (1976) in St Louis have described a cell-derived chemotactic factor that seems to be a glycoprotein.

Gowans: So you can collect your material by just lining up masses of polymorphs against aggregated gammaglobulin on a glass surface?

Zigmond: Yes. The trouble is that the material may not be highly defined. Wilkinson in particular has shown that a number of different denatured proteins may be chemotactic (Wilkinson & McKay 1971). Wilkinson has also shown that lymphoblasts exhibit chemotaxis (Russell et al 1975).

Gowans: Do they move towards the same range of things?

Zigmond: I think Wilkinson used casein and showed that lymphoblast chemotaxis was inhibited by colchicine. I think it remains true that many factors shown to be chemotactic for neutrophils and macrophages are not chemotactic for lymphocytes.

References

Atherton A, Born GVR 1972 Quantitative investigations of the adhesiveness of circulating poly-morphonuclear leucocytes to blood vessel walls. J Physiol (Lond) 222:447-474

Bessis M, Burté B 1965 Positive and negative chemotaxis as observed after the destruction of cells by u.v. or laser beams. Tex Rep Biol Med 23:204-212

Higgs GA, McCall E, Youlten LJF 1975 A chemotactic role for prostaglandins released from polymorphonuclear leucocytes during phagocytosis. Br J Pharmacol 53:539-546

Russell RJ, Wilkinson PC, Sless F, Parrott DMV 1975 Chemotaxis of lymphoblasts. Nature (Lond) 256:646-648

Spilberg I, Gallacher A, Mehta JM, Mandell B 1976 Urate crystal-induced chemotactic factor. Isolation and partial characterization. J Clin Invest 58:815-819

Wilkinson PC, McKay IC 1971 The chemotactic activity of native and denatured serum albumin. Int Arch Allergy Appl Immunol 41:237-247

Dynamic theory of leucocyte adhesion

PETER D. RICHARDSON

Division of Engineering, Brown University, Providence, RI 02912

Abstract   Leucocyte chemotaxis in *in vitro* chambers is analysed by applying the theory of transient diffusion to the chemotactic agent. This illustrates that use of physical principles can help in the interpretation of such experiments. A fluid-dynamical background is sketched for analysis of the dynamics of leucocyte adhesion in flow through blood vessels: this background includes the flow field induced around a particle in a velocity gradient. Discussion of the dynamics of leucocyte adhesion centres on estimation of the wall collisional rate of leucocytes from the bloodstream and leads to questions concerning measurement of collision efficiency. Specific experiments to elucidate factors remaining in question are suggested.

The adhesion of leucocytes to vessel walls has received much attention. This attention has been dominated by investigation of chemotactic agents, and much has been done to catalogue agents and their inhibition in experiments with Boyden chambers.   Zigmond and Hirsch developed an alternative technique more recently which involves observation of movement in response to a stripe of a chemotactic agent deposited on a slide. Rather little has been done to investigate the effect of fluid-mechanical and diffusional dynamic factors in the circulation (or in test chambers) on leucocyte adhesion.

The purpose of this paper is to redress the balance a little by discussion of diffusional and fluid-mechanical factors. First, it is interesting to review some chamber experiments on chemotaxis with the theory of transient diffusion applied to them. Second, in moving towards a theory of leucocyte adhesion, it is desirable to consider some relevant background work in fluid mechanics and diffusion. Third, the interpretation of this in terms of leucocyte motion and adhesion is discussed. Finally, the questions of what may be determined by further experiments and computation, and how this may be interpreted in specific terms for the leucocyte, are reviewed.

313

© *Excerpta Medica 1980*
Blood cells and vessel walls: functional interactions
(Ciba Foundation symposium 71) p 313-335

THE ZIGMOND AND HIRSCH STRIPE CHAMBER:
INTERPRETATION IN TERMS OF TRANSIENT DIFFUSION

The viewpoint of the fluid-dynamicist is different in character from that of, say, a biochemist. The latter may consider that a process is understood if the steps in going from a chemical initiator to a selected terminus are known; the representation is often in the form of a chain of reactions, sometimes branched, with kinetics known. There is a pathway with direction from cause to effect. Fluid dynamicists are interested primarily in what a fluid is doing in a chosen region of space; they believe they know all the basic causes and effects and want to know what the local balances will lead to as values for parameters such as velocity or species concentration.

One extreme case in fluid dynamics is hydrostatics, which describes fluids at rest. The rest is macroscopic only; on the microscopic scale Brownian motion still occurs, of course, and diffusion with it. Several tests for leucocyte chemotaxis use chambers in which there is no steady motion but where there is transient diffusion as a result of the procedure of assembly and where a gradient of concentration of a chemotactic factor is established. Quantitative aspects of this concentration gradient seem to be largely ignored. In this section the spatial and temporal variations of concentration of a chemotactic factor are discussed and compared with the original experiments.

The chemotaxis test chamber with hydrostatic use chosen for this discussion was described by Zigmond & Hirsch (1973). A stripe of a protein solution is drawn on a slide and allowed to dry. A small drop of cell suspension (containing the cells which will react to the chemotactic factor) is placed near the centre of the stripe and a coverslip is gently lowered onto it, spreading the cell suspension smoothly between the slide and the coverslip. The edges of the coverslip are sealed. The preparations are observed in a microscope for an hour or more, and locomotion is evaluated by scoring the percentage of cells in a visual field which display a locomoting morphology. Cells close to the stripe begin moving towards the stripe sooner than those further away. The physical arrangement is illustrated in Fig. 1a.

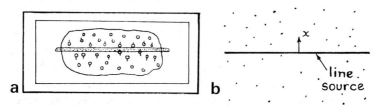

Fig. 1. Zigmond and Hirsch stripe chamber for leucocyte chemotaxis. (a) Physical arrangement, (b) idealization for theory.

If the stripe is a line source (cf. Carslaw & Jaeger 1959) of a chemotactic agent we must expect some delay before significant concentrations are achieved far from the stripe in the chamber. The transport of the chemotactic agent occurs by diffusion, and the leucocyte motion is so slow that it has very little effect on the diffusion. If the cell population is relatively sparse it is unlikely that cells nearer the stripe will consume so much of the agent that they will starve those further away. It is probably fair to ignore the leucocytes in calculating the diffusion of the agent.

On this basis, a physical idealization of the chamber can be developed. At time $t = 0$, a long thin source of a diffusible substance is placed at $x = 0$ in a uniform stationary medium (Fig. 1b).

The local concentration c of the diffusing factor varies with position x (distance from the line source) and time t. This variation should satisfy the diffusion equation:

$$\partial c/\partial t = D(\partial^2 c/\partial x^2) \tag{1}$$

where D is the diffusivity. Boundary conditions for this differential equation include the condition that far from $x = 0$, $c = 0$. This assumes that nothing has been smeared into the chamber when the drop of cell suspension is spread out. On the stripe itself, idealized as a line at $x = 0$, we may have an exhaustible source of strength Q per unit length of stripe (the case with a simple chemotactic agent), or possibly a sustained concentration $c = c_0$ (which would be one example of a chemotaxigenic agent which sustained the concentration through reaction). The solutions for both of these cases are well known in chemical and in thermal diffusion (Carslaw & Jaeger 1959), and take the forms:

For the exhaustible line source

$$c = (Q/2\sigma \sqrt{\pi Dt}) \exp(-x^2/4Dt) \tag{2}$$

where σ is the solubility,
and for the sustained concentration c_0

$$c = c_0 \operatorname{erfc}(x/2 \sqrt{Dt}) \tag{3}$$

where erfc (.) is the complementary error function (Abramowitz & Stegun 1964). The typical solution curves for these functions are sketched in Fig. 2a,b.

It is not known what concentrations c or concentration gradients $\partial c/\partial x$ are required to cause directed leucocyte motion. If some threshold values are required, then cells far enough from an exhaustible source may never experience enough to activate them. It is not known whether a finite time is

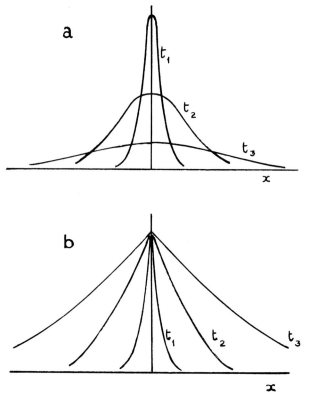

Fig. 2. Transient distributions of concentration in a line-source chamber as a function of distance from the line, at different times. (a) Exhaustible line source, (b) sustained line concentration.

required for cells to develop directed leucotactic motion after stimulation. It is not known how much variability in thresholds or activation delay times there is between members in a population of leucocytes. It seems difficult to attempt a comparison with experiments. However, in equations (2) and (3) the space–time factor is separate, with $(x/2\sqrt{Dt})$ or its square appearing. In Fig. 3 of their paper, Zigmond & Hirsch (1973) have given space–time data on leucocyte movement scores. In their Fig. 3b, they score the furthest positions where at least 50% of the cells are moving towards the source. It seems worthwhile to re-plot this data on coordinates x vs. $t^{1/2}$; if the data provide a straight line it would imply that the diffusion delay time in reaching x is the cause of the delay in chemotactic response there. Such a graph of the data is shown here in Fig. 3. The scatter from the original data persists, but the upper limit (the deepest penetrations away from the stripe) now appear close to a straight

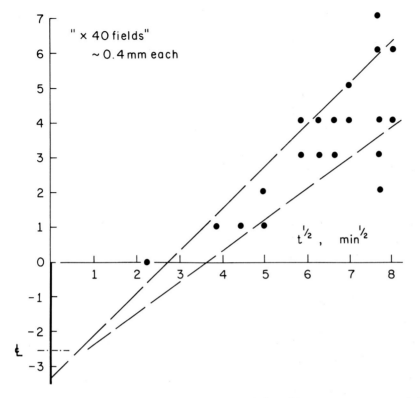

FIG. 3. Zigmond and Hirsch data on space–time relation of leucocyte movement replotted in coordinates suggested by the theory of transient diffusion.

line. One feature of this line is that it does not pass through the origin of the graph, but has an offset. This could be due to a delay time in activation, but is more likely to be due to the fact that the counting of × 40 fields was made from the edge of the deposited stripe rather than its centre. If the midpoint of the stripe is taken as a better origin (which would be reasonable), then the upper limit line extrapolates quite closely to this.

If we assume that the chemotactic agent had a diffusivity of 0.5×10^{-5} cm^2/s, and combine this with the slope measured for the upper limit line in Fig. 3 so that the argument of the complementary error function in equation (3) can be calculated, the result indicates that the leucocytes reach the scored state (at least 50% moving towards the stripe) when the concentration is 6% of the maximum concentration c_0 and the concentration gradient is about 16% of that near the origin.

The variability indicated by the points lying below the upper limit line could arise because the same points correspond to weaker stripes (biological variability) or because of response to a different agent with a smaller diffusivity.

Experiments could be designed to clarify the possibly different effects of concentration c and concentration gradient $\partial c/\partial x$. For example, if the stripe was made as a ring rather than a line, points inside the ring would have different combinations of c and $\partial c/\partial x$ from points outside.

The lesson to be gleaned from this discussion is that a more absolute and quantitative result may be obtained by adequate consideration of physical principles, and modifications of tests may help to resolve issues otherwise left in question. Zigmond subsequently has sought to take account of some of these effects by using another chamber design incorporating two linear wells with a narrow thin bridge between them; the wells serve as sources of chemotactic agents and a linear concentration gradient develops across the leucocyte-bearing bridge after a while (Zigmond 1977).

FLUID-DYNAMICAL PERSPECTIVE

The first point to make clear is the viewpoint and perspective of a fluid-dynamicist on flow problems. The essence of this viewpoint is a quantitative representation of the flow field; a flow is considered known when the local velocity is known throughout the space of interest. If the flow varies with time, then the fluid dynamicist wants to know the local velocity in the space of interest, serially in time. The matters of cause and effect are handled as distinct terms in differential equations which represent the phenomena involved. The distribution of velocity in the space of interest must be such that the terms in the differential equations balance each other out at every point. The equations represent simple physical principles, including the conservation of matter, the conservation of momentum, the conservation of energy and the conservation of individual chemical species. It is typical that the balance of different terms in any one equation varies in different regions in the space of interest. Thus a pressure gradient may serve mainly to overcome viscous drag near a stationary solid surface and to accelerate the fluid well away from that surface.

The second point to be made is that the equations which are generated to represent real problems are often very difficult to solve. The equations are generally easier to solve if some terms are dropped. To drop terms means that what is left is only an approximation, but this can work well. Sometimes it is possible to drop one type of term in one region of the flow, and a different term in another region of the flow, and somewhat match the two where the regions adjoin each other. This is all heuristic, and it is at this stage that the die is cast for each method to be a better or a worse approximation. Values of dimensionless numbers, such as the Reynolds number, are used to help to choose methods. The process is perhaps analogous to a combination of gene manipulation and surgery. The results depend on the appropriateness of the perceptions applied in generating the solutions, and rival schools of thought can struggle over the same problems for a long time before one perception emerges as superior. To an outsider the flurry of different studies can be baffling. The most attractive situation is one where a particular case has been formulated clearly in a straightforward geometry with distinct boundary conditions, a solution has been calculated, corresponding experiments have been carried out—and the results agree.

The third point is that blood flow provides many difficulties to fluid-dynamicists for calculation. Most progress in developing the theory of fluid flow has come with simple fluids such as gases and simple liquids flowing over fixed walls (or in fixed ducts) which have geometries that can be described with just a few parameters. Flow of blood in compliant vessels does not fit these conditions. Some aspects of blood flow in the large vessels can be analysed well (e.g. Shapiro 1977) if blood is regarded as a homogeneous fluid and the individual red cells are ignored. This viewpoint is often called macrorheological. As part of this viewpoint, individual cell types which are relatively dilute in the blood, e.g. white cells and platelets, are considered to be like solute species in an otherwise uniform liquid. On the other hand, if a theoretical model of the flow of blood in small vessels is to have any prospect of realism, the existence of the individual cells must be incorporated. Indeed, it is necessary to consider also the mechanical properties of the cell membranes —the factors that govern cell deformability. The viewpoint that focuses on the fluid dynamics around individual cells or aggregates is called micro-rheological.

There are flows, or events within flows, which call for both macrorheological and microrheological theory to account for what occurs. The two viewpoints are then applied simultaneously. This is not a new exercise for fluid dynamicists (Goldsmith & Mason 1968), but it is an area where the most effective perceptions and manipulations are still being worked out for blood flows. One issue not yet fully resolved by matching solutions at the macrorheological and microrheological levels is the prediction of the effective viscosity of whole blood. Values are known for this at the macrorheological level as a function of haematocrit, for example. At the microrheological level it is known that red cells are deformable under shear and that the membrane can rotate steadily around the cytoplasm, but the distribution of pressure and shear around a cell under these conditions is still not known.

These considerations lead to recognition of the need to discuss the flow pattern around a cell carried in a shear field.

FLOW AROUND A CELL IN A SHEAR FIELD

Let us suppose that a cell is carried in a continuum flow with a velocity gradient Γ locally. We can write a shear flow Reynolds number for the particle, based on its radius a and the shear rate (velocity gradient):

$$R_s = a^2 \Gamma / \nu \qquad (4)$$

The flow field around a sphere or long rod, for example, is affected quite strongly by the Reynolds number even when the latter is still smaller than unity (Poe & Acrivos 1975). If we focus on the cell and hold it stationary in view we can draw the streamlines we would see around it at a small R_s value (Fig. 4). The cell itself rotates (in this illustration, clockwise). This motion is entirely passive. The rotation rate is approximately one-half of the shear rate imposed on the surrounding fluid, i.e. it is about $\Gamma/2$.

The flow induced around the particle deserves particular attention. Immediately surrounding the rotating body is a closed streamline region, a blanket of liquid. This region is not penetrated by a convective flow, but substances can obviously enter or leave by diffusion. At the margin of this region the streamlines come to two stagnation points, where different streamlines come into contact (seen to the left and the right of the body in Fig. 4). Laterally beyond these points there are regions where the streamlines turn around. From a more distant perspective the scene looks as if the fluid is a layer about as thick as the body radius which approaches the body e.g. from the left, above the mid-plane of the flow, for the region on the left 'collides' with the body and is pushed away, below the mid-plane of the flow for the region on the left of the body.

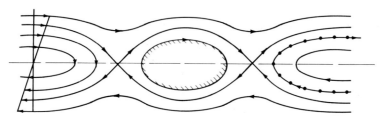

Fig. 4. Streamlines around a cell (hatched) when it is in a steady shear field.

A shearing flow field that is carrying particles has some properties that are due to the presence of the particles. One important property is that the motion of the particles can facilitate diffusion across the flow. The mechanism for this is convective and can be seen readily enough from Fig. 4. With the closed streamline motion around the body, fluid is quickly carried through a lateral distance exceeding twice the radius and can proceed further by normal diffusion. The lateral deflection of the 'colliding' flow beyond the two stagnation points is a direct convective facilitation of transport and may contribute more than the closed-streamline flow around the body itself. The process is illustrated on a streamline on the right-hand portion of Fig. 4. Dots on the streamline represent the successive positions of a large molecule (or small particle) at uniform intervals of time. A lateral transfer of about $2a$ in size occurs during a 'collision'.

The diagram is drawn for a single major body rotating in the fluid. If there are many bodies in a volume of liquid, each separated from the others well enough for the flow around one not to interfere with the flow around the others, then 'collisions' for diffusible species will be more frequent and transport will be much enhanced. This will be noticed most strongly for the species of higher molecular weight because the normal diffusivity (associated with the Brownian motion) decreases as the molecular weight increases. The flow around a cell in a shear field gives about the same lateral transfer to all molecules or particles and the facilitation (effective diffusivity compared with molecular diffusivity) will be larger for larger molecules. In blood flows the facilitation factor can be as much as 10^2 for platelets, for example. This depends on the shear Reynolds number.

When the number-density of the particles in the liquid increases so that their flow fields mutually interfere, the flow field is harder to represent. Even with a steady shear field, the flow field around any one particle is unsteady, with other particles coming and going continuously. All the evidence points to persistence of the effects seen clearly in the flow around an isolated particle, in the sense that augmented diffusion still occurs. When the particles are near-identical in size, a situation representative of red cells in blood, one can investigate self-diffusion. Eckstein et al (1977) did experimental studies on this with particle volume fractions up to about 0.5 and found that facilitated diffusion persisted. Their studies were done with solid particles, and values measured with these particles in blood flows should be used only with reservations. The reason for caution is that blood flows do not show the same rate of increase of effective viscosity with particle volume fraction (haematocrit) as do slurries of solid particles—and some of the increase in effective viscosity is due to enhanced diffusivity of the particles.

Red cell deformation and rotation in a shear field has been demonstrated by Fischer & Schmid-Schönbein (1977).

SHEAR FLOW IN A BLOOD VESSEL

Flow near the wall in a blood vessel has a velocity profile, e.g. Fig. 5, where the implications for particle translation over finite intervals of time are well appreciated. What may not be so well appreciated is that the particles also rotate (see right-hand part of Fig. 5), and the rotation is fastest where the velocity gradient is largest (usually near the wall). Thus the microrheological effects of a particle in a shear flow must be considered around cells in vessels larger than capillaries—a factor to be reckoned in developing a theory for leucocyte motion and adhesion in blood vessels.

Chemotactic factors generally originate in extravascular tissue. They enter the bloodstream by diffusion through vessel walls. The transport through extravascular space to the vessel may be assisted by Starling flow. As the chemotactic factor diffuses into the blood flow itself it is swept downstream in a continuous stream. This stream lies against the vessel wall, which continuously feeds the factor into the stream, and the stream is usually called a concentration boundary layer. For typical substances identified as chemotactic factors (Gallin & Wolff 1975) diffusivity is small compared with the kinematic viscosity of blood, and therefore the Schmidt number is large. Diffusion in laminar motion into a flow with large Schmidt on Prandtl numbers is one of the problems that has been analysed well (e.g. Meksyn 1961, Evans 1962) and, provided that the stream-wise variation of shear rate at the wall is known, concentration profiles can be worked out in good detail. The typical consequence is illustrated for a single tube or vessel in Fig. 6, where the downstream development of the concentration boundary layer is illustrated by the radially inward growth of contours of constant concentration.

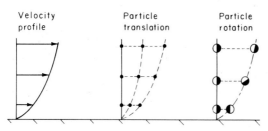

FIG. 5. Macrorheology and microrheology of flow near a vessel wall: velocity profile and particle rotations.

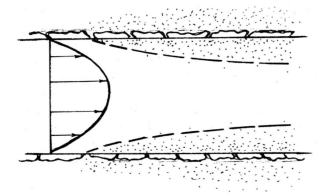

FIG. 6. Diffusion of a chemical agent into the flow in a vessel: development of a concentration boundary layer. Dashed lines indicate edges of boundary layers in the flow.

The portions of the microcirculation where leucocyte adhesion is observed do not consist of single tubes. Rather, they involve a microvascular bed where blood flows from capillaries into venules which increase in calibre with progress downstream. It is easier for a chemotactic factor to spread all across a stream in a capillary than in a venule, so leucocytes may first be affected in capillaries and then make their way to the venules, where a boundary layer of additional factor is found. This may not be appropriate to those circumstances, specially contrived, where thermal injury is caused by an optically heated dye being injected transiently into a vessel to facilitate the effect, with perfusion with blood being restored soon afterwards (Grant 1973).

DYNAMIC THEORY OF LEUCOCYTE ADHESION

The first issue to try to settle is whether leucocytes can exert any migratory effects towards a vessel wall while they are in the bloodstream. If the leucocytes have no means of knowing of the effective stickiness of endothelial cells until they physically encounter the cells, the question is moot. However, if the leucocytes are affected by a chemotactic factor through diffusion in the bloodstream, the question should be considered. The prospects for migratory effects occurring in response to a chemotactic factor are dim, because of the rotation of the cells. This rotation has two effects that militate against migratory behaviour. The first effect is that with the closed streamline region around a rotating cell the factor will be transported all the way round the cell and will diminish the gradient of concentration across the cells. The second effect is that, because of the cell rotation in the flow, the cell would have to keep rotating its efforts at migratory work in synchrony with the rotation externally imposed. This is an unlikely talent for the cell. Therefore it will be assumed that there is no migratory behaviour of leucocytes in the blood-stream.

The next point to consider is characterization of the flow. Let us suppose that the typical vessel of interest has a diameter of 30–300 μm, and that the velocity distribution is parabolic. If the mean velocity is U, the shear rate at the wall is $8\ U/d$. The shear Reynolds number experienced by a cell close to the wall is

$$R_s = (8\ U/d)a^2/\nu \tag{5}$$

where a is the cell radius. A typical value for R_s, with $a = 5$ μm, is 10^{-8}. If the leucocyte count is 5000/mm^3, the average spacing of leucocytes if uniformly distributed in blood in bulk is about 60 μm.

The effective diffusivity can be assessed using the data of Eckstein et al (1977).
From their Fig. 2, with a particle volume fraction of 40%,

$$D/a^2\Gamma \sim 0.025$$

and taking $a = 5$ μm, $\Gamma = 80$ s^{-1}

$$\text{we obtain } D = 50\ \mu\text{m}^2\,\text{s}^{-1}$$

We can next estimate the collision rate of leucocytes with the wall. If n is the number-density of the particles of interest, then the flux J, which is the mean number of particles crossing (or colliding with) a unit area of a plane per unit time, can be written

$$J = -D(\partial n/\partial z) \tag{6}$$

A considerable uncertainty arises in estimating $\partial n/\partial z$. A crude estimate is given by $\Delta n/\Delta z$, where Δn is the midstream value for n ($5 \times 10^{-6}/\mu$m^3) minus the effective value at the wall (zero if they all stick), and Δz is some fraction of the radius—say 50 μm. Then $J = 5 \times 10^{-6}\,\mu$m^{-2}s^{-1}. It should be emphasized that this is an approximation that could be below or above reality, for various reasons. A 100 μm length of a 100 μm diameter vessel has a surface of about 3×10^4 μm^2, so the collision rate in such a segment would be one cell about every 6.25 s. Cells that collide with the wall will either adhere or return to the stream. The average fraction that adheres is called the collision efficiency. Even when chemotactic effects are strong it seems unlikely that the collision efficiency is 100%: it has been remarked many times that the adhesion of leucocytes to endothelial cells is patchy, with the leucocytes sometimes skipping over patches that are not particularly adhesive. If a leucocyte collides with such a patch the chance of its adhering seems relatively low. Some spatial selectivity of adhesive sites has been observed also with platelets (Richardson 1979): there were sites in a flow chamber which platelets repeatedly preferred for adhesion well above the expectation from random choice. On the other hand, *in vitro* experiments suggest that a large fraction of leucocytes can exhibit adhesive and chemotactic behaviour. It must be pointed out that there is no clear indication of typical values for the collision efficiency under chemotactic conditions.

At this point it is useful to consider some experimental results that may help to guide analysis. Atherton & Born (1972) have recorded roll-by counts of granulocytes for fixed observation points in a preparation. Their Fig. 14 indicates that the roll-by count increases roughly linearly with venule diameter. For a venule diameter of 100 μm the count is roughly 16/min. The steady roll-by count will be equal to the total attachment rate upstream of the observation point, minus the number of cells which have remained stationary and out of sight. If the flux discussed above for a 100 μm vessel applied in this case, and if the collision efficiency was 100%, a roll-by of 16 cells/min would imply a catchment length of about 167 μm. This length is small compared with the visual field, and it is likely that major changes in the rolling count would have been noticed. There are two possibilities to consider: (1) the collision efficiency is far below 100%, so the catchment length required is much longer, and (2) catchment occurs over a limited range rather far upstream, and the traffic of leucocytes simply passes through the field of view.

One factor which complicates assessment of these possibilities is the anatomy of the microcirculation itself. This involves two aspects. The first is that with larger vessel sizes it is not possible to see all the adherent leucocytes because of the optical interference of the red cells in the bloodstream. The second is that on the venous side it is often hard to find uninterrupted segments of a length great enough to make roll-by counts at considerable separation. G. Schmid-Schönbein (personal communication, 1978) has pointed out that when an adhesive leucocyte emerges from a capillary into a larger vessel the nearest red cell behind the leucocyte tends to leap-frog around it into the larger vessel, deforming as it does so, and pushing the leucocyte into contact with the endothelium of the larger vessel in the process. The leucocyte then tends to continue its adhesive way rather than join the free flow. Born et al (1978) remarked, however, that the combined frequency of cells in two venules of similar calibre was higher just above their confluence than in the single venule just below.   They also noted that there were no significant differences in frequency counts of rolling granulocytes at two points separated by distances of 400–3000 μm along junctionless segments of venules. However, individual leucocytes have not been followed to determine the mean length of the rolling path in such preparations, and correspondingly the dynamic interchanges with the free stream have not been quantified.

In summary, consideration of the dynamics of blood flow in vessels makes it possible to estimate the opportunities for leucocyte adhesion in terms of the frequencies of random collisions with endothelium.   Observations have not yet caught up with the needs of theory through sufficient observations of stream-wise changes of roll-by frequencies, mean length of rolling path, systematic recording of leucocyte count for each experimental animal used, selection of junctionless segments of vessel for observation, and spatial control of the inflammatory background.

PROSPECTS FOR FURTHER EXPERIMENTS TO QUANTIFY ASPECTS
OF LEUCOCYTE ADHESION AND CHEMOTAXIS

Without leucocyte adhesion there is no migration, in response to chemotaxis, through the vessel wall.   Adhesion is the important first step. Some sort of activation of leucocytes as they pass through capillaries may be insufficient to explain their persistent rolling in venules.   Experiments *in vitro* are not able to provide the carpet of endothelial cells in an observable fluid-mechanically-controllable system.   Unless this situation changes, only the microcirculation appears likely to yield more information about the dynamics of adhesion to endothelial cells.   However, a basic weakness for quantitative study is that the spatial distribution of concentration of chemotactic agents is unknown in the microcirculation preparation.

Besides a follow-up of the deficiencies in experimental data obtained so far, remarked on at the end of the previous section, other steps can be taken to improve quantification of adhesion so that collision efficiencies and so on can be assessed.   Briefly, these include use of arterioles with inflammation strong enough to cause leucocyte adhesion (the advantage is that one can find relatively long, unbranched arterioles), and measurement of leucocyte incorporation in platelet-dominated mural thrombi (the growth of such mural thrombi is related to the blood flow rate [Begent & Born 1970, Richardson 1973], so incorporation of leucocytes is an indication of the collision efficiency of leucocytes encountering the thrombus during its growth).   The problem of observing the roll-by count of leucocytes on the walls of vessels too large to permit full visual observation may be resolved by use of focused ultrasound, but the technology for this is not yet developed far enough.

The question of the strength of the force holding the leucocyte to the endothelial wall despite the rapid passage of the free stream is interesting. Knowledge of this force would allow the bond strength which is made and unmade as the cells roll along to be estimated, and this bond strength could be compared with what is known of the chemistry of the adhesive bond. The fluid dynamics of a single particle at or near a wall in an otherwise uniform fluid has been studied theoretically, and this has been done also for a dilute suspension (Tözeren & Skalak 1977), but the presence of red cells is likely to make a significant difference. This is an area where fluid dynamicists need to make more studies.

It could be useful to use chambers to interrogate leucocytes about their response to shear forces while they are also responding to chemotactic agents. If a chamber is assembled so that diffusion of a factor can occur, as in the Zigmond and Hirsch stripe chamber, and if a shear field is established after some time has elapsed under hydrostatic conditions, the response of leucocytes already in various stages of response to the chemotactic agent could be observed. Observations could be made on the shear range required to detach cells; the shear sufficient to move the cells a cell diameter or so from their previous positions, and whether the cells resume locomotion in the appropriate direction when the shear is removed; and the shear which is the threshold to produce any alteration at all in behaviour. A typical chamber in which this could be attempted is the plate and cone. The chemotactic agent would be placed at the central region and allowed to diffuse radially outwards through the cell suspension between the plate and cone. After a finite wait such as 30–60 min, the cone would be rotated to produce a uniform shear for some finite period. With this configuration, any cell movement is most likely to occur azimuthally, so cells will stay at the same concentration and concentration gradient. A slight modification of the protocol would allow a different experiment: a chemotactic agent could be deposited in a straight stripe on the plate along a diameter of the chamber and, after a motionless delay to allow some time for diffusion, steady shear would be applied for a finite period, so that the shear could serve to change the concentration distribution by convective mixing even if the shear rate is gentle enough not to move cells. With the shear removed, observations of cell locomotion could help to resolve whether rates of change of concentration in time are significant in affecting leucocyte behaviour compared with concentration gradients in space. Published evidence suggests that some types of cells are affected by the temporal rate of change of a stimulus, e.g. as shown by the rapid release of insulin by pancreatic β cells in response to a temporal square-wave loading of glucose concentration (Curry 1971).

DISCUSSION

The time appears ripe for a coordinated effort in both theory and experiment to be made so that a better understanding of the physical processes of leucocyte adhesion and locomotion can be achieved. A surge of interest by fluid dynamicists in particle-bearing flows, the development of *in vitro* chambers and of *in vivo* protocols which allow physical interpretations to be made, and advances in microrheology of blood cells, all provide a favourable setting. A key to progress will be the development of a dialogue between the various disciplines so that sharper questions can be asked and pertinent techniques and theories improved.

ACKNOWLEDGEMENT

This work was supported in part by NIH Grants HL-11945 and HL-22338.

References

Abramowitz M, Stegun IA 1964 Handbook of mathematical functions (NBS Appl Math Ser 55) U.S. Govt. Printing Office, Washington DC

Atherton A, Born GVR 1972 Quantitative investigations of the adhesiveness of circulating polymorphonuclear leucocytes to blood vessel walls. J Physiol (Lond) 222:447-474

Begent N, Born GVR 1970 Growth rate in vivo of platelet thrombi, produced by iontophoresis of ADP, as a function of mean blood flow velocity. Nature (Lond) 227:926-930

Born GVR, Planker M, Richardson PD 1978 Influence of blood flow on granulocyte adhesion in venules. J Physiol (Lond): 289:76-77 p

Carslaw HS, Jaeger JC 1959 Conduction of heat in solids, 2nd edn. Oxford University Press, Oxford

Curry DL 1971 Insulin secretory dynamics in response to slow-rise and square-wave stimuli. Am J Physiol 221:324

Eckstein EC, Bailey DG, Shapiro AH 1977 Self diffusion of particles in shear flow in a suspension. J Fluid Mech 79:191-208

Evans HL 1962 Mass transfer through laminar boundary layers, 7. Further similar solutions to the b-equation for the case B = 0. Int J Heat Mass Trans 5:35-57

Fischer T, Schmid-Schönbein H 1977 Tank tread motion of red cell membranes in viscometric flow: behaviour of intracellular and extracellular markers. Blood Cells 3:351-365

Gallin, JI, Wolff SM 1975 Leukocyte chemotaxis: physiological considerations and abnormalities. Clin Haematol 4:567-607

Goldsmith HL, Mason SG 1968 The microrheology of dispersions. In: Eirich FR (ed) Rheology: theory and applications. Academic Press, London, vol 4

Grant L 1973 The sticking and emigration of white blood cells in inflammation. In: Zweifach B et al (eds) The inflammatory process, 2nd edn. Academic Press, New York, vol 2, ch 7

Meksyn D 1961 New methods in laminar boundary layer theory. Pergamon Press, Oxford

Poe GG, Acrivos A 1975 Closed-streamline flows past rotating single cylinders and spheres: inertia effects. J Fluid Mech 72:605-623

Richardson PD 1973 Effect of blood flow velocity on growth rate of platelet thrombi. Nature (Lond) 245:103-104

Richardson PD 1979 Platelet adhesion in a flow chamber. Proc 7th New Engl Bioeng Conf, Pergamon Press, New York, p 57-60
Shapiro AH 1977 Steady flow in collapsible tubes. Trans Am Soc Mech Eng J Biomech Eng 99:126-147
Tözeren A, Skalak R 1977 Stress in a suspension near rigid boundaries. J Fluid Mech 82:289-307
Zigmond SH 1977 Ability of polymorphonuclear leukocytes to orient in gradients of chemotactic factors. J Cell Biol 75:606-616
Zigmond SH, Hirsch JG 1973 Leukocyte locomotion and chemotaxis. J Exp Med 137:387-410

Discussion

Born: In the last few weeks we have been trying to find out how the adhesiveness of rolling granulocytes—that is, the adhering cells that are still moving under the shear—to vessel walls is affected by haemodynamic changes. One primitive approach is to count the rolling cells above and below a confluence of two small venules. Some of the cells arriving at such a confluence find themselves without a wall to adhere to and are bound to 'fall off' into the streaming blood. For them one has to make the best possible correction on the basis of geometrical considerations [See Fig. 1(Richardson).]

We recently provided evidence (Born et al 1979) indicating that the rolling can be affected by local haemodynamic effects. In rats anaesthetized with pentobarbitone sodium (60 mg/kg), exteriorized mesenteries were superfused with Krebs bicarbonate solution at 37 °C. In venules small enough (diameter 15–33 μm) for all rolling granulocytes to be countable, the combined frequency of the cells was higher in two venules just above their confluence than in the single venule just below. The difference, which was highly significant in three experiments in which the frequencies were high, but not in one in which it was low, is less likely to be due to a change in the properties of the granulocytes or the endothelium than in the properties of the blood flow.

Secondly, in junctionless segments of venules, there were no significant differences in the frequencies of rolling granulocytes when these were determined under apparently constant conditions, on the same side, at two points separated by distances of 400–3000 μm, in 17 experiments using the cheek pouch of anaesthetized hamsters. If rolling depended on a change affecting granulocytes as they passed through capillaries before entering venules, the latter might be expected to contain more rolling granulocytes upstream than downstream. Instead, the results support our proposition that rolling follows random collisions of circulating granulocytes with the walls of the venules in which it is observed.

Vane: What is the interval in the flow between the points at which you counted? How long is the period in which the leucocyte might be losing any conditioning it could have picked up in the capillary?

Born: The time difference was from about 1 to 3 s. It is, of course, conceivable that in that time some change in the surfaces of the granulocytes disappeared again. For these observations we chose the longest unbranched stretches of venules we could find to give the cells the greatest chance of being sheared off the walls.

Vane: But if the cell isn't stuck does it flash by?

Born: Yes. These venules had mean flow velocities of 100–1000 μm/s. Granulocytes that roll do so at about one-hundredth of that velocity, and in proportion to it in any given venule.

Richardson: You may be talking about something like a 10-s duration in the microscope viewing field, which is not very long. If the cell sticks, it has already made up its mind; if it doesn't, it is in such a fast flow that it has gone out of the field and may stick somewhere else where you don't know about it.

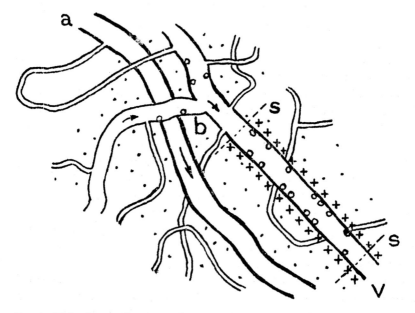

Fig. 1. (Richardson). Sketch of microcirculation with arteriole *a*, venule *v*, bifurcation of venules *b*, and rolling polymorphonuclear leucocytes in venule *v*. Natural chemotactic factor can diffuse throughout extravascular space (dotted) and affect cells in capillaries as well as in venules. An ideal way of investigating collision efficiency in a venule is to find a long section *s*–*s* without branches and apply a chemotactic agent (or localized damage) along the venule wall (crosses) so that changes in rolling numbers can be observed along that section.

Howard: The vessel diameters we are discussing are fairly small. The places where leucocytes or granulocytes are sticking are not much larger in diameter than the cell itself. The capillary venules may have diameters of less than 10 μm or so, or perhaps one or two cell diameters, though they may also be larger. You said that the collision efficiency in some vessels was undoubtedly low. That presumably means that a cell can approach the endothelial surface through augmented diffusion or simple diffusion and collide, making an instantaneous attachment which is not efficient. This will tend to bring it into the vicinity of the wall and it seems to follow that in small vessels cells will always tend to move down the walls. Their diffusion will be constantly limited by their incompletely efficient contact and it is as if they have a low affinity bond with the wall.

Richardson: This problem also interests people who worry about platelets, because if platelets are 'marginated', giving a higher concentration in the flow closest to the wall instead of a uniform concentration all the way across the stream, this might account for thrombosis when something triggers it off. The degree of margination depends on the length of the vessel that is available after some big disturbance, such as a junction. Goldsmith & Mason (1967) reported that sometimes the flow of cells would be tens of thousands of microns downstream before they noticed a strong tendency for a skimming layer to come to its asymptotic state. It can happen, but often it takes some time of bouncing motion for it to show itself.

Howard: Isn't another way of describing inefficient collision to say that the cell is spending more than the average time near the margin?

Richardson: If the cell is spending more time near the wall than it is expected to spend and it still doesn't adhere, then you will find a low collision efficiency. The efficiency will be lower than you would calculate if you assumed that the cells were randomly distributed across the flow and moved to the wall according to some normal diffusion–collision rate. Augmented diffusion can do something else. If the cell spends some time near the wall and is not adhering, in a sense that is not very useful. If another cell in the stream would do better, facilitated diffusion would help to bring that other cell to interrogate the wall more quickly.

Gowans: Where the dimensions of the wall are more or less the same as those of a white cell, keeping the white cells on the move must be a problem. To what extent is the skimming phenomenon a device that fine blood vessels use to rid themselves of white cells and keep the red cells flowing?

Born: Any such effect would presumably depend, *inter alia,* critically on the size of the blood vessel. The evidence seems to be that the change during inflammation which has this glue-like effect on granulocytes does not occur in capillaries. Even in inflamed areas, both the leucocytes and the red cells squeeze through the capillaries one by one. There is no hold-up. Skimming occurs in larger vessels or at branchings with particular relationships between the geometrical configuration and the haemodynamics.

Richardson: Some strange situations can arise where small vessels join together at a 'T' and at one side of the 'T' there is no flow of cells—there is a flow through just two arms of the 'T'. At the end of the vessel with the stationary cells there is a leucocyte with a whole stack of red cells behind it. There is no real flow through there, except some plasma. The leucocyte just spins, without adhering. It is like a rotating plug. All sorts of phenomena like that can be seen in the microcirculation.

Weiss: At that right-angled branch would plasma and white cells be preferentially skimmed out of the flow?

Born: Yes.

Weiss: That is exactly the arrangement of the vasculature in the periarterial lymphatic sheath of the white pulp of the spleen. The central artery goes barrelling right through with the right-angled branches. That might be one mechanism that brings lymphocytes in plasma to one part of the spleen and allows blood with a high haematocrit to go out to the red pulp.

Richardson: Rheologists are trying to find out how red cells that come to a junction in a vessel decide which way to go. It seems to depend in part on the history: if a lot go one way, they increase the effective resistance to flow down that branch and they slow down along that branch, while the other branch, which has fewer red cells in transit is flowing faster for the same pressure difference, and then the arriving red cells start distributing to the other side (Yen & Fung 1977). There may even be a bi-stable situation where one line will fill for a while, then the other. If you have microcirculations that you can look at with a large enough field you can watch what is going on in parallel branches.

Butcher: You have given us a particular collision frequency for a standard situation. How does this collision frequency relate to the rate of flow of the blood and the diameter of the vessel?

Richardson: The effect of flow rate is rather small. The diffusivity, which is really what dominates the issue, gets to an asymptotic value. In other words if the shear rate increases or decreases beyond a certain magnitude there is no significant change in the effect on diffusivity. It gets to a terminal value.

Butcher: But would the flow rate affect the efficiency of collisions, if not their frequency?

Richardson: Yes. It could also affect the contact time. This is where the question of being preconditioned or not preconditioned may be critical. As I was trying to point out with the platelets and the aggregation delay time, the contact time seems to be important for cells when they are deciding whether to adhere, especially if they are not preconditioned by prior exposure to activating agents before they strike the vessel wall.

The question of diameter is more complicated. I picked a diameter which seemed to represent a vessel size in the range that is experimentally accessible and meaningful for this sort of study. Anything bigger will be more unobservable in terms of the total amount going on. I expressed the rate per square micron, so as the segment length decreases a smaller area will be available but also the rate is likely to go down somewhat, depending on a number of factors.

Butcher: So the collison frequency per unit area of wall will go down.

Richardson: Yes, but not as fast as vessel size.

Butcher: Is anything known about the inside diameter of capillaries or post-capillary venules *in vivo*?

Richardson: The architecture covers a range of sizes. The tricky part is that branches are frequent. In other words the entry of the capillaries into the postcapillary venules is much more frequent than the exit of capillaries from comparably-sized arterioles.

Simionescu: Postcapillary venules vary from one tissue to another, measuring from 10 to 15 μm up to about 30 μm (inner diameter). These venules are succeeded by muscular venules of about 30 to 100 μm diameter.

Gowans: There seem now to be two possibilities about the mechanism of emigration through venules. Firstly, emigration is initiated by changes which affect the endothelium of the venules exclusively. But secondly you have introduced the idea that the leucocyte is preconditioned in the capillary and is therefore different when it arrives at the venule and it sticks. What is the evidence for the second view?

Richardson: The evidence is only circumstantial. In the microcirculation small vessels are embedded amongst larger vessels and therefore if there is some coherent source of some factor which is going to diffuse in the extravascular space, it will have an opportunity to present itself to the capillaries. In fact it may even be able to diffuse into those a little more readily in the sense of getting across the flow, because the blood moves in the capillaries at a slower rate than in the postcapillary venules.

Vane: The concept of cells being conditioned as they are squeezed through the capillaries is attractive, especially if the capillary wall is making an active substance such as prostacyclin. Those cells would have to squeeze through in close contact with the prostacyclin and the deformability of the red cell membrane might be changed, or the stickiness of the leucocytes or any of the cells that have to squeeze through. The concept is attractive but there is no evidence for it yet.

Gowans: Can one get at this experimentally?

Vane: One could use prostaglandin synthetase inhibitors to see whether they change the condition of the cells as they come out of the capillaries.

Born: Can one build a bridge between Peter Richardson's contributions and those of Joe Hall? What is known about the flow of lymphocytes in vessels from which they are extracted?

Hall: Very little, I think. The extraction rates that have been calculated are very unreliable (Hall 1967)—something like 10% under basal conditions, but even to put a figure to it is probably presumptuous.

Marcus: Is it fair to say that your calculations are actually being made for rigid tubes? If we consider that stimulated platelets are producing thromboxane A_2, probably the most powerful vasoconstrictor yet described, one can envision the end of the blood vessel closing down. If one abolishes thromboxane production by aspirin medication, this lead-in time would become longer. Thus, one should take into consideration the role of blood vessel reactivity to biological substances produced by the contents therein. For example, serotonin itself causes vasoconstriction although we know it is not essential for haemostasis.

Richardson: Yes, but variations in flow arise in the microcirculation even without that sort of thing. Natural shimmerings arise just by changes in the effective haematocrit in different parts of parallel vessels.

Vessel distension arises from the nature of the vessels and the balance of pressure and the elasticity of the vessel wall. The whole scene shimmers (i.e. the capillary flows start, accelerate, slow, stop, start again), for that reason. But with some of the approximations that are already being introduced, such as taking the walls as having uniform properties, and the local effective haematocrit as varying with time, some of these phenomenological features are already explained.

Vane: I have only looked at rolling leucocytes for a few minutes but what impresses me is that a cell sticks and jumps, then sticks again and jumps a bit, and sometimes it sticks completely. I have never seen one sticking and then starting to go through. Do you see this when you look at them for a long time?

Born: Yes, you can see it very clearly in inflamed vascular beds. There are beautiful films of this.

Gowans: These events can be observed very nicely in the rabbit ear chamber.

Richardson: But even there you can run into another problem: as you change the strength of the field illumination you may change the number of leucocytes seen.

It is really desirable to try to design experiments which can give more quantitative answers to some of these questions. Sometimes a framework like the analysis here can lead to experiments which give some of the numbers that would be helpful.

References

Born GVR, Planker M, Richardson PD 1979 Influence of blood flow on granulocyte adhesion in venules. J Physiol (Lond) 289:76-77P

Goldsmith HL, Mason SG 1967 The microrheology of dispersions. In: Eirich FR (ed) Rheology: theory and applications. Academic Press, London, vol 4:85-250

Hall JG 1967 Quantitative aspects of the recirculation of lymphocytes: an analysis of data from experiments on sheep. Q J Exp Physiol Cogn Med Sci 52:76-85

Yen RT, Fung YC 1977 Red blood cell at cross road, which way would it go? Trans Am Soc Mech Eng J Biomech Eng 99:71-73

Final general discussion

Ford: We have all been talking about the adhesion of lymphocytes to high endothelial cells in terms of complementary surface receptors but in my view we have seen no evidence that this interaction is mediated by receptors.

I should clarify what I understand by 'receptor' by giving two examples. The B lymphocyte inserts antibody molecules into its cell membrane. The corresponding antigen binds to this antibody and a train of events may follow. This is undoubtedly a surface receptor. I would also be willing to include the structure on human T lymphocytes that binds sheep erythrocytes although we do not understand the biological relevance of this.

Even allowing this broad definition I am not convinced that we are necessarily dealing with receptors. Suppose that the lymphocyte covers itself with an Araldite type of glue and the high endothelial cell secretes the 'hardener', then the lymphocyte would stick to the first surface it hits— usually the endothelial cell. The 'hardener' might operate enzymically or by ion transfer.

Gowans: The problem of adhesion is clearly very complex and I agree it may be a mistake to think of 'receptors'. What is the 'receptor' on glass that enables macrophages to stick to it so firmly, for example?

Vane: Peter Richardson showed another receptor on glass where different platelets would go on sticking to the same spot.

Gowans: The snail's trace left behind by the first one?

Richardson: Quite often we can hang a whole chain of other platelets on one platelet. If the flow is disturbed so that the thing waves around we can see it very clearly. There is an anchor platelet and a tremendous number of other platelets on a string behind.

Ford: I think we should keep in mind the many possibilities. In Judith Woodruff's hands the lymph node sections work optimally after fixation in

strong glutaraldehyde but lymphocytes do no adhere even after mild glutaraldehyde treatment. This underlines the asymmetry between the lymphocyte
and endothelial cell. It seems to discount two possibilities: (1) that each cell
type displays the same receptor, with linkage accomplished through a bivalent
ligand, and (2) that the cells have mutually complementary receptors of
basically similar composition.

Howard: An important immunological analogy suggests an alternative
view. The cytotoxic T lymphocyte sticks to the relevant antigenic monolayer
when the monolayer is fixed with low enough concentrations of
glutaraldehyde, but the lymphocyte receptor can never be fixed.

Ford: Some metabolic activity may be demanded for the optimal function
of such an antigenic monolayer. However, I would concede that the
conditions for adherence between the cytolytic T cell and its target are asym-
metric although it is dependent on surface receptors.

Woodruff: This kind of comparison doesn't seem right. We are talking
about a section of a cell on the one hand and an intact cell which is a sphere on
the other. It seems not unreasonable to expect that they will have different
physical requirements for adhesion. The finding that certain lymphoid
populations exhibit poor HEV binding activity suggests that those cells which
do bind express surface molecules which mediate HEV recognition and it is in
this sense that the term receptor is used.

Davies: I don't think the problem is simply semantic. The concept of
receptors proposes some positive structure, whereas Bill Ford's 'Araldite'
analogy proposes that there is adhesiveness without any recognition of
structure *per se*—it depends upon effluvia, if you like, and this is not the same
as having something on the surface which facilitates contact.

McConnell: If we look at the experimental evidence for things moving
through vessel walls it is clear that for polymorphs it is quite easy to have
chemotactic agents, including active complement components, which induce
margination of cells before their emigration from vessels. Exactly the same
could apply to lymphocytes—lymph nodes might secrete lymphocyte chemotactic factors to produce initial margination of cells.

Williams: Sally Zigmond clearly has receptors with the peptides, since she
can demonstrate the binding.

I would like to say something about carbohydrate in recognition and this
may be relevant to the fact that fixing of lymph node sections promotes
binding of lymphocytes to HEV but fixing of lymphocytes abolishes binding.
You would get fixing knocking out one part of a recognition system and not
the other if you had a protein recognizing carbohydrates. The fixing wouldn't
affect the carbohydrate side but it could very easily affect the protein side.

Recognition like this occurs: for example, the human receptor for sheep erythrocytes appears to be a lectin because if glycopeptides are made from human erythrocytes, they inhibit that rosetting (Boldt & Armstrong 1976). That recognition appears to be like a lectin recognizing carbohydrate.

Mechanisms of this type are also relevant in this meeting with respect to the role of carbohydrate in lymphocyte recirculation. Experiments (Ashwell & Morrell 1977) on control of degradation of serum glycoproteins suggest why removal of sialic acid with neuraminidase led initially to the confusing results on the role of sialic acid in recirculation of lymphocytes. If the sialic acid is taken off certain serum glycoproteins, exposing galactose sugars, then this carbohydrate is bound by a lectin in the liver. That is probably why the lymphocytes go to the liver when the sialic acid is taken off. The lectin is quite specific and doesn't bind to all carbohydrate structures with exposed galactose. For instance, if you take the sialic acid off transferrin carbohydrates or immunoglobulins, these won't bind. A recognition system like this can be modified in time. You can start off with cells that haven't had the sialic acids put on and then put them on in the course of a differentiation series. Thymocytes may do funny things *in vivo* because they haven't got completed carbohydrates.

Woodruff: On the question of neuraminidase treatment of lymphocytes it has also been shown that serum contains antibodies which react with lymphocytes after sialic acid is removed; this could lead to accumulation of lymphocytes in the liver (Winchester et al 1975).

Williams: Where do the lymphocytes go in the liver?

Woodruff: From what we have seen by autoradiography they are in the hepatic sinuses.

Gowans: When platelets adhere after the addition of ADP what sticks them together?

Born: Fibrinogen and calcium are essential.

Gowans: But what sticks them together?

Born: Probably, first, calcium bridges between negative charges on the platelet surface and the carbohydrate moiety of fibrinogen. On the fibrinogen these charges are almost certainly those of *N*-acetylneuraminic acid. And secondly, interactions between specific sequences of the fibrinogen molecules with complementary receptor sites on the platelets. The calcium bridges initiate adhesion by providing coulombic interactions which are the strongest but quite non-specific. Specificity comes in through weaker receptor interactions.

Humphrey: Do you mean that the rearrangement could take place in the absence of calcium?

Born: What do you mean by rearrangement?

Humphrey: You said that calcium formed the bridge and then the rest of the strengthening of adherence depended on other things.

Born: Calcium is part of the fibrinogen link. The receptor part of fibrinogen is becoming known mainly through competitive effects of various fractions of the molecule.

Vane: What you are talking about is platelet–platelet stickiness. What about platelet adhesion?

Richardson: With the flow chamber I described in my paper here, I tried running ADP down one side, with the other side as the control. The adhesion rate of platelets to the glass doesn't seem to change with, say, 10^{-6}M-ADP present, but the rate of adhesion of platelets to those platelets that adhere to the glass increases. In fact aggregates grow on adherent platelets faster in the presence of ADP than in its absence.

There is also some trickery with fibrinogen binding. When I put platelets from an afibrinogenaemic patient through the flow chamber I get no adhesion at all on glass. If I coat half the slide with fibrinogen I get adhesion. If I put the effluent through another chamber with plain glass I don't find any adhesion. In other words, it doesn't look as though any of the platelets that have gone through the first chamber have been able to pick up fibrinogen, take it down into the next chamber and use it for adhesion to the glass there.

Gowans: Do you now know why platelets stick to glass?

Vane: No, but it sounds as if fibrinogen is involved. Is calcium involved too?

Richardson: Yes, and I can get them to stick in the absence of fibrinogen. I can also coat a slide with thrombin and get platelets from the afibrinogenaemic patient to stick.

Weissman: Most thymocytes are known to have agglutinability by a peanut lectin which binds to galactose residues, but the small population of postmature thymocytes does not. The galactose residues are obscured, apparently, when a terminal sialic acid is added. That fits with the kind of maturation scheme mentioned earlier.

If a lymphocyte is putting out a very specific Araldite hardener and a particular HEV is putting on its surface a particular Araldite precursor, and if the expression is limited to the cell membranes in both of them, that is a receptor–ligand interaction as far as I am concerned. Measuring how big the molecules are and whether they are attached to the membranes is not critical.

Vane: That wouldn't fit the definition of a pharmacological receptor which you can antagonize with something and prevent the stickiness.

Weissman: We still find that out if we start characterizing the molecules that are involved in this interaction.

Williams: Immunoglobulin wouldn't fit that either, would it? You said you would have to be able to antagonize the receptor. You couldn't do that with surface immunoglobulin on the B cell.

Vane: Each discipline uses the term receptor in its own way.

McConnell: It is not just a semantic problem. If you agreed with Bill Ford's view, you would ask whether high endothelial cells actually secrete something which causes lymphocytes to become 'sticky'. Can you find tumours of endothelial cells that are releasing a product that alters lymphocyte traffic, migration and so on?

Weissman: If he finds tumours of endothelial cells that are involved in this interaction we'll both be happy.

Zigmond: Is a change in the presence of a specific molecule involved? Or is there a change in membrane fluidity or some other parameter, with no change in the molecular composition of the membrane? One doesn't necessarily have to envisage that there is a new molecular species there. In one case you might speak of a receptor and in the other of an activity.

Gowans: We started to discuss lymphocyte chemotaxis because we wanted to know what happened after adherence to the surface of the high-walled endothelium. What provokes lymphocytes to move across the vessel wall once they have stuck to the endothelium?

Ford: Could their movement be facilitated by a pressure gradient?

Richardson: It is feasible but the pressure differences available in these locations seem extremely small. The rates of movement would be difficult to account for, I suspect.

Zigmond: P.C. Wilkinson has done one experiment which not only shows lymphocyte activation or enhanced motility but also a directional response. There is other work on enhanced migration through filters but that may or may not be chemotaxis.

References

Ashwell G, Morell AG 1977 Membrane glycoproteins and recognition phenomena. Trends Biochem Sci 2:76-78

Boldt DH, Armstrong JP 1976 Rosette formation between human lymphocytes and sheep erythrocytes. Inhibition of rosette formation by specific glycopeptides. J Clin Invest 57:1068-1078

Winchester RJ, Fu SM, Winfield JB, Kunkel HG 1975 Immunofluorescent studies on antibodies directed to a buried membrane structure present in lymphocytes and erythocytes. J Immunol 114:410

Summing up

IRVING L. WEISSMAN

Department of Pathology, Stanford University Medical Center, Stanford, California 94305

Twenty years ago Jim Gowans and his colleagues found that lymphocytes recirculate through lymphoid tissues, entering by a special type of postcapillary high-endothelial venule. At that time he contrasted the pathway that lymphocytes take through these venules with inflammation-induced movement of other types of cells. He also demonstrated that lymphocytes were immunospecific elements in the immune response and were entirely responsible for all of the antigen-specific aspects of the immune response that we knew about then. From these studies it seemed fairly clear that the particular interaction that is the focus of this meeting—the small lymphocyte-HEV interaction—was responsible for the histogenesis and maintenance of antigen-reactive lymphoid tissues. It provided a nice rationale for how a large number of lymphocytes, each expressing its particular antigen-specific receptor, could file by and meet antigen at the site of antigen drainage. Jim Gowans then moved on to look at particular questions of localized recirculation and local immunity in the gut, which he summarized for us in this conference.

How far have we got since then? In particular, how far have we got as demonstrated by this meeting? We have been concerned with three general topics here. The first concerned the general interactions between vascular endothelia and elements in the blood, both plasma proteins and cells. For a major part of the meeting we then tried to define a specific kind of interaction between the vascular endothelium and a cellular element, the lymphocyte. The third topic that came out in an unexpected way was the role of various non-lymphocytic cells which end up in lymphoid tissue and which apparently are also involved in the establishment and maintenance of specific architectural domains within lymphoid tissue. Some of these cells may be involved directly in the generation of immune responses within those tissues.

© *Excerpta Medica 1980*
Blood cells and vessel walls: functional interactions
(Ciba Foundation symposium 71) p 343-347

Now let's see what kinds of general correlations between studies might be made. As to the general interactions between vascular epithelium and blood components, it has been demonstrated for us that capillary endothelia apparently have several mechanisms for transmitting, admitting, or transporting plasma proteins. These have well-defined morphological correlates and they may relate to the size of the protein constituents, as well as their charge distributions which may be involved in facilitated transport of these molecules. The transcytotic vesicles described by Dr Simionescu are of special interest here. For example, it is possible that the electron micrographs of Marchesi and Gowans showing lymphocytes apparently going through the cytoplasm of postcapillary venule endothelial cells could be due to very large transcytotic vesicles rather than exclusively to cell movement through intercellular spaces. To my mind that question has not yet been conclusively resolved.

Part of the conference that I didn't understand very well had to do with endothelial cells being fairly active participants in the maintenance of cells and platelets at a distance, especially in terms of endothelial cells being involved in the elaboration of opposing agents which are involved in platelet aggregation and haemostasis. As I understand it, there are specific endoperoxidases in different tissue endothelia, which may define the ratio of prostacyclins to thromboxanes. If that is true it would make us think of these vascular endothelial cells as interesting differentiated sets of cells which in a particular organ may express both morphological differences and enzymic differences that subsume different functions.

That leads us to something that I am more familiar with—the question of lymphocyte–high endothelial interactions and the specificity of these interactions. It is clear from several studies that there is local recirculation of lymphocytes to local sites—e.g. specific recirculation to the gut compared to the lymph nodes. As I understand it, entry of cells into these sites is antigen-independent. Once an immune response is set up you get local immunity, and at least a portion of that local immunity is due to restricted egress of cells from those sites, resulting in local build-up of effector cells and their proliferating precursors. Local recirculation may involve some sort of recognition between lymphocyte and high-endothelial venules in these particular sites. However, that recognition would only lead to cells binding at a particular place. We don't know whether other factors play a role in the migration of cells through these endothelia. There is perhaps an important clue in the studies on the development of local recirculation pathways. We are told that in the adult sheep there is a very clear and specific local recirculation, gut to gut and peripheral node to peripheral node, but that in the fetal sheep this difference has not yet developed. It is therefore a rather late event in the maturation of the organism. One should ask: what are the signals that influence the maturation of either subtypes of lymphoid cells or subtypes of endothelial cells in these particular regions?

It is my opinion that lymphocyte–high-endothelial venule interactions are in fact due to lymphocyte receptor recognition of endothelial cell ligands. That was most importantly demonstrated by Judith Woodruff's very interesting and, at the time, shocking finding that one could demonstrate this interaction *in vitro*, allowing a quantitative assessment of it. It has been demonstrated at this meeting that lymphocyte–HEV interactions show genetic variations, organ-specific variations and species variations. As the interaction must take place in the fetal lamb to allow lymphocyte recirculation, it appears that recirculation itself is independent of exogenous antigen administration. This shoots down a major hypothesis of at least two groups of people who feel that virgin lymphocytes themselves are non-recirculating, while memory lymphocytes are recirculating. That is now a very difficult position to hold.

Next, I was going to tell you that there may be a common thread between the various cell surface receptor and ligand interactions we have discussed today. I pushed Sally Zigmond hard on whether chemotactic factors are active in the fluid phase or only on surfaces because I wanted to develop the idea that chemotaxis of leucocytes can tell us something about lymphocyte–HEV interactions. That is, if a local concentration of a particular ligand—in this case a chemotactic factor—is put on a surface, a cell recognizing the factors by cell-surface receptors would walk up a concentration gradient. That still remains (to me) a strong possibility, and there is perhaps a real similarity between that kind of interaction (leading to leucocyte mobility) and lymphocyte movements through HEV walls. I suppose you could call that a factor rather than receptor–ligand interactions, but again that is perhaps only a semantic difference.

The other part of the thread again relates to transcytosis and asks whether there are also very specific molecules involved in this type of transport. Thus the maintenance of membrane concentration gradients of particular ligands might allow for movement of molecules or movement of membrane moieties. A prime example of a kind of endothelial molecule that could be highly variable from organ to organ would be the sulphated glycoprotein of very high molecular weight that was described by Andrews and Ford in lymph node HEV.

An interesting point that came from Bede Morris was that there is apparently a large-scale recirculation of lymphocytes through non-lymphoid organs. Several people showed that recirculation can be increased through different organs by antigenic exposure; in skin by painting with various contact-sensitizing agents, and through surgically anastomosed allogeneic kidneys. What is really surprising about these findings is how little is known about entry sites and movements through those tissues. One of the first things that was found when lymphocyte recirculation was discovered was the entry pathway; that finding opened a new field for the study of cell interactions. It would be nice to find out in greater detail what kinds of endothelia are involved in recirculation through non-lymphoid tissues, and to see whether there are specific interactions in those models also.

The third topic was the role of non-lymphoid cells in the maturation of lymphocytes, in the organization of lymphoid tissue, and in the generation of cell interactions leading to an immune response. The two major maturational microenvironments for lymphocytes, the bone marrow and thymus, apparently have in common at least one morphological class of cells—those with long dendritic processes in the bone marrow and the dendritic cells in the thymus (which we think are epithelial cells). What is interesting, at least in the thymus, is that different MHC markers are expressed by these cells, depending on their location. The implication is that in different layers of the thymus these dendritic cells subsume different kinds of maturational tasks, resulting in the maturation of different sub-classes of T lymphocytes. Weiss has proposed that bone marrow dendritic cells may also form maturational microenvironments. However, it is likely that these two types of dendritic cells are really quite different. For example, it is believed that the thymic dendritic cells are epithelial and derived from pharyngeal endoderm, while there is no evidence that adventitial cells in the bone marrow are of endodermal origin. Perhaps our criteria for characterizing thymic dendritic cells as epithelia are wrong. Certainly they are interesting dendritic cells, and they are not bone-marrow-derived in chimeras.

Another type of dendritic cell that was described was the interdigitating dendritic cell found in T cell regions of spleen and lymph nodes. There was some very interesting evidence from J. Humphrey that such cells may be derived from non-sessile precursors—the 'veiled' cells in afferent lymphatics. Although that point is not yet clear it is an interesting possibility. If it is a cell that ends up in a T cell region, and if it is a cell that is antigen processing, one would expect that it would have the kinds of markers that T lymphocytes and not B lymphocytes might recognize and respond to, at least in the context of antigen presentation.

Yet another kind of cell appears to be in the afferent lymphatics—efferent from one lymph node, afferent to another, and perhaps afferent to the first—and that is the kind of cell that Bede Morris finds in antigen-activated sites.

This provides us with an entirely new view of lymphoid tissue architecture. In antigen-independent states our current understanding is that cells enter the lymph node through the high endothelial venule, break up into two independent paths (the so-called B and T paths) and, in the absence of antigen, filter through and go out. Now Bede Morris says that there is another site of entry. Antigen-activated lymphocytes enter via afferent lymphatics and localize in germinal centres (structures which are present only in antigen-activated tissue). It seems likely, although not proven, that cells in germinal centres aid in the development of antigen-activated effector cells of both T and B cell lineages by direct differentiation or by recruitment of other cells. Thus one can sketch two traffic zones—the antigen-independent zone and the antigen-dependent zone. The problem with this formulation is the problem that Jonathan Howard and I brought up. That is, where does sensitization really occur? If antigen-sensitization occurs in the periphery, as Bede Morris's model implies, you have the problem of what is the function of a lymph node. Why is this constant file of a large number of cells going through the node rather than allowing peripheral antigen to be both the sensitizing stimulus and the site of the immune response? We are going to have to come to grips with that issue in the future. The evidence for peripheral sensitization is positive in some circumstances, and negative in others. If it turns out that peripheral sensitization is common, it may relate to the recirculation through non-lymphoid tissues that Bede Morris and others talked about. Perhaps only particular subsets of lymphocytes go to particular places (because of their interaction with specialized venules in those places) and therefore antigen recognition need not result in the generation of an immune response. Or perhaps some sites are excluded from this traffic, e.g. the central nervous system. In the latter case one may be able to explain why peripheral sensitization to myelin basic protein does not occur unless one deliberately injects these antigens in sites outside the CNS.

Unfortunately, we must conclude that most of the major questions first raised by Gowans are still unresolved. With all these questions unanswered I would like to invite Jim to come back and resolve some of these problems when he finishes his vacation at the MRC.

Epilogue

LEON WEISS

School of Veterinary Medicine, University of Pennsylvania, Philadelphia, PA 19104

Some facets of Dr Gowans' scientific achievements are readily understood. Others are difficult to convey. His work demonstrating the recirculation of small lymphocytes from blood to lymph through lymph nodes and that showing the immunological competence of lymphocytes by means of graft rejection and the development of graft-versus-host disease are understood and constitute the keystone of cellular immunology. Some 20 years after Dr Gowans' first paper on the subject, we recognize the recirculating pool of lymphocytes as a tissue, as much an entity as the more physically coherent tissues. We are beginning to discover, moreover, that lymphocytes control not only the proliferation and differentiation of plasma cells but of other haemopoietic cell lines and of certain non-haemopoietic tissues as well. Dr Gowans' definition of the recirculating pool of immunologically competent lymphocytes is leading us to an understanding of the immune system that constitutes a considerable expansion of our ideas of immunity—to a system of regulation that rivals the endocrine and nervous systems.

What is difficult to convey in Dr Gowans' work is the nature of the study of haematology in the forties and the fifties. In the decades that preceded his work intuition reigned and investigators were persuasive in proportion to their persistence and articulateness: things were what one said they were. Isaiah Berlin, in a lovely essay, speaks of the hedgehog and the fox, where the fox knows many things and the hedgehog knows one big thing. Study of the haemopoietic, the immune and the reticuloendothelial systems, at the time when Dr Gowans began his work, was being done in the world of hedgehogs and foxes, of lumpers and splitters, of monophyleticists and polyphyleticists.

What Dr Gowans did was to send Humpty Dumpty back through the looking glass, to dismiss proof by intuition and magic and to put science in their place. His work re-established the experimental foundations of

© *Excerpta Medica 1980*
Blood cells and vessel walls: functional interactions
(Ciba Foundation symposium 71) p 349-350

microscopic anatomy and histology, which had become mechanically descriptive. He showed that the immune system was not an arbitrary assemblage of cells but the morphological expression of functional interactions of diverse cell types.

When I consider the two levels of Dr Gowans' achievement—the enduring contributions on the recirculation of small lymphocytes and the immunological competence of lymphocytes, and the significance and the power of this work in revitalizing a major scientific discipline and bringing it into the main stream of science—I know of no achievement in modern cell biology that equals it.

I should like to thank Bill Ford for suggesting this meeting, Ruth Porter for planning and carrying it out, and James Gowans for chairing it.

Index of contributors

*Entries in **bold** type indicate papers; other entries refer to discussion comments*

Andrews, P. 52, **211,** 227, 228, 229

Born, G. V. R. 36, 54, 55, 58, **61,** 73, 74, 76, 77, 91, 93, 94, 95, 96, 99, 100, 104, 108, 165, 193, 208, 257, 283, 286, 308, 309, 329, 330, 331, 332, 334, 339, 340

Butcher, E. C. 59, 229, 260, **265,** 282, 283, 284, 285, 286, 332, 333

Cahill, R. N. P. 124, 141, 144, **145,** 157, 158, 159, 160, 161, 162, 164, 165, 194

Davies, A. J. S. 14, 92, 124, 144, 157, 158, 162, 193, 234, 235, 236, 239, 258, 261, 285, 286, 310, 338

Ford, W. L. 35, 103, 104, 109, 125, 161, 164, 166, 190, 208, **211,** 227, 228, 230, 231, 235, 237, 259, 261, 262, 285, 297, 298, 337, 338, 341

Gowans, J. L. **1,** 13, 34, 55, 56, 73, 92, 93, 95, 99, 100, 103, 106, 107, 108, 109, 110, **113,** 122, 123, 124, 125, 126, 158, 159, 162, 163, 190, 191, 192, 193, 205, 206, 207, 208, 209, 228, 231, 233, 234, 235, 237, 261, 262, 263, 282, 286, 296, 297, 298, 308, 310, 311, 331, 333, 334, 337, 339, 340, 341

Gryglewski, R. J. 74, 90, 92, 193, 206, 227

Hall, J. G. 53, 58, 122, 163, 164,165, 166, **197,** 205, 206, 207, 208, 209, 233, 234, 238, 257, 258, 296, 334

Heron, I. **145**

Hopkins, J. **167**

Howard, J. C. 52, 58, 76, 95, 96, 100, 103, 122, 141, 142, 161, 162, 165, 190, 209, 228, 229, 232, 233, 234, 235, 239, 240, 284, 331, 338

Humphrey, J. H. 19, 36, 93, 125, 190, 191, 194, 206, 233, 240, 258, 263, **287,** 296, 297, 298, 310, 339, 340

Kuttner, B. J. **243**

Lachmann, P. **167**

McConnell, I. 35, 57, 94, 122, 142, 159, 164, 165, **167,** 190, 191, 192, 193, 194, 207, 236, 238, 240, 261, 262, 285, 338, 341

Marcus, A. J. 17, 18, 51, 76, 77, 91, 92, 93, 282, 283, 334

Moncada, S. **79**

Morris, B. 53, 54, 55, 57, 58, 92, 100, 104, 106, 107, 108, 122, 123, 124, 125, **127,** 140, 141, 142, 143, 144, 158, 159, 162, 163, 164, 165, 166, 191, 192, 193, 207, 208, 234, 235, 237, 238, 260, 296, 297, 298, 309

Owen, J. J. T. 18, 33, 99, 110, 159, 160, 161, 162, 230, 240

Poskitt, D. C. **145**

Richardson, P. D. 34, 35, 56, 73, 96, **313**, 330, 331, 332, 333, 334, 337, 340, 341

Simionescu, M. **39**, 52, 53, 54, 55, 56, 57, 58, 59, 99, 207, 228, 333
Steer, H. W. **113**
Stoddart, R. W. **211**

Trevella, W. **127**
Trnka, Z. **145**

Vane, J. R. 36, 73, **79**, 91, 92, 94, 95, 96, 99, 100, 108, 165, 191, 192, 193, 207, 228, 286, 309, 310, 330, 333, 334, 337, 340, 341
van Ewijk, W. **21**, 34, 35, 36, 37, 109, 190, 193, 260, 262

Weiss, L. **3**, 13, 14, 15, 16, 17, 18, 19, 52, 58, 74, 93, 94, 99, 100, 107, 109, 142, 162, 228, 237, 259, 332, **349**
Weissman, I. L. 16, 34, 57, 103, 106, 123, 125, 140, 141, 143, 144, 157, 161, 190, 208, 209, 227, 229, 233, 234, 236, 237, 240, 260, **265**, 281, 283, 297, 307, 308, 340, 341, **343**
Williams, A. F. 15, 122, 124, 125, 160, 161, 235, 259, 260, 261, 262, 263, 309, 338, 339, 341
Woodruff, J. J. 108, 110, 126, 228, **243**, 257, 258, 259, 260, 261, 262, 263, 283, 284, 286, 338, 339

Youdim, M. B. H. 75, 94

Zigmond, S. H. 95, 192, 227, 228, 240, **299**, 308, 309, 310, 311, 341

Indexes compiled by Peter Hatton

Subject index

acids *see* **specific acids**
adherence 243-255, 257-263, 265-280, 283-286, 288, 297, 323-327, 329, 337-341
ADP 63, 64, 66, 67, 69, 70, 74, 75, 87, 91, 92, 95, 99, 339
afferent lymphatics 114, 119, 131, 137, 140, 143, 149, 165, 173, 175, 176, 191, 192, 199, 206, 207, 208, 293, 294, 297, 347
aggregation, platelet *see* **platelets,** aggregation
AMP 69, 76, 81, 82, 83, 95, 257, 308
antibodies 22, 28, 33, 36, 55, 93, 94, 100, 120, 124, 138, 140, 159, 168, 178, 207, 217, 227, 233, 238, 262, 292, 337
antibody response, suppression of 183-184
antigen(s) 7, 28, 57, 58, 105, 113, 114, 115, 116, 117, 120, 123, 127, 136, 137, 140, 141, 143, 144, 146, 153, 155, 156, 157, 158, 159, 160, 161, 164, 167, 170, 171, 172, 173, 175, 178, 179, 180, 181, 182, 183, 184, 185, 186, 187, 188, 190, 191, 192, 194, 198, 204, 206, 207, 209, 231, 232, 233, 234, 235,

236, 238, 239, 240, 261, 268, 269, 287, 292, 293, 294, 298, 337, 338, 343, 344, 345, 346, 347
antigen Ia 28, 33, 34, 35, 36, 37, 99, 100-103, 187, 190, 293
antigen-reactive cells 179-188
antigen-sensitive cells *see* **memory cells**
antigen-stimulated lymph nodes 176-179, 190, 275
anti-thrombotics 80, 81, 82, 83, 85, 86, 88, 90
aplastic anaemias 16, 17
arachidonic acid 70, 71, 80, 81, 85, 86, 87, 88, 94, 193
arteriolar haemostasis 65
arterioles 39, 42, 43, 47, 63, 65, 67, 68, 228, 326, 330
Arthus reaction 202, 209
aspirin 83-84, 86, 91, 100, 191
atherogenesis 62-63
ATP 69, 70, 76, 293

B lymphocytes 6, 14, 16, 21, 22, 23, 24, 25, 26, 27, 29, 30, 34, 101, 102, 103, 104, 114, 117, 119, 120, 122, 123, 125, 135, 140, 141, 147, 150, 152, 155, 159, 160, 161, 163, 170, 171, 173, 176, 178, 231, 233, 235, 236, 237, 238, 248, 259,

266, 287, 292, 297, 298, 337, 346, 347
binding *see* **receptors**
binding, genetic variability of 275-278, 282, 284-286
binding, leucocyte-endothelium 323-327
blood flow, chemotaxis and 319-323, 332, 334
Birkbeck granules 36, 293, 294, 296
bone marrow 4, 5, 7, 8, 9, 10, 11, 13, 14, 15, 16, 17, 18, 19, 22, 34, 100, 103, 113, 158, 159, 166, 232, 236, 248, 275, 291, 346

cats 90, 91
capillaries 40, 41, 42, 43, 44, 45, 46, 47, 53, 55, 92, 93, 96, 110, 127, 154, 194, 199, 207, 209, 247, 325, 330, 331, 333
charcoal haemoperfusion 85
chemotaxis 225, 299-306, 308-311, 313-327, 338, 341, 345
chickens 80, 278, 285
chlorpromazine 75-76, 77
contact sensitivity 197-200, 209, 240
defence mechanisms
lymphoid system 235-236, 237

delayed-type hypersensitivity *see* hypersensitivity
dendritic cells 13, 16, 17, 22, 28, 292, 346
diacytosis *see* transcytosis
differentiating lymphocytes 101
dogs 85, 86

efferent lymphatic cells 53, 114, 115, 128, 129, 130, 131, 138, 148, 149, 150, 151, 158, 165, 166, 173, 174, 175, 176, 177, 179, 180, 184, 190, 203, 297, 347
eicosapentaenoic acid 87, 88
electron microscopy 22, 23, 28, 29, 30, 31, 35, 40, 44, 56, 59, 66, 190, 200, 201, 202, 207, 212, 214, 215, 221, 254, 293, 330, 344
elephants 166
endocytosis 11, 227
endoperoxides *see* prostaglandin
endothelial cells 8, 10, 14, 19, 24, 26, 31, 35, 36, 46-50, 51, 52, 55, 56, 57, 59, 62, 67, 81, 87, 93, 94, 95, 99, 100, 103, 105, 106, 107, 108, 109, 110, 126, 146, 159, 178-179, 184, 213, 344, 345, 346; *see also* high-endothelial cells
endothelial venules *see* venules
endothelium 18, 19, 25, 31, 32, 35, 36, 39-41, 49, 50, 52, 53, 54, 55, 56, 57, 58, 59, 67, 68, 92, 93, 94, 95, 99, 100, 103, 107, 108, 127, 128, 130, 145, 146, 190, 211, 225, 228, 229, 253, 286, 325, 343, 344
  arterial 58, 62, 63
  circulation and 9, 10, 11, 12
  structure of 40-43, 44, 50, 53, 54, 55, 56, 57, 211-213
  transport across 10, 40, 42, 43, 44-46, 49, 50, 53, 55, 56, 57, 58, 59, 105, 108, 109, 110

eosinophils 5, 7, 8, 11, 14
epithelial cells 28, 34, 36, 37, 39-40, 102, 103, 346
erythroblasts 5, 6, 7, 14
erythrocytes 12, 35, 69, 70, 92, 266, 282, 291, 292, 337, 338
extracorporeal circuits 82, 85, 96

fats, bone-marrow of mice 9, 16, 17, 19
  see also polysaturated and polyunsaturated fatty acids 193
fetus, sheep *see* sheep
fibroblasts 5, 7, 9, 16, 17, 261
fluorescent polysaccharides *see* polysaccharides
formylmethionyl *see* peptides

genetic variability *see* binding
germinal centres, lymph nodes 159, 259, 288, 347
  formation of 137, 138, 140, 141, 143, 144
glucose dependence 218
glycoprotein 102-103, 104, 221
glycosaminoglycan 221
Golgi apparatus 212, 213, 214, 215, 225, 228, 244
gradient, chemotaxis in 299, 300, 301, 302, 303, 305, 307, 308, 318, 320, 341
granules *see* Birkbeck granules
granulocytes 10, 17, 18, 146, 149, 199, 282, 325, 326, 329, 331
guinea pigs 62, 100, 190, 238, 240, 284, 293
gut-associated tissue 118, 122, 123, 125, 129, 136, 158, 205, 212, 343

haemodialysis 85
haemolysis 74, 75, 76, 77, 95, 96, 173
haemoperfusion *see* charcoal
haemopoietic microenvironment 4-17

haemostasis 61-71, 83-84, 93, 334, 344
hamsters 309, 329
heparan sulphate *see* sulphate
HEV (high-endothelial cell venules) *see* venules
high-endothelial cells 35, 126, 192, 213, 215, 225, 227, 228, 229, 244, 245, 337-341, 343, 344, 345
high-endothelial venules *see* venules
homing, lymph node and venules role in 103, 244-245, 247, 248-249, 252
hormones 34, 90, 99, 239-240
hypersensitivity, delayed-type 202, 206

immune response 104, 113, 116, 128, 136-138, 140, 141, 142, 143, 145, 154, 159, 160, 161, 165, 192, 230, 232, 233, 234, 235, 236, 237, 240, 298, 347, 349, 350
immunoblasts 203, 268
immunoglobulin 23, 34, 118, 119, 120, 127, 135, 140, 154, 159, 170, 202, 233, 267, 341
immunoglobulin A 117, 118, 119, 120, 122, 123, 124, 146, 147, 148, 153, 157, 165
  internal 118
  surface 114, 115, 117, 118, 119, 120, 123, 129
immunoglobulin-bearing cells 148, 153, 170, 205
immunoglobulin D 161
immunoglobulin E 7;
immunoglobulin G 55, 58, 153, 165, 171, 173, 288
immunoglobulin M 55, 56, 58, 119, 128, 129, 153, 161, 288
  negative 208
  positive 202, 203, 205, 208, 249
  surface 120, 138, 140, 171, 202, 204, 208

inflammation 107, 108, 143, 191, 206, 207, 209, 212, 239, 240, 331, 332, 334, 343
interdigitating cells 36, 37, 292-293, 294, 295, 296, 346
intestine, small 117, 118, 120, 123, 126, 151, 153, 154

Langerhans cells 34, 103, 293, 294, 296
leucocytes, polymorphonuclear 6, 299-306, 307-309, 313-327, 330, 331, 332, 333, 334, 335; see also binding
liver 153, 154, 207, 238, 262, 267, 289, 298, 339
locomotion, chemotaxis in 299, 300, 301, 304, 314, 315, 316-318, 328, 334
lymph see more specific headings
lymphocytes see B lymphocytes, T lymphocytes
lymphoid microenvironment 21-32, 33-37
lymphomas, thymic see thymic lymphomas

macrophages 6, 7, 10, 13, 14, 17, 19, 27, 32, 36, 57, 100, 108, 114, 186, 191, 193, 199, 205, 207, 240, 282, 287-295, 296-298
polysaccharide uptake by 288-290
marrow see bone marrow
megakaryocytes 17, 18, 19, 84
membrane, permeability of see permeability
memory lymphocytes 123, 155, 160, 167, 231, 232, 233, 234, 235, 236, 237
mice 4, 5, 7, 8, 9, 12, 14, 15, 22, 23, 24, 25, 26, 27, 28, 30, 33, 34, 36, 41, 42, 46, 47, 48, 49, 103, 105, 122, 124, 130, 141, 142, 148, 160, 162, 164, 206, 209, 212, 213, 233, 234, 238, 239, 248, 249, 266, 267, 275, 277, 278, 279, 281, 283, 284, 288, 289, 290, 291
nude 14, 23, 24, 239

microvilli 22, 24, 25, 26, 29, 30, 31, 32, 35, 36, 193, 221, 225, 257, 259, 260, 286
migration, chemotaxis and 326
lymphocytes of 4, 10, 11, 22, 24, 25, 26, 32, 35, 36, 52, 103, 104, 106, 107, 108, 109, 110, 125, 128-138, 143, 151-154, 207, 212, 225, 228, 230, 232, 239, 243, 244, 255, 259, 260, 265, 286, 344
migration, differential 287-295, 297, 300; see also recirculation and traffic
mucopolysaccharides 213, 220
murine binding see thymic lymphomas

N-formylmethionyl peptides see peptides
nodes, antigen-stimulated see antigen-stimulated
nodes, lymph 21, 23, 24, 25, 26, 27, 32, 35, 53, 58, 104, 106, 108, 110, 113-120, 125, 126, 128, 129, 130-138, 140, 142, 147, 149, 151, 152, 153, 154, 155, 158, 161, 162, 163, 164, 165, 190, 193, 194, 199, 203, 204, 206, 212-215, 225, 227, 229, 230, 234, 235, 236, 238, 240, 243, 244, 248, 254, 258, 259, 261, 266, 267, 268, 269, 270, 271, 272, 274, 275, 276, 277, 278, 279, 282, 285, 291, 292, 294, 296, 297, 298, 344, 345, 347
cell output of 133, 137, 138, 147-148, 152
nude mice see mice

octopi 235
oxazolone 198-204, 206, 207, 208, 209, 234

paracortex 211, 227
parenchyma 22, 23, 26, 254, 289

peptides, N-formylmethionyl 300, 301, 302, 303, 304, 306, 309, 338
periodic acid-Schiff-positive cells 6, 17
peripheral lymph 105, 108, 124, 129, 131, 136, 137, 141, 149, 160, 165, 198, 199, 200, 201, 202, 203, 205, 206, 207, 235, 238, 266, 267, 268, 269, 270, 271, 274, 281, 296, 297, 298
permeability 39, 42-44, 49-50, 52, 55, 57, 59, 99, 104-106, 108, 109, 209
Peyer's patches 114, 117, 118, 119, 120, 123, 124, 125, 126, 136, 244, 266, 268, 269, 270, 271, 272, 275, 279
PGI₂ see prostacyclin
phagocytes 7, 37, 57, 100, 143, 190, 192, 297, 298, 310
pigs 68, 88, 162, 165, 166, 293
platelets 18, 54, 55, 61-71, 73-77, 92-93, 94, 95, 96, 97, 193, 282, 331, 333, 334, 337, 339, 340, 344
activation time 66-71
aggregation 61-71, 73-77, 83-84, 91-92, 96, 257
in vitro 66, 73, 75, 82, 94
in vivo 64, 66-70, 82
prostacyclin and 79, 80, 81, 82, 83, 84, 85, 87, 90, 91, 92, 94
polarity, cells 302-306
polymorphonuclear leucocytes see leucocytes
polysaccharides 288-292
polysaturated fats 87
polyunsaturated fats 87, 88
postcapillary venules see venules
prostacyclin 51, 52, 73, 74, 77, 79-89, 90, 91, 92, 94-95, 96, 99, 177, 178, 193, 333, 344
prostaglandin(s) 51, 70, 71, 85, 86, 88, 100, 176-179, 191, 310, 334
endoperoxides 79-84, 86, 87, 88, 89, 95, 96, 344
prostaglandin E₁ 208

prostaglandin E₂ 177, 178, 179, 191, 192, 193
pulmonary circulation 85, 99, 154

rabbits 5, 7, 9, 18, 68, 80, 81, 84, 85, 86, 89, 93, 105, 190, 207, 267, 278, 293, 301, 334
radioactive sulphate see sulphate
rats 5, 8, 15, 29, 36, 46, 80, 85, 86, 102, 103, 105, 108, 110, 116, 117, 118, 122, 124, 125, 130, 133, 141, 148, 149, 160, 161, 162, 164, 166, 179, 206, 212, 213-225, 230, 245, 246, 247, 248, 249, 255, 261, 278, 279, 282, 296, 297, 329
reactivities, differential of lymph cells 136, 138, 140, 141, 143, 144
reception, chemotaxis in 303-305
receptors, lymphocyte-venule binding 31, 32, 35, 57, 146, 227, 237, 266-280, 282-286, 337-341, 343, 344, 345
in vitro detection 31, 246-248, 345
see also binding and venules
species specificity, lack of 249-250
recirculation, lymphocytes 4, 23-29, 32, 35, 71, 104-105, 106-122, 122-125, 127-128, 130-131, 133, 134, 141, 145-156, 157-166, 179, 185, 190, 193, 194, 199, 205, 206, 208, 225, 230, 233, 234, 235, 236, 237, 243-247, 265, 268, 285, 343, 344, 345, 346, 347, 349
see also migration and traffic
recognition, cellular 244, 279, 280, 338-339
regulation, lymphocyte traffic of 131, 132, 238

reticular cells 5, 7, 8, 9, 10, 13, 14, 16, 18, 19, 28, 32, 37, 102, 193, 296

scanning see electron microscopy
sensitivity see contact sensitivity and hypersensitivity
serotonin 71, 75-76, 93, 99, 334
sheep 105, 117, 119, 127-137, 140-144, 145-156, 157-166, 168, 179, 180-183, 191, 192, 193, 194, 199, 200-204, 205-206, 208-209, 233, 234, 238, 293, 296, 339, 345
fetus 145-156, 157-166
shutdown, cell and lymphocytes 167-188, 191, 192, 194, 199, 230, 232, 238, 338
skin 129, 198-200, 205, 222, 223, 224, 232, 238, 239, 240, 289, 293, 294, 298
spleen 5, 7, 13, 14, 16, 21, 32, 35, 74, 104, 125, 142, 153, 154, 155, 161, 164, 166, 228, 233, 236, 238, 249, 259, 266, 267, 278, 287, 289, 290, 291, 292, 294, 332, 346
stem cells 4, 5, 7, 8, 14, 15, 16, 17, 19, 110, 113
stromal cells 5, 7, 26
sugars 221, 222
sulphate 218, 219, 220, 221, 222, 224, 225, 227, 228, 229, 262
heparan 220
incorporation 224-225
radioactive 215, 217, 219, 227
suppressor lymphocytes 4, 6, 9, 13, 14, 114, 142, 170, 179, 231, 232, 233, 235
surface immunoglobulin see immunoglobulin

T lymphocytes 6, 14, 16, 21, 22, 23, 24, 26, 28, 29, 30, 100, 101, 102, 103, 104, 119, 120, 124, 125, 140,

141, 142, 147, 150, 151, 152, 153, 154, 155, 156, 161, 163, 170, 171, 186, 190, 231, 232, 233, 235, 236, 237, 238, 239, 248, 259, 266, 267, 274, 275, 281, 287, 293, 294, 295, 297, 298, 337, 338, 346, 347
thoracic duct 117, 118, 119, 122, 123, 125, 126, 142, 147, 150, 151, 158, 160, 161, 164, 165, 179, 221, 223, 232, 235, 245, 246, 247, 248, 251, 252, 255, 259, 266
thrombosis see anti-thrombotics
thromboxane A₂ 63, 69, 70, 77, 79-80, 82, 83, 84, 87, 91, 92, 94, 95, 100, 191, 192
thymic lymphomas, binding to HEV 271-275, 280, 283
thymocytes 22, 29, 30, 31, 35, 141, 268, 340
thymus 7, 22, 28, 29, 30, 33, 34, 36, 37, 100, 101, 102, 110, 125, 140, 147, 155, 158, 159, 160, 161, 213, 233, 248, 262, 268, 272, 274, 275, 281, 292, 294, 296, 346
thrombogenesis 62, 63-65
traffic, lymphocytes 35, 128-129, 186, 187, 194, 202, 204, 206, 208, 211, 231, 296; see also migration and recirculation
transcytosis 46, 50, 51, 52, 56, 58, 219, 345
trypsin 222, 223, 224, 228, 251, 252

vasodilatation 80, 85-86
veiled cells 292, 294, 297, 346
venules, high-endothelial 31, 35, 36, 54, 55, 65, 67, 104, 105, 106, 108, 109, 115, 119, 126, 130, 190, 192, 199, 207, 208, 211-221, 224, 225, 227, 228, 229, 230,

244-245, 246-255, 257-263, 265-280, 281-286, 288, 295, 329, 330, 331
see also **binding** and **receptors**
emigration through 333
fixatives, alteration by 251, 253-255, 257, 258, 259, 261,

262, 266
node homing, function in 103, 244-245, 247, 248-249, 252-268
organ-specific adherence to 268-271, 275, 278-279, 284

postcapillary 35, 41, 42, 43, 44, 46, 47, 50, 107, 110, 187, 211, 259, 260, 269, 333, 343, 344
radioactivity and 220, 225
virgin cells 124, 147, 148, 155, 157, 160, 161, 231, 236, 275